VIDEO MINING

THE KLUWER INTERNATIONAL SERIES IN VIDEO COMPUTING

Series Editor

Mubarak Shah, Ph.D.
University of Central Florida
Orlando, USA

Other books in the series:

VIDEO REGISTRATION
Edited by Mubarah Shah, Rakesh Kumar; ISBN: 1-4020-7460-3

MEDIA COMPUTING: COMPUTATIONAL MEDIA AESTHETICS
Chitra Dorai and Svetha Venkatesh; ISBN: 1-4020-7102-7

ANALYZING VIDEO SEQUENCES OF MULTIPLE HUMANS: ***Tracking, Posture Estimation and Behavior Recognition***
Jun Ohya, Akita Utsumi, and Junji Yanato; ISBN: 1-4020-7021-7

VISUAL EVENT DETECTION
Niels Haering and Niels da Vitoria Lobo; ISBN: 0-7923-7436-3

FACE DETECTION AND GESTURE RECOGNITION FOR HUMAN-COMPUTER INTERACTION
Ming-Hsuan Yang and Narendra Ahuja; ISBN: 0-7923-7409-6

VIDEO MINING

edited by

Azriel Rosenfeld
University of Maryland
College Park, MD, U.S.A.

David Doermann
University of Maryland
College Park, MD, U.S.A.

Daniel DeMenthon
University of Maryland
College Park, MD, U.S.A.

KLUWER ACADEMIC PUBLISHERS
Boston / Dordrecht / New York / London

Distributors for North, Central and South America:
Kluwer Academic Publishers
101 Philip Drive
Assinippi Park
Norwell, Massachusetts 02061 USA
Telephone (781) 871-6600
Fax (781) 681-9045
E-Mail: kluwer@wkap.com

Distributors for all other countries:
Kluwer Academic Publishers Group
Post Office Box 322
3300 AH Dordrecht, THE NETHERLANDS
Telephone 31 786 576 000
Fax 31 786 576 474
E-Mail: services@wkap.nl

Electronic Services <http://www.wkap.nl>

Library of Congress Cataloging-in-Publication Data

Video Mining / Edited by Azriel Rosenfeld, David Doermann, and Daniel DeMenthon
p.cm.
Includes bibliographical references and index.
ISBN: 1-4020-7549-9 (alk. paper)
1. Data Mining. I. Rosenfeld, Azriel. II. Doermann, David. III. DeMenthon, Daniel. IV. Series.

Printed on acid-free paper.

Printed in the United States of America.

Contents

Series Foreword

Traditionally, scientific fields have defined boundaries, and scientists work on research problems within those boundaries. However, from time to time those boundaries get shifted or blurred to evolve new fields. For instance, the original goal of computer vision was to understand a single image of a scene, by identifying objects, their structure, and spatial arrangements. This has been referred to as *image understanding*. Recently, computer vision has gradually been making the transition away from understanding single images to analyzing image sequences, or *video understanding*. Video understanding deals with understanding of video sequences, e.g., recognition of gestures, activities, facial expressions, etc. The main *shift* in the classic paradigm has been from the recognition of static objects in the scene to motion-based recognition of actions and events. Video understanding has overlapping research problems with other fields, therefore *blurring* the fixed boundaries.

Computer graphics, image processing, and video databases have obvious overlap with computer vision. The main goal of computer graphics is to generate and animate realistic looking images, and videos. Researchers in computer graphics are increasingly employing techniques from computer vision to generate the synthetic imagery. A good example of this is image-based rendering and modeling techniques, in which geometry, appearance, and lighting is derived from real images using computer vision techniques. Here the *shift* is from *synthesis* to *analysis followed by synthesis*. Image processing has always overlapped with computer vision because they both inherently work directly with images. One view is to consider image processing as low-level computer vision, which *processes* images, and video for later analysis by high-level computer vision techniques. Databases have traditionally contained text, and numerical data. However, due to the current availability of video in digital form, more and more databases are containing video as content. Consequently, researchers in databases are increasingly applying computer vision techniques to analyze the video before indexing. This is essentially *analysis followed by indexing*.

Due to MPEG-4 and MPEG-7 standards, there is a further overlap in research for computer vision, computer graphics, image processing, and databases. In a typical model-based coding for MPEG-4, video is first *analyzed* to estimate local and global motion then the video is *synthesized* using the estimated parameters. Based on the difference between the real video and synthesized video, the model parameters are *updated* and finally *coded* for transmission. This is essentially *analysis followed by synthesis, followed by model update, and followed by coding.* Thus, in order to solve research problems in the context of the MPEG-4 codec, researchers from different video computing fields will need to collaborate. Similarly, MPEG-7 is bringing together researchers from databases, and computer vision to specify a standard set of descriptors that can be used to describe various types of multimedia information. Computer vision researchers need to develop techniques to automatically compute those descriptors from video, so that database researchers can use them for indexing.

Due to the overlap of these different areas, it is meaningful to treat *video computing* as one entity, which covers the parts of computer vision, computer graphics, image processing, and databases that are related to video. This international series on *Video Computing* will provide a forum for the dissemination of innovative research results in video computing, and will bring together a community of researchers, who are interested in several different aspects of video.

Mubarak Shah
University of Central Florida

Orlando

Preface

The goal of data mining is to discover and describe interesting patterns in data. This task is especially challenging when the data consist of video sequences (which may also have audio content), because of the need to analyze enormous volumes of multidimensional data.

The richness of the domain implies that many different approaches can be taken and many different tools and techniques can be used, as can be seen in the chapters of this book. They deal with clustering and categorization, cues and characters, segmentation and summarization, statistics and semantics. No attempt will be made here to force these topics into a simple framework. In the authors' own (occasionally abridged) words, the chapters deal with video browsing using multiple synchronized views; the physical setting as a video mining primitive; temporal video boundaries; video summarization using activity and audio descriptors; content analysis using multimodal information; video OCR; video categorization using semantics and semiotics; the semantics of media; statistical techniques for video analysis and searching; mining of statistical temporal structures in video; and pseudo-relevancy feedback for multimedia retrieval.

The chapters are expansions of selected papers that were presented at the DIMACS Workshop on Video Mining, which was held on November 4-6, 2002 at Rutgers University in Piscataway, NJ. The editors would like to express their appreciation to DIMACS and its staff for their sponsorship and hosting of the workshop.

Azriel Rosenfeld
David Doermann
Daniel DeMenthon
College Park, MD
April, 2003

Chapter 1

EFFICIENT VIDEO BROWSING

Using Multiple Synchronized Views

Arnon Amir, Savitha Srinivasan and Dulce Ponceleon
IBM Almaden Research Center
650 Harry Road, CA 95120
arnon@almaden.ibm.com

Abstract People can browse text documents very quickly and efficiently. A user can find, within seconds, a relevant document from a dozen retrieved items listed on a screen. On the other hand, browsing of multiple audio and video documents could be very time-consuming. Even the task of browsing a single one-hour video to find a relevant segment might take considerable time. Different visualization methods have been developed over the years to assist video browsing. This chapter covers several such methods, including storyboards, animation, slide shows, audio speedup, and adaptive accelerating fast playback. These views are integrated into a video search and retrieval system. A synchronized browser allows the user to switch views while keeping the context. The results of a usability study about audio speedup in different views are presented.

Keywords: Video retrieval, multimedia browsing, video streaming, synchronized views, audio time scale modification (TSM), fast playback, video browser, visualization techniques, storyboard, moving storyboard (MSB), animation, slide show, audio speedup, adaptive accelerating, usability study, TREC Video Track, navigation, hierarchical taxonomy, movieDNA.

Introduction

In the last several years we have witnessed a significant growth in the digital video market. Digital video cameras become ubiquitous with the proliferation of web cameras, security and monitoring cameras, and personal hand-held cameras. Advances in video streaming technology, such as MPEG-4, and the penetration of broadband Internet home connections allow home users to receive good quality live video streams. The home entertainment market is growing rapidly with game computers, DVD players/recorders and set-top boxes. All these trends indicate a promising future for digital video in a variety of applications, such as entertainment, education and training, distance learning, medical and technical manuals, marketing briefings, product information, operation guides and more.

As the amount of video-rich data grows, the ability to find and access video data becomes critical. Users should be able to quickly and efficiently find the information they look for. In the past ten years there has been a major research effort to develop new video indexing, search and retrieval methods which would allow for efficient search in large video repositories. It evolved as a multidisciplinary effort, which includes a wide range of research topics from computer vision, pattern recognition, machine learning, speech recognition, natural language understanding and information retrieval (see, e.g., [Aigrain *et al.*, 1996; Bach *et al.*, 1996; Jones *et al.*, 1996; Chang *et al.*, 1997; Gupta, and Jain, 1997; Del Bimbo, 1999; Wactlar *et al.*, 1999; Adams *et al.*, 2002; Chang *et al.*, 2003]).

Search and browse are tightly coupled operations. Sometimes it is easier to search, especially for specific entities, like a specific company, a person, etc. In other cases, when the topic is generic, such as Chinese food, or modern houses, it is more efficient to browse through a well crafted taxonomy. Most often, an initial search can take the user to a good starting point for browsing. For example, in the case of looking for a computer accessory, the user may search for the computer model (known, specific), and browse from there to the accessories list. An opposite example is searching for a Ford model car on eBay. A query to the entire database might suffer from high ambiguity of those words. Instead, a user might first navigate through the categories to Toys & Hobbies, then to Models, and then initiates a search for the car model within this category. In this case the browsing provides context to the search.

Extracting semantic information from image, audio and video is a complicated task. The state of the art might be well represented in the NIST TREC Video track (see Chapters 8 and 11, and [Over and Taban, 2001; Smeaton and Over, 2002]). Even manual cataloging and annotation of multimedia is as difficult as it is important. It requires considerable human labor, but is limited to what is considered relevant by the annotator at indexing time, rather than the specific user needs at query time [Enser, 1993; Armitage and Enser, 1997]. Manual annotation of a single image, to be useful, must be done at three different levels; "pre-iconography," "iconography" and "iconology," using special lexicons [Shatford, 1985]. Still, this provides only a partial solution to the real needs of image and video retrieval. Despite limitations and cost, manual annotation is still considered one of the best available ways to index videos and images. The MPEG-7 standard provides a unified description scheme for annotating multimedia content, including audio, video, etc. [Pereira, 1997]. As such, it widely supports multimedia indexing, searching and browsing needs. However, the standard does not provide all the algorithms and tools for populating those indexes. Many of those are still to be developed.

Searching speech transcripts of videos using the familiar metaphor of free text search has been studied in several projects [Hauptmann and Witbrock, 1997; Jones *et al.*, 1996; Wactlar *et al.*, 1999; Srinivasan and Petkovic, 2000; Garofolo *et al.*, 1999]. First, automatic speech recognition (ASR) is applied to the audio track, and a time-aligned transcript is generated. The indexed transcript provides direct access to semantic information in the video. A word inverse index provides a simple way to search and retrieve query words. Another popular choice is dividing the transcript into short segments (100-200 words long), and indexing those using statistical document indexing techniques (e.g., OKAPI). The ASR transcripts can be further processed to extract keywords, phrases (N-Grams) and topics. Those are similar to standard statistical text analysis (e.g., [Baeza-Yates and Ribeiro-Neto, 1999]), with some special consideration of the word error rate in ASR transcript [Jones *et al.*, 1995].

The main advantage of speech indexing, compared to image content, is the representation. A speech index is naturally built of words, which are a natural choice for query formulation as well. Image indexing, however, requires to extract semantic labels from the visual content and then represent them by concepts [Adams *et al.*, 2002], keywords, multinets [Naphade *et al.*, 1998] or other semantic representations (e.g., see [Chang *et al.*, 2003]). There is no clear answer to the problem of representation of visual content.

Browsing of visual content is much easier and faster than browsing audio and speech. A single page of results may contain a dozen of matches, with thumbnails and textual information, and could be glanced over in seconds. Browsing of speech and audio, however, requires the user to listen to one audio stream at a time. Furthermore, audio is a temporal signal which requires time to be played. Hence visual information is much more efficient for browsing than audio and speech.

The audio and video are complementary modalities. The theme of our work in the CueVideo project is "Search the speech, browse the video" [Amir *et al.*, 2003]. The audio track and the video track of a video are synchronized and time-aligned. It is therefore possible to search the speech and browse the video, or images, thus implementing these two coupled operations using the two complementary modalities of the video data. Visual browsing may use static keyframes, animation, fast playback, keywords and transcripts from the speech, storyboards, mosaics and video summaries and more.

While search and browse are tightly coupled operations, the focus of his chapter is about audio and video browsing. We differentiate between three levels of browsing, depending on the scope :

- Browsing a large collection of videos.
- Browsing a ranked list of videos, such as a query result.
- Browsing a single video to find relevant segments

At the highest level, navigation through a preprocessed hierarchical taxonomy is often used. At this level there is no need to access the document itself. This process requires a highly semantic cataloguing of the videos into those classes, and would work for multimedia documents as well as for text documents.

Browsing of search results is a slightly different task, where all the video are provided in a single list, sorted either by the match rank or by other criteria. The user looks for one or more relevant videos. A brief description of each video is provided to assist the user with the browsing process, showing how the document is related the query. When dealing with a list of multimedia documents, other than text, the description of each document and its relevancy to the query are not as simple to present as for text. Representing in a single table the audio and visual contents of one hour of video and their relationship to the query is quite a challenge.

At the finest level, a selected video document is being browsed for information. People are very efficient at reading text documents, and they can quickly find the relevant part of a document. However, when it comes to the browsing of one hour of video, this could be a very time-consuming effort.

We will mainly focus on the last two levels, as those seem to be the most challenging ones, causing a bottleneck in many existing applications. The comparison to text browsing helps to understand the user's expectation and behavior patterns, and to define the desired functionality.

1.1 Video Browsing - Related Work

Video browsing seems to receive somewhat less attention than search, or what is often called content-based image and video retrieval. However, the browsing efficiency problem arises whenever a system is built over a large collection of videos. Below we provide several examples of such systems.

The Informedia research project [Hauptmann and Witbrock, 1997; Christel *et al.*, 1998; Wactlar *et al.*, 1999] has created a terabyte digital library where automatically derived descriptors for video are used to index, segment and access the library contents. The system offers several ways to search and to view the search results, such as poster frames, filmstrips and skims (video summaries). The filmstrip view reduces the need to view each video paragraph in its entirety by providing storyboards for quick viewing [Hauptmann and Witbrock, 1997]. The most relevant subsections of the video paragraph are displayed as key scenes and key words are clearly marked.

The Collaborative and Multimedia group at Microsoft Research accomplished considerable work in multimedia browsing, studying both the technical aspects and the user behavior [He *et al.*, 1998; Omoigui *et al.*, 1999; Li *et al.*, 2000]. This work includes video summarization, audio speedup, collaborative annotation, and the usage patterns of such technologies. The interaction of a user, or multiple users, with the media content is at the core of this line of work. The results are supported by multiple usability and case studies.

Chang et al. developed the WebClip system for the search and browse of video for education [Chang *et al.*, 1998] and tested it in several K-12 schools. Their main focus was visual indexing and retrieval. The system includes most of the various components described above and includes content-based and semantic image retrieval, object segmentation, motion trajectories, textual annotation and advanced browsing capabilities such as a client video editor. Indeed they report on the difficulties of automatically extracting semantic knowledge from images, and state in regards to content-based indexing that videos have an advantage over images. The reported client-server implementation requires high communication bandwidth over a local network.

Yeung et al. [Yeung *et al.*, 1995] proposed the story graph as an efficient way to capture the temporal relationships between shots and scenes. Shots with similar visual content are grouped together into a single node in a graph, and directional edges connect between nodes with consecutive shots. In a situation comedy, where most of the shots occur at only a handful of different places, this typically result in a compact graph that captures an entire episode.

Aner and Kender [Aner and Kender, 2002] developed a compact video representation and summarization technique. First each shot is represented with a mosaic, and then the mosaics are matched and grouped according to the different setups. A single mosaic is selected to represent each scene. Related scenes from different episodes are then grouped together and can be efficiently accessed. This was successful demonstrated on several situation comedy and sports programs.

Video skimming, or summarization, is another common technique to support efficient video browsing. The main challenge is how to automatically select the most important or most representative video segments which will compose the skimmed video. Different algorithms use various combinations of shots detection, speech recognition, face detection and recognition, motion detection, pauses, salient, music and sound effect detection, text overlay, and video production models to come up with good skims [Smith and Kanade, 1995; Christel *et al.*, 1998]. Other methods compose collaborative skims, based on the most frequently played segments of the video, or personalized skims, depending on a query or on user profile [Tseng and Lin, 2002]. In general, skims could be used as teasers, as an alternative to watching the full video, or for search and browse purposes.

The NIST TREC Video Track provides a benchmark for video indexing, search and retrieval. In its second year, 2002, there were two Search tasks, denoted Manual Search and Interactive Search. The difference between them can be defined by a single word: browsing. In the Manual task, the user composes a query to search for a given topic, but is not allowed to interact with (browse) the results. The Interactive task continues from that point and allows the user to browse, refine the query, and provide relevance feedback. The results of the Interactive runs were significantly higher than those of the Manual runs, for all the systems, with often twice as high or higher MAP scores. Striving for fast and efficient browsing, some of the participating groups developed powerful user interfaces for video and shot browsing. Two such examples are presented in Chapter 8 and Chapter 11 of this book, and in [Browne *et al.*, 2002].

As video browsing is a time consuming task, it is suggested that multiple views of the video content could help reducing the time. In the rest of this chapter we describe several such views, for both visual and audio content. When integrated together in a single synchronized browser, those views form a coherent video browsing interface which is very simple and intuitive to use. More views can be added into this interface, such as the ones mentioned above. A usability study was performed to assess the effect of different views on content comprehension by users.

1.2 Shot Boundary Detection

A key to efficient video visualization is accurate detection of shot boundaries. A shot is a continuous sequence of frames as captured by a camera. It is often being represented using a single keyframe in a storyboard. Hence it is important to accurately detect the shot boundaries. These boundaries, or changes between shots, are created during the editing phase of a video production. There are multiple types of shot boundaries, from cuts, or abrupt changes, to dissolves, which make gradual changes and through many other editing effects, some of which have unique artistic design. The aim of a shot boundary detection (SBD) algorithm is to find all the shot boundaries, including the abrupt ones and the gradual ones. Extensive work has been done on shot boundary detection. The reader is referred to several surveys and comparisons [Brunelli *et al.*, 1996; Lienhart, 2001] and the TREC benchmark [Over and Taban, 2001; Smeaton and Over, 2002].

The video content is processed using the SBD program from the *IBM CueVideo Toolkit* [Amir *et al.*, 2001]. This is a single pass algorithm that operates on uncompressed frames. Keyframes are selected and extracted as the shot boundaries are detected and are saved as JPEG files.

The algorithm uses a 512 bin three-dimensional color histograms (3-bits per channel) in RGB color space to compare pairs of frames. Histograms of 60 frames around the current frame are stored in a moving-window buffer to allow comparison between multiple pairs of frames. Frame differences at one, three, five, and up to thirteen frames apart are computed around the middle frame. Statistics of frame differences are computed in the moving window and used to compute adaptive thresholds. A different threshold is computed for each frame distance. In general, the correlation between frame pairs is lower as the frames are further apart from each other. Hence the threshold for further pairs should be higher. The program is completely automatic and does not accept any user sensitivity-tuning parameters. An example of frame differences and their corresponding thresholds is shown by the three upper graphs in Figure 1.1.

A state machine is used to detect and classify the different shot boundaries. The thirteen states include In-shot, Cut, After-Cut, Dissolve, an so on. At each frame a state transition is made from the current state to the next state, based on rules and on the multiple frame comparisons. An appropriate operation is taken as the state transition occurs (e.g., report

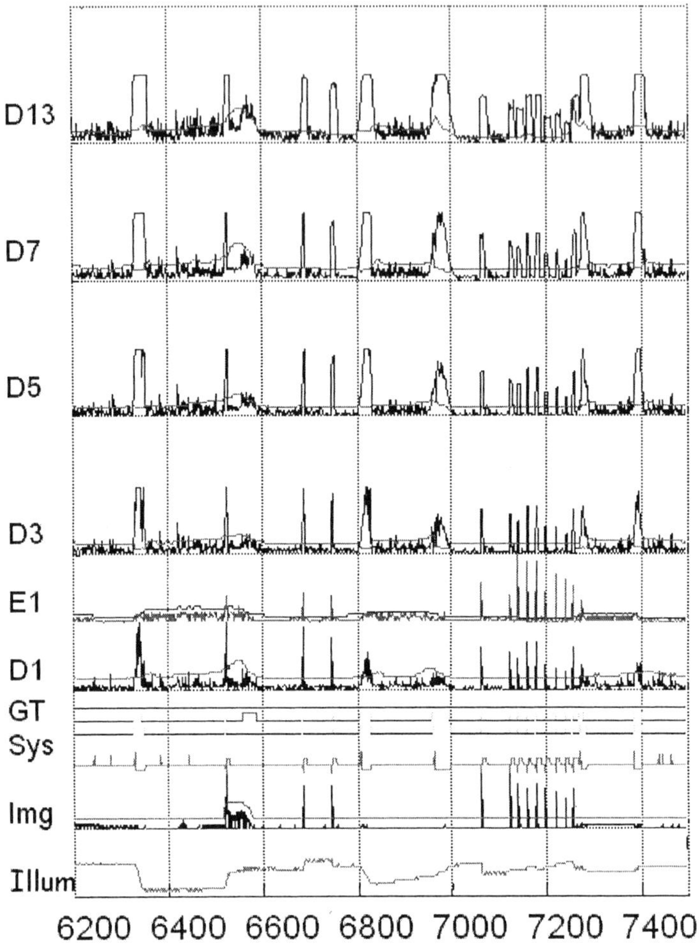

Figure 1.1. This example, captured from the SBD debugging program, represents 1200 frames (40 seconds). Color differences between frames for frame offsets equal to $1, 3, 5, 7, 12$ are shown by $D1, D3, D5, D7, D13$ respectively. Graph $E1$ represent edges difference, Sys corresponds to the machine state during the process, and GT shows two graphs, the ground truth and the algorithm output. Errors are marked in red color. Twelve cuts and six dissolve are present, all but one are correctly detected.

a shot, save a keyframe to file). Rank filtering in time/space/histogram is used at various different points along the processing, to handle high noise levels and low quality videos.

Evaluation of the SBD algorithm was carried at the NIST TREC Video Track conference, in years 2001 and 2002. Table 1.1 summarizes the evaluation of the current system on the data sets of both years.

	Video Data Set			All		Cuts		Gradual		Gradual Fr.	
Year	Videos	Hrs	#Shots	Rc	Pr	Rc	Pr	Rc	Pr	Rc	Pr
TR-01	42	5:48	3006	.96	.92	.99	.98	.89	.79	.66	.90
TR-02	18	4:51	2090	.88	.83	.93	.87	.76	.72	.57	.89

Table 1.1. Shot boundary detection results of the CueVideo SBD system on both TREC-01 and TREC-02 video data test sets.

The system performance on TREC-02 data is noticeably lower than on TREC-01 data set. This is due to the old videos which were used in 2002. It is very noticeable in all other participating systems as well.

In order to find the main causes of errors, a manual verification of all insertion errors reported by NIST evaluation has been done for two of the eighteen benchmark videos (36553.mpg and 08024.mpg). The breakdown of the sixty insertion errors, as summarized in Table 1.2, shows that 55% were due to false detection while the other 45% were correctly detected events, and were either reported in a wrong way or missed in the NIST evaluation ground truth.

Cause of insertion error	freq.	%
Fast motion	12	20%
Fast zoom	5	8%
Illumination changes	4	7%
Long grad reported as two	4	7%
MPEG video error	1	2%
others	7	12%
Short/long grad mismatch	12	20%
FOI reported as two (FO+FI)	7	12%
Missing cuts in ground truth	5	8%
Fade in at video start	2	3%
Fade out at video end	1	2%
Total	60	100%

Table 1.2. Shot boundary detection insertion errors on TREC-2002 data set - a breakdown by cause. The first group corresponds to false detections, the second group to inappropriate reporting of correctly detected events, and the third one is of events not reported in the NIST ground truth.

1.3 Time Scale Modification of Audio Signals

1.3.1 Introduction

Efficient video browsing also requires efficient audio browsing. One approach is to speedup the audio, also known as Time Scale Modification (TSM) of audio signals. Audio TSM techniques allow speeding up or slowing down audio without noticeable distortion. High quality time scaling of speech signals is achievable mostly because large portions of human speech signals have a quasi-periodic nature. The basic idea in most algorithms accounted below is to skip pitch periods when speeding up, and duplicate them when slowing down. By deleting (or inserting) small audio segments, the total playing time changes. When performing time scaling on musical signals, one cannot rely on a quasi-periodic nature, especially when more than one musical instrument is playing. However, when the time scaling factor is not too extreme, some of the methods described below do achieve acceptable quality.

For speech signals, time scaling aims at speeding up or slowing down the signal, such that it will sound as if the original speaker was recorded again saying the same thing faster or slower. This means the TSM algorithm should affect the speaking rate only, while preserving other perceived aspects such as timbre, voice quality and pitch. Malah [Malah, 1979] laid the foundations and mathematical background for time scaling of speech signals, as part of the general formulation of Time-Domain Harmonic Scaling (TDHS). Valbret et. al. [Valbret *et al.*, 1992] demonstrated a simple Time-Domain, Pitch Synchronous Overlap Add (TD-PSOLA) technique for time scaling. Over the years, these methods were improved in terms of robustness, speech quality and computational efficiency. Roukos and Wilgus [Roukos and Wilgus, 1985] presented the Synchronized Overlap Add (SOLA) method which was used as the basis for most modern speech TSM algorithms. Verhelst and Roelands [Verhelst and Roelands, 1993] introduced an improved variant to the SOLA titled "Waveform Similarity Overlap Add" (WSOLA). This algorithm is described in the next section. Recently, the new ISO MPEG-4 audio standard ([ISO/IEC, 1999], part 3 subpart 1) has adopted an algorithm titled PICOLA (Pointer Interval Controlled Overlap Add) for time scaling, as an optional algorithm applicable to all MPEG-4 audio coding schemes.

The method used to perform time scaling in our application is based on the Waveform Similarity Overlap Add (WSOLA) algorithm by Verhelst and Roelands [Verhelst and Roelands, 1993], with 20 msec sections of the input signal. The effect of audio TSM on comprehension is discussed in Section 1.8.

An illustration of the WSOLA technique for speeding up audio is presented in Figure 1.2. We use 20 msec triangle windows, shown in the figure, to segment and weight the sampled audio signal. The first graph corresponds to windows over the input audio, and the second one corresponds to windows over the time scaled output audio (the sampled audio signal itself is not shown in the figure). In WSOLA, the output signal is constructed by taking small 20 msec sections of the input audio waveforms from different time points, and "pasting" them together to form the output waveform. The triangle windows help weight these short waveforms such that by overlap–adding them to the already constructed output waveform, the continuity of the waveform is preserved.

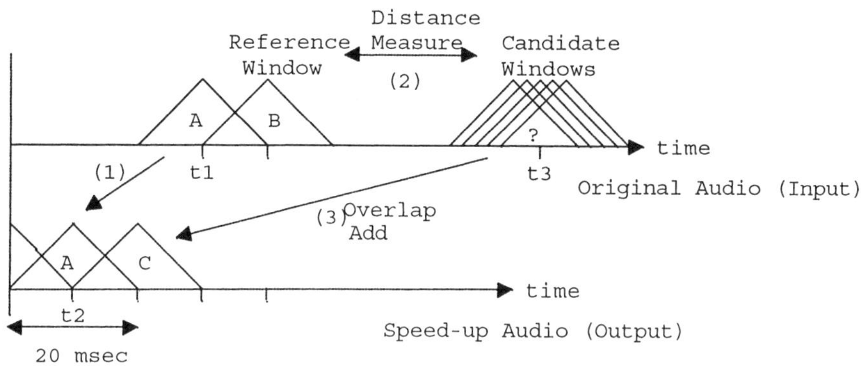

Figure 1.2. Illustration of WSOLA method for speeding up audio

For example, assume that window A in Figure 1.2 was previously copied, (Step 1), from the input time point t_1 to the output time point t_2, and overlap–added to the already constructed output waveform until time t_2. The time t_1 is ahead of t_2 since we are speeding up. We now look for the next window C to add to the output waveform. To assure a time and frequency continuity of the constructed waveform, the best selection would be to copy window B (the reference window), to location C. However, since we are speeding up, a window must be selected from a time point further ahead. We therefore search for the proper window in the vicinity of the new input time point t_3, among several candidate windows. The time point t_3 is calculated in advance, as a function of the time scaling factor.

Step 2 involves calculating a distance measure between the reference window B and each of the candidate windows, and choosing the candidate window which is closest to the reference. The normalized cross-correlation, a common distance measure in audio signal processing, is used to select the best candidate. In Step 3, the selected candidate window is copied to location C, and is overlap–added with the already constructed output waveform. The degree of freedom expressed in the fact that there is more than one candidate window to choose from assures that the output signal preserves the possible quasi-periodic nature of the input signal.

The WSOLA algorithm can be implemented in such a way that during time scaling of an audio signal, one can change the scaling factor and hear the result immediately. This allows automatic or manual adjustment of the speedup or slowdown factors according to the played context. In recent years, several TSM plug-ins were made commercially available for popular computer media players. However it is inevitable that TSM will become an integrated part of media players, supporting playback of both local files and streaming media.

1.4 Storyboards, Moving Storyboards and Animation

A **storyboard** is a static representation of the visual content of a video. The storyboard is composed of one or more pages of representative still frames called keyframes, or thumbnails, which are (automatically) selected from the video frames based on time intervals, on a segmentation of the video into shots, or other selection criteria. One example is shown later in Figure 1.5. Due to its large popularity we will save the reader from extra details, which may be found in any video analysis textbook, e.g., [Maybury, 1997].

A **moving storyboard (MSB)** is a slide show, composed of representative keyframes, fully synchronized with the original audio track. Each keyframe is shown for the duration of the video segment it represents. Typically one keyframe per shot is extracted and is displayed for entire duration of the associated shot. In other cases, a long shot might be represented by several keyframes.

The MSB is useful for browsing videos over low bit-rate connections. In the case of video from the classroom, for example, the static keyframes capture the speaker's slides at a much higher quality than that of a low-bitrate streaming video. In such applications, the missed motion might be of only secondary importance to the user.

Slide shows are easy to compose in all major streaming formats, including Apple Quicktime, SMIL (e.g., with Real Networks), Macromedia Flash, etc. By default, the Quicktime player assigns to the keyboard arrow keys the functions of next-frame/previous-frame. Skipping one keyframe in an MSB corresponds to skipping one video shot, making it easy to browse. The required bitrate is typically 25-35 Kbps, which meets standard modem bandwidth. It is 30-50 times smaller than the original, 1.2-1.5Mbps MPEG-1 video.

MSBs are best suited for news, education and commercial clips where the combination of audio with still images conveys most of the content. However, they are not suitable for summarizing high motion events such as a tennis match, a car race or a dance performance, where motion is very relevant in conveying content.

The length of the MSB is solely governed by length of the audio track, which by default is the same as the original video. An MSB can be further compressed to save time by speeding up the audio. This topic is discussed in detail in Section 1.3.

An **animation** is a very fast (salient) slide show of the keyframes. Each keyframe is being displayed for a fixed duration (typically 0.6sec). It allows fast keyframe browsing of long videos on a small screen, where a storyboard is inadequate. Instead of scanning large tables of keyframes, the keyframes change in one window. A time bar and basic play control make it less tedious than a regular Storyboard. Its duration is much shorter than the original video, making it a very fast view. However, both the Storyboard and the salient slide show are unsuitable for certain domains, like education and training, where most of the information is found in the audio track.

1.5 Adaptive Accelerating Fast Playback

Adaptive accelerating fast playback is a very fast video playback (without audio). In a regular fast playback of digital video, frames are skipped at a fixed rate, independent of the video content. Some set-top box products allow up to 64× speedup ratios and more. However, at fast speeds, short video events might either be missed or become just too short to be noticeable. Adaptive fast playback methods address this problem by adjusting the speedup ratio according to the video content. The speed might depend on various properties of the video, such as the amount of object motion, camera motion, the shot length and more.

The method described here depends on the video segmentation into shots. It starts at a slow, 2× speedup rate at the beginning of a shot. After 0.5 seconds into the shot, it gradually accelerates to 30× faster or more, and continues at the fastest target speed throughout the rest of the shot. It resets back to the low rate at the beginning of the next shot. This ensures that short shots are not missed. At a linear acceleration, the frame selection criteria is provided by this simple iterative rule

$$t_1 = 1 \tag{1.1}$$

$$t_n = ((t_{n-1} * d < n) \;\&\; (t_n - t_{n-1} < T_{max}))\;?\; n \;:\; t_{n-1} \tag{1.2}$$

$$s_n = 1 - (t_n == t_{n-1}) \tag{1.3}$$

where n, d, T_{max} are the frame number in the shot, the acceleration slope and the maximal rate, respectively. An example is shown in Figure 1.3, using typical values, $d = 1.25$ and $T_{max} = 30$. The ramp-up period gives the user enough time to fixate on some of the shot content, after which the user is better able to follow the increasingly fast playback of the rest of the shot. The average speedup could be very fast, however users' reaction after watching this fast playback is that it does not "feel" as fast as the numbers tell. Indeed, it does not have those abrupt fast frames that are typical to most other fast forward methods. Other rate adaptation criteria could be used in combination with this method, e.g., to control T_{max} according to object motion.

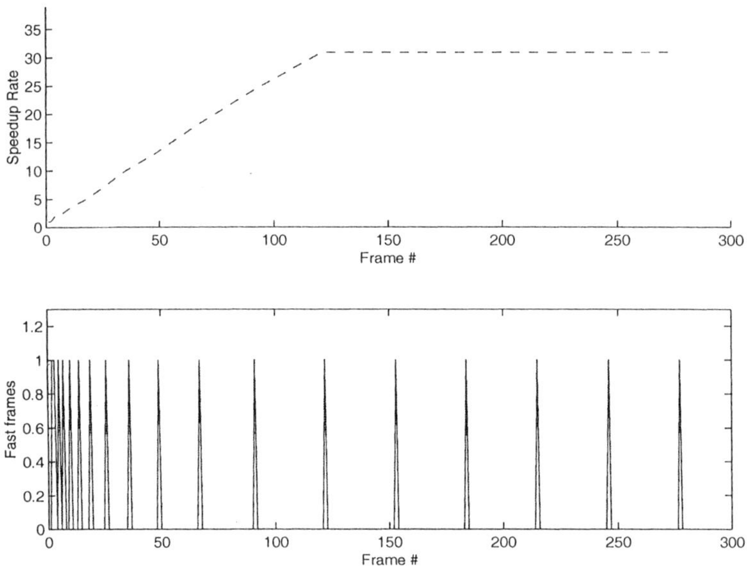

Figure 1.3. The frame skipping policy for adaptive fast forward. The speedup ratio and the selected frames are plotted in the upper and lower graphs, respectively. This policy restarts at the beginning of each shot.

The adaptive accelerating fast playback method does not require much computational load in computing the frames to be presented. It requires however the shots to be detected. In our implementation with streaming media the fast version was encoded as a new video stream and stored along with other views. A list of the frame numbers of corresponding shots in both the original and the fast version is stored to allow synchronized switching between the two. For a non-streaming local video file, this list is enough to allow decoding of the required frames from the original video file at playback time. The concept of synchronized views concept is introduced in the next section.

1.6 Streaming Synchronized Views

Having multiple views of the same video, such as animation, slide show, adaptive fast playback, fast audio, video skims and others, one would like to be able to switch between views without loosing the context. The context is defined as the current position in the current view, and in order to keep the context the new view must take from the point left by the previous view. Hence, even these views might have different duration and playback rates, the user experience would be as the views are synchronized.

Some video players provide functions like fast forward and fast backward. Advanced players, sometimes found in video editing tools, allow to skip to the next shot, next scene, and other browsing functions. The reader might rightfully consider some of the views as additional browsing functions, that should look and feel the same as ones already in a player. However, most if not all of these commercial players still require the video to be accessed as a local file, to be rapidly accessed at any arbitrary frame numbers, and manipulated in real time during playback time. This limits the use of such advanced players in video retrieval applications, where typically a large repository of videos is accessed remotely over a network such as the Internet.

Applying a similar real-time manipulation approach to streaming media is rather complicated. Frames cannot be arbitrarily accessed but are rather streamed in consecutive order, from a desired starting point. A client implementation of fast playback would require much faster streaming of the original stream. It requires very high communication bandwidth and is impractical with existing networks.

A second approach is to implement the real-time stream manipulation on the server, generating a modified stream on the fly and streaming it to the client at a regular bitrate. This is not as simple as it might sound either. Most video compression techniques encode individual video frames based on other neighboring frames. This limits the ability to stream arbitrarily selected encoded frames. The server, however, cannot be loaded with any complicated computation that would need to be performed independently per each client. This would hurt the server scalability. High scalability is a major concern in video streaming servers, as those are required to serve as many simultaneous clients as possible at minimal cost.

A third approach is to keep each of the individual views as a separate stream, and switch between streams at playback time. For example, in order to provide the user with three different audio speedup rates, three different versions are preprocessed and stored on the server. When a different speed is selected by the user, the client just switches to another stream. This approach requires only standard communication bandwidth and no extra computational load on the server. However, it requires more storage for all the views, and is limited to only those preprocessed views (e.g., three audio speeds). These tradeoffs were studied in the context of usage patterns of audio speedup technology [Omoigui *et al.*, 1999].

We decided to use the third approach. The main technical challenge with this approach is how to keep the context when switching between views. The desired process would have the following steps:

1 Pause the current view

2 Get the current position

3 Close the current stream

4 Compute corresponding position in the selected view

5 Open the selected view

6 Seek to the corresponding position

7 Play the selected view

Ideally all of this process would have done by the streaming server, who holds the streams of the two views. However current streaming technology requires some extra effort to resolve step 4 and step 6.

The first issue is illustrated in Figure 1.4, showing the server-client architecture of our video search and retrieval system. The web page is composed by the application server and served by the HTTP server to the client. The web page contains an embedded player, which connects to the video streaming server and plays the video. This web page is shown in Figure 1.5 (left). The user may switch to a different view at any time by pressing one of the buttons below the player. The computation is not supported by the streaming server. It could have been done on the application server. However, since there is no link between the application server and the streaming server, the result should have to go through the client. This would cause extra communication between the client and the application server that would delay the switching process.

Our approach is to perform the entire computation on the client, using Javascript code in the HTML page. This code contains the data required for the computation, including the relative speedup rates of the different streams and corresponding frame numbers between shots of non-linear views. Hence there is no longer need to contact the application server for switching between views.

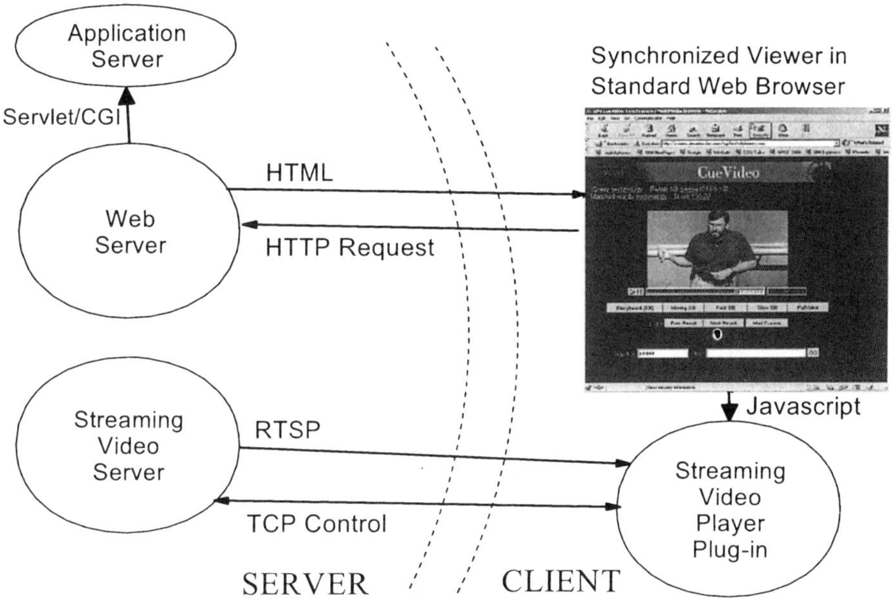

Figure 1.4. Server client architecture of the synchronized media browser with multiple views.

The second issue corresponds to the actual seek process. We used the APIs of the Real Media player to implement the seek. In this player, a seek can only be performed while in Play mode. However, after initiating a Play request it takes some time before the player starts to play, and this time varies depending on the connection speed and other variables. Hence a subsequent seek command is likely to arrive too soon and be ignored. The seek algorithm uses a timer to wait for the player to reach its Play mode, and only then invokes the seek request. This overcomes the problem, however a better design of the client API-s would make a better solution. More details can be found in [Srinivasan *et al.*, 2001].

While all the streaming views play in the embedded player of Figure 1.5 (left), the storyboard is displayed in a separate window (Figure 1.5 (right)). A user may select one of the keyframes and it would take him to the beginning of that shot in the video or the selected view.

A Mail Current button allows users to capture any point in a video and send it as a URL via email, keep it as a bookmark, embed it as a hyperlink in a document, etc. The mail window is shown in Figure 1.5 (left). This is useful in collaborative environments, distance learning, etc.

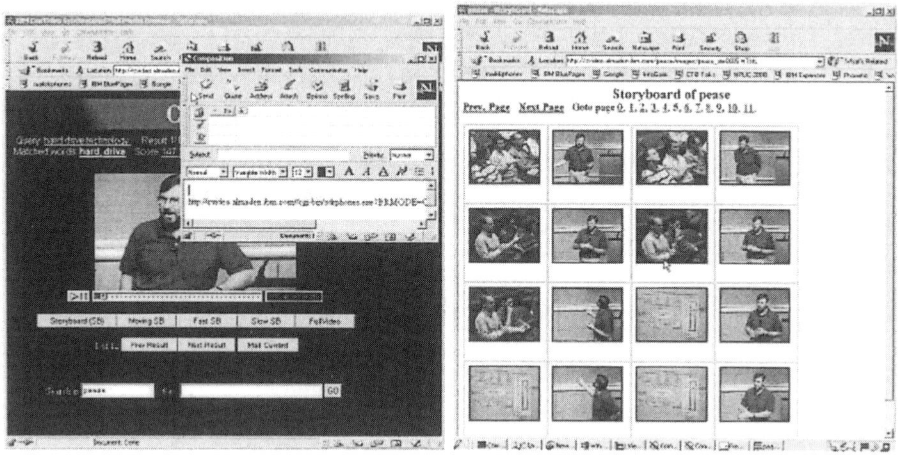

Figure 1.5. left: synchronized media browser with multiple views. The buttons below the player allow to switch between views, to scroll through the search results, and to generate a URL link of the current point in the current video and mail it (see the popup mail window). Right: one of the views is a typical storyboard. Selecting a frame starts the player at the beginning of the corresponding shot.

1.7 Browsing Multiple Videos: MovieDNA

Navigating large collections of linear data is a big challenge, because it is impossible to directly apprehend a linear media type (video, music) at one glance. Indirectly, such a one-glance assessment of linear data can be achieved through an abstraction of the data. However, unlike the case of a single document, where storyboards and other views are found very efficient, when multiple videos are to be browsed to find the relevant ones, it is desired to be able to navigate quickly between the videos without spending much time on diving into each one of them.

We designed and built a visualization and navigation tool for large video collections, called movieDNA, which provides an overview of video data through an abstraction and permits easy navigation in the data as well. Currently there are not many tools or approaches that support navigation through large amounts of video data that are easy to use and keep the user within context. We achieved this by focusing on the three fundamental questions of navigation in any type of information space:

1 Where am I?

2 Where can I go?

3 Where is X?

The movieDNA is based on our work on Context Lenses [Dieberger and Russell, 2000]. Context Lenses were inspired by earlier work by Hearst [Hearst, 1995]. The term movieDNA was chosen because our visualization shows some visual similarities to DNA prints.

The movieDNA and hierarchical brushing can serve as navigation tool either for a single long video, or for a larger collection of videos. The movieDNA approach is general enough so that it can be applied to any kind of linear data, like video, music, event logs, etc. Here we focus on its application to the domain of video data. One example is a set of videos retrieved as a result of a query, as illustrated in Figure 1.6. It shows three videos, represented by the three blocks on the left, having a vertical time line. Each block contains a small matrix. Lines represent video segments (10 minutes in this example). The labeled columns represent attributes of those videos. Those could be derived from keywords of the query, keywords and topics which are automatically derived from speech [Ponceleon and Srinivasan, 2001], or may be selected by the user. A cell's color tells about the relevancy of the (column) feature to the (row) video segment.

Users brush through the video by moving the mouse cursor over the movieDNA. Every line in the DNA triggers display of a fold-out window to the right of the corresponding line. At the first hierarchical level each video is a single selectable unit. At the next level, the selected video expands into a longer DNA, providing finer details of video segments. At the last and finest level, a keyframe is shown along with other metadata associated with this video segment.

The aggregated DNAs of the hierarchical movieDNA aim to convey as much information about the movieDNA of the second level as possible. Currently our aggregation function is relatively straightforward: we aggregate several (5) DNA lines into one line by counting occurrences of features and visualizing the count as gray levels. This way, if a feature occurs in several of the 5 aggregated lines of the DNA this feature shows up strongly in the top level DNA (meaning: it will show strong information scent). Logical extensions of this model would consider weights of features, relevance rankings, etc. In addition to the aggregation step, the aggregated DNAs are drawn at a smaller scale.

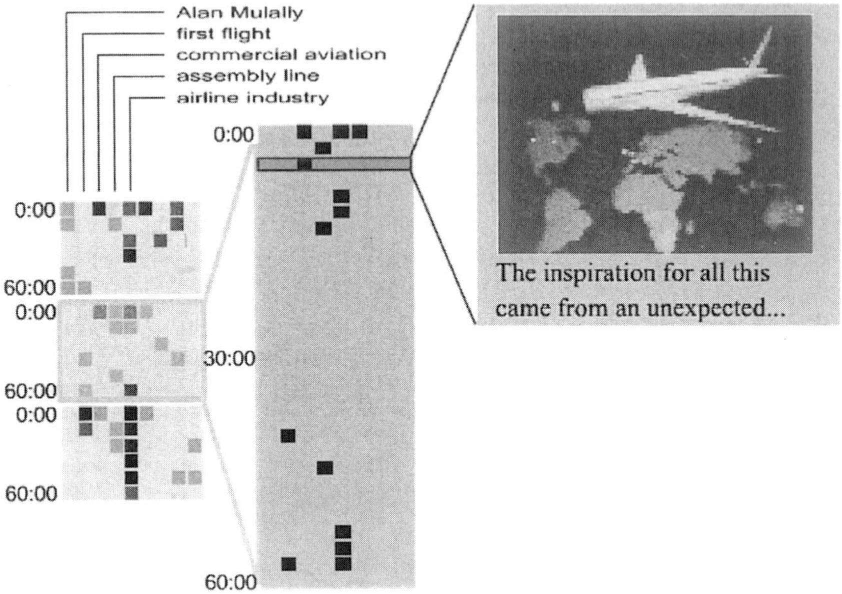

Figure 1.6. The hierarchical movieDNA allows efficient navigation and browsing through many hours of video. Users brush through the video by moving the mouse cursor over the DNA.

A movieDNA may display an arbitrary number of features for the entire video collection. In our current implementation the aggregated DNAs show the same number of features as the second level DNAs. However, this is not an iron cast rule. Just as we aggregated several segments into one, it might be feasible to aggregate several related features into one, in order to achieve an even more compact representation on the first level. Such a design has to ascertain that the first level contains enough information scent to allow users to navigate.

1.8 Usability Study

A usability study was conducted to evaluate the effectiveness of different views to video browsing. We study how audio TSM affects speech comprehension in several aspects. The main motivating questions were

- TSM effect on comprehension How does the audio speedup affect speech comprehension? It is expected that the level of comprehension will degrade when play speed accedes some "optimal" speed.
- Views effect on comprehension Is video better understood than speech alone? In other words, is there any difference between play-

ing the full video (in regular or fast speed) compared to playing the audio only, without the images, at the same speed? How is the MSB doing relative to these two extreme cases?

1.8.1 Usability Study Procedure

We selected three different videos (broadcast news, a technical talk and a company technical training video). Eight short (10 seconds) clips and eight long (30 seconds) clips were extracted from each video. The 48 segments were selected to contain mostly speech (in English), and to include complete sentences. Each of the clips was then modified by two factors; speed and media. Each clip was transformed to four different speeds, selected from one of these two (interlacing) sets of speeds: $[0.6944, 1.0000, 1.4400, 2.0736]$ and $[0.8333, 1.2000, 1.7280, 2.4883]$. For each speed, the clip was generated in three views (media forms): full video, MSB, and audio-only. This resulted in twelve variations of each original clip, or a total of 576 video/MSB/audio clips (1.3 GByte).

The study includes 24 human subjects, 17 men and 7 women, ages between 20-50. Only 10 subjects were native English speakers, while the other 14 have English as second language (see Table 1.4).

Each subject listens to one variation of each of the 48 original clips. The clips order, speed and media are mixed and reordered for a group of 12 subjects in a balanced way, so that

1 each subject listens to all combinations of eight different speeds and three different media the same number of times,

2 each of the 576 clips is watched once, and

3 the playing order of the 48 clips for each subject is different, and is balanced across users.

There was no training stage before the test. A subject plays the 48 media clips one by one using the graphical user interface shown in Figure 1.7. After watching and listening to a clip, the subject completes an interactive three questions form, shown in Figure 3.

1 Speed Assessment In the first question the user is asked how comfortable he/she feels with the clip speed (the actual speed is not revealed to the user). The five discrete grades are between -2 and 2, and are labeled from "much too slow" to "much too fast", respectively. The "optimal" value is 0 ("acceptable").

2 Subjective Comprehension In the second question the user is asked about his comprehension of the speech. The grades are from 1 to 5, and are labeled from "not at all" to "completely", respectively.

3 Objective Comprehension In the last question, the user is asked to type and summarize the main points in the clip. After all the subjects have done with the test, the summaries were manually graded with a value between 1-5 (similar to question 2).

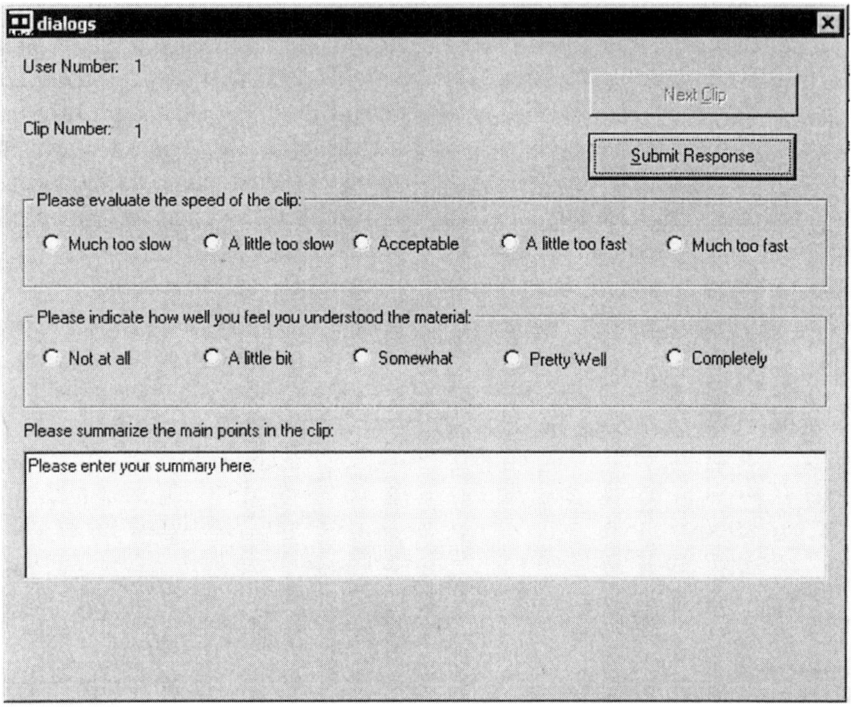

Figure 1.7. User interface used for the usability study. It first plays the clip in a separate window, then closes the player and opens this form for typing.

The time it takes for a subject to grade 48 clips is between an hour and an hour and a half. Unlike [Omoigui *et al.*, 1999], here the subjects were not put under any time pressure to complete a task, and could not select the speedup rate, nor the media type. The collected data was then sorted in the form of a 1152 rows by 7 columns table. Each line corresponds to one clip observed by one user and includes the four clip factors and the three user answers. In the rest of this section we discuss the main results that are derived from this data.

User's Speed Assessment. The speed assessment results are shown in Figure 1.8. The user feeling about the speedup ratio is plotted against the actual speedup ratio. The graphs show the average speed assessment over all clips from the same video (a), and over all clips from the same media type (b). Apparently, all graphs fit a linear model very well. It suggests that the user speedup assessment increases linearly with the logarithm of actual speedup ratio, as defined by Equation 1.4

$$S_{user}(S) = a\ log\frac{S}{S_0}, \tag{1.4}$$

where S denotes the actual speedup ratio, a is a constant which depends on the video and its domain, and S_0 denotes the *natural speed* of the video and is defined as the speedup ratio at which the linear model cross zero. This is the preferred speedup ratio for the video for the average user.

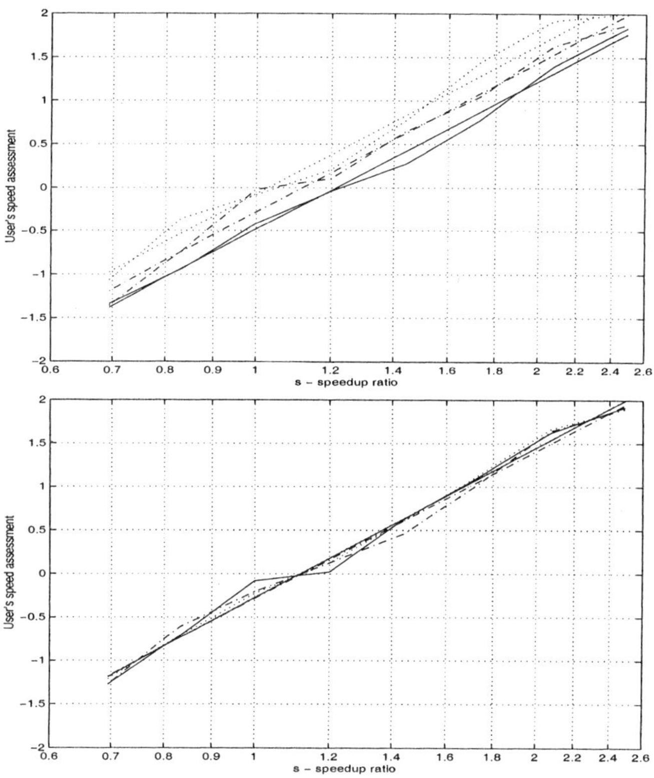

Figure 1.8. The average user assessment of audio speedup is plotted against the actual speedup ratio. Each measurement graph is accompanied by a straight line, representing the logarithmic speedup perception model (Equation 1.4). (a) The three graph pairs correspond to news (solid), training (dashed) and technical talk (dotted), each has a different natural speed. (b) The three graph pairs correspond to media types; audio (solid), MSB (dashed) and video (dotted), showing no significant difference (on average).

The estimated (LMS) model parameters for the graphs are summarized in Table 1.3. The video with fastest natural speed of 1.22 is News, and the slowest is of 1.03 is found for the technical talk. Hence there is no single "average" speedup factor that will be optimal for all types of videos. However, if one would first speedup each video to its natural speed, and then repeat this study with those "normalized" videos, then all three would produce the same speedup assessment graph. In other words, it suggests that for any given video there is a fixed natural speed that can be used to "normalize" the video by speeding it up at this rate, S_0, to produce a unified speedup assessment across videos with the natural speed being 1.0 after this normalization.

Domain	a	S_0	RMS error
News	2.46	1.22	0.05
Talk	2.48	1.03	0.16
Training	2.49	1.12	0.12

Media	a	S_0	RMS error
Audio	2.49	1.12	0.08
MSB	2.49	1.12	0.03
Video	2.44	1.12	0.08

Table 1.3. Model parameters for the graphs in Figure 1.8, using the model of Equation 1.4.

The graphs in Figure 1.8(b) shows the average speed assessment over all audio clips, MSB clips and video clips. As before, the "optimal" speedup is where the graph cross zero. Apparently, there is no measurable difference between the three graphs, and the three linear models applied to the three graphs are essentially the same (see Table 1.3). It means that the average user speed assessment is independent of the media. However, it does not project on an individual user. As shown in Table 1.4, for some users there is a significant advantage for one type of media over the other. Some people are better with the visual information while other are better with the audio only view. More studies need to be done to further explore these differences. However, from a design point of view, it is desired to provide the user with the flexibility to choose whatever view type he likes to use.

User Speech Comprehension. The speech comprehension results are shown in Figure 1.9. Graph (a) shows the average user subjective recall over all clips from the same video, for the three different videos, news, technical talk and training. Graph (b) shows the average user objective recall over all clips from the same video, for the same three videos. Apparently there is a significant difference between the comprehension of the talk to the other two videos. It is evident in both subjective and objective comprehension. Interestingly, for the news clips and for the training clips, comprehension remains at a high level up to speedup of

User ID	Preferred speedup ratio All	Audio	MSB	Video	Gen.	Age grp	English level	Rank A	M	V
11	1.4754	1.4322	1.5331	1.4610	M	1	2	3	1	2
01	1.4110	1.5238	1.3816	1.3275	M	4	native	1	2	3
19	1.3547	1.3795	1.3649	1.3198	M	1	native	1	2	3
22	1.2780	1.2886	1.2810	1.2645	M	4	2	1	2	3
18	1.2499	1.1486	1.2828	1.3183	F	2	native	3	2	1
14	1.2491	1.3941	1.1620	1.1911	M	3	native	1	3	2
13	1.2483	1.2347	1.2879	1.2222	M	2	2	2	1	3
23	1.2214	1.1795	1.2230	1.2617	F	1	native	3	2	1
15	1.2159	1.2240	1.1918	1.2320	F	1	native	2	3	1
16	1.2146	1.1903	1.2580	1.1954	M	1	2	3	1	2
06	1.2027	1.0842	1.2315	1.2924	M	1	1	3	2	1
20	1.1932	1.2087	1.2741	1.0968	M	3	1	2	1	3
04	1.1768	1.1767	1.2089	1.1447	M	2	1	2	1	3
12	1.1575	1.0663	1.0968	1.3093	M	1	1	3	2	1
03	1.1544	1.2234	1.1219	1.1179	M	2	3	1	2	3
08	1.1342	1.2293	1.1046	1.0688	M	2	1	1	2	3
24	1.1278	1.1665	0.9470	1.2700	M	3	native	2	3	1
10	1.1228	1.0822	1.1093	1.1770	F	1	2	3	2	1
17	1.1216	1.2075	1.1082	1.0491	M	1	3	1	2	3
21	1.1001	0.9119	1.1659	1.2224	M	2	1	3	2	1
07	1.0681	1.1329	0.9099	1.1616	F	1	native	2	3	1
09	1.0527	1.0607	1.0080	1.0893	M	6	1	2	3	1
02	1.0454	1.1695	0.9453	1.0214	F	3	native	1	3	2
05	1.0357	1.0270	1.0861	0.9940	F	1	native	2	1	3

Table 1.4. Preferred audio, MSB and video speedup ratios for all subjects, sorted by average speedup ratio (All). There is no consensus about a single, widely preferred media among video, MSB and audio. Different people perform better with different views. Hence it is desired to provide users with several choices to select from.

1.4. That is, although the users would prefer to play it slower, at the natural speed, they still don't miss anything at this rate. The difference between graph (a) and graph (b) suggests that while subjects feel uncomfortable with the higher speed, they still do not loose the content until about $1.7x$.

1.9 Summary

Video browsing has an important role in any video retrieval system. Search and browse are two tightly coupled functions in most information systems, from traditional databases, through unstructured textual information and in particular for multimedia documents such as speech and video. This chapter covered some of the techniques for supporting

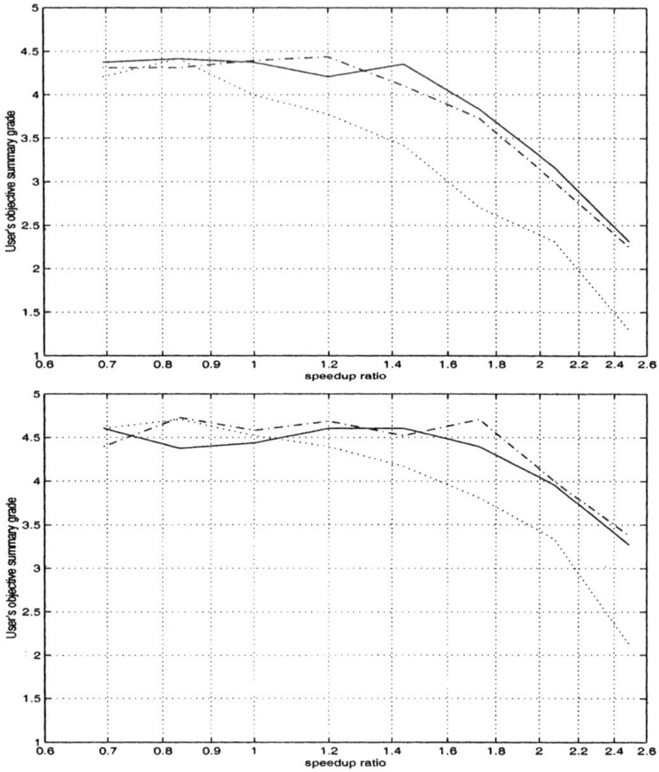

Figure 1.9. (a) Subjective comprehension decreases with the speedup ratio after 1.5×. (b) Assessed comprehension decreases with the speedup ratio after 1.7×. The three graph pairs correspond to the three domains: news (solid), training (dashed) and technical talk (dotted), the later appears to be more difficult to some of the subjects.

fast and efficient browsing of multiple videos and within videos. The main properties of several views are summarized in Table 1.5.

	Visual			Typical speedup
View	Static (images)	Dynamic (motion)	Audio	ratio
Full video (w/o TSM)		+	+	1 – 2×
Video Skim		+	+	2 – 20×
Slide show (w/o TSM)	+		+	1 – 2×
Adaptive Fast Playback		+		5 – 30×
Animation	+			10 – 40×
Storyboard, mosaic	+			NA

Table 1.5. Several views and their properties.

All streaming views are made synchronized by a synchronized video browser, allowing the user to switch between views while keeping the context. This requirement imposes several engineering challenges with current streaming technology. We believe that in the near future these problems will be addressed by the rapidly evolving multimedia technology for the Internet and for the home entertainment market.

The NIST TREC Video Track is the *de facto* benchmark in video retrieval. It provides a benchmark for video retrieval. The year 2002 results for Manual and Interactive search show a very significant improvement in retrieval MAP score that was achieved by browsing of the search result. While the video track continues to promote new research in visual indexing, it also recognizes the power of browsing as a major vehicle for information retrieval in video collections. Until the huge challenge of visual indexing is resolved, browsing will continue to play a very dominant role in video information retrieval.

References

B. Adams, A. Amir, C. Dorai, s. Ghosal, G. Iyengar, A. Jaimes, C. Lang, C-Y. Linx, A. Natsev, M. Naphade, C. Neti , H. J. Nock, H. H. Permuter, R. Singh, J. R. Smith, s. Srinivasan, B. L. Tseng, A. T. V.z , D. Zhang, "IBM Research TREC-2002 Video Retrieval System", in TREC-11 Video Track, Maryland Nov. 20-22, 2002.

P. Aigrain, H. Zhang and D. Petkovic, "Content-Based Representation and Retrieval of Visual Media: A State-of-the-Art Review", in Multimedia Tools and Applications, Vol. 3, pp. 179-202, Kluwer Academic Publishers, 1996.

A. Amir, G. Ashour and S. Srinivasan, "Towards Automatic Real Time Preparation of On-Line Video Proceedings for Conference Talks and Presentations", in Video Use in Office and Education, Thirty-Fourth Hawaii Int. Conf. on System Sciences, HICSS-34, Maui, January 2001.

A. Amir, S. Srinivasan and A. Efrat, "Search The Speech, Browse The Video - A Generic Paradigm for Video Collections", in EURASIP Journal on Applied Signal Processing, Special Issue on Unstructured Information Management from Multimedia Data Sources, EURASIP JASP 2003:2 (2003) pp. 209-222.

A. Aner and J. R. Kender, "Video Summaries through Mosaic-Based Shot and Scene Clustering", Proc. European Conference on Computer Vision, Denmark, May 2002.

L. H. Armitage and P. G. B. Enser, "Analysis of user need in image archives. Journal of Information Science 23(4), 1997, 287-299.

J. R. Bach et al., "Virage image search engine: An open framework for image management", in Proceedings of SPIE Storage and Retrieval for Still Images and Video Databases IV, Vol. 2670, IS&T/SPIE, February 1996.

R. Baeza-Yates and B. Ribeiro-Neto, *Modern Information Retrieval*, Addison-Wesley, 1999.

R. M. Bolle, B. L. Yeo and M. M. Yeung, "Video Query: Research Direction", IBM Journal of Research and Development, March 1998.

P. Browne, C. Czirjek, C. Gurrin, R. Jarina, H. Lee, K. Mc Donald, A. F. Smeaton and J. Ye, "Dublin City University Video Track Experiments for TREC 2002", in TREC-11 Video Track, Maryland Nov. 20-22, 2002.

R. Brunelli, O. Mich and C. M. Modena, "A Survey on Video Indexing", IRST Technical Report 9612-06.

K. E. Chang, G. Goh Sychay and W. Gang, "CBSA: content-based soft annotation for multimodal image retrieval using bayes point machines", Circuits and Systems for Video Technology, IEEE Transactions on ,Volume: 13 Issue: 1 , Jan 2003 Page(s): 26 -38.

S.F. Chang, W. Chen, H. J. Meng, H. Sundaram and D. Zhong, "VideoQ: An Automated Content Based Video Search System Using Visual Cues", in Proceedings of MM'97, pp. 313-324, ACM Press, November 1997.

S. Chang, A. Eleftheriadis and R. McClintock, "Next Generation Content Representation, Creation and Searching for New Media Applications in Education", Proceedings of IEEE, Special Issue on Multimedia Signal Processing, Part One, pp. 884 - 890, May 1998.

M.G. Christel, M.A. Smith, C.R. Taylor and D.B. Winkler, "Evolving Video Skims into Useful Multimedia Abstractions", in Proceedings of CHI '98 (Los Angeles CA, April 1998).

A. Del Bimbo, **Visual Information Retrieval**, Morgan Kaufmann, San Francisco, CA 1999.

Dieberger, A. and Russell, D.M., Context Lenses - Document Visualization and Navigation Tools for Rapid Access to Detail, ACM CHI 2000.

P.G.B. Enser, Query analysis in a visual information retrieval context. Journal of Document and Text Management, 1(1), 1993, 25-52.

J. Garofolo, C. Auzanne, and E.Voorhees, "The TREC Spoken Document Retrieval Track: A Success Story", in E. Voorhees, Ed., NIST Special Publication 500-246: The Eighth Text REtrieval Conference (TREC 8), Gaithersburg, Maryland, November 1999.

A. Gupta, and R. Jain, "Visual information retrieval", in Communications of the ACM 40, 5 (May 1997), Pages 70 - 79.

A. Hauptmann and M. Witbrock, "Informedia: News-on-demand multimedia information acquisition and retrieval", in Intelligent Multimedia Information Retrieval, chapter 10, pp. 215–240. MIT Press, Cambridge, Mass., 1997.

L. He, A. Gupta, S. A. White and J. Grudin, "Corporate Deployment of On-demand Video: Usage, Benefits, and Lessons", MSR-TR-98-62, Microsoft Research, Redmond, WA 98052.

Hearst, M.A., TileBars: Visualization of terms distribution information in full text information access. In Proc. of the ACM SIGCHI Conference on Human Factors in Computing Systems, p. 59-66, Denver, CO, May 1995.

ISO/IEC 14496: Information technology - coding of audio-visual objects (final draft international standard), 1999.

G. J. F. Jones, J. T. Foote, K. S. Jones and S. J. Young, "Video Mail Retrieval: the effect of word spotting accuracy on precision", in Proceedings of ICASSP 95, volume 1, pp. 309-312, Detroit, MI.

G. J. F. Jones, J. T. Foote, K. S. Jones and S. J. Young, "Retrieving Spoken Documents by Combining Multiple Index Sources", in Proceedings of SIGIR 96, pp. 30-38, Zurich, Switzerland.

R. Lienhart, "Reliable Transition Detection in Videos: A Survery and Practitioner's Guide", International Journal of Image and Graphics, Vol. 1, No. 3 (2001) 469-486.

F.C. Li, A. Gupta, E. Sanocki, L.W. He and Y. Rui "Browsing Digital video", CHI-00, pp. 169-176.

D. Malah, "Time-domain algorithms for harmonic bandwidth reduction and time scaling of speech signals", IEEE Transactions on acoustics, speech and signal processing, ASSP-27(2) pp.121-133, April 1979.

M. T. Maybury, *Intelligent Multimedia Information Retrieval*, MIT Press, 1997.

M. Naphade, T. Kristjansson, B. Frey, and T. S. Huang, "Probabilistic multimedia objects (multijects): A novel approach to indexing and retrieval in multi-media systems," in Proceedings of IEEE Int. Conf. on Image Processing, Chicago, IL, Oct. 1998, vol. 3, pp. 536-540.

N. Omoigui, L. He, A. Gupta, J. Grudin and E. Sanocki, "Time-compression: systems concerns, usage, and benefits", CHI-99, pp. 136-143, 1999.

P. Over and R. Taban. "The TREC-2001 video track framework", in 10th Text Re-trieval Conference (TREC-10), Gaithersburg, Maryland, USA, November 2001.

F. Pereira, "MPEG-7: A standard for content-based audiovisual description", Proc. of Int. Conference on Visual Information Systems (VISUAL'97), San Diego, CA, Dec. 1997.

D. B. Ponceleon and S. Srinivasan, "Structure and Content-Based Segmentation of Speech Transcripts", ACM SIGIR 2001, 404-405.

S. Roukos and A. M. Wilgus, "High quality time-scale modification for speech", IEEE Int. Conference on Acoustics, Speech, and Signal Processing, ICASSP-85, pp. 493-496, 1985.

S. Shatford, "Analyzing the Subject of A Picture: A Theoretical Approach", Library of Congress, Cataloging and Classification Quarterly, Vol. 6, 1985.

A. Smeaton and P. Over, "The TREC 2002 Video Track Report", in 11th Text Re-trieval Conference (TREC-11), Gaithersburg, Maryland, USA, November 2002.

M. A. Smith and T. Kanade, "Video Skimming for Quick Browsing Based on Audio and Image Characterization", Carnegie Mellon University, Technical Report CMU-CS-95-186, July 1995.

S. Srinivasan, D. Ponceleon, A. Amir, B. Blanchard, D. Petkovic, "Engineering the Web for Multimedia", in S. Murugesan and Y. Deshpande (editors), Web Engineering: Managing Diversity and Complexity in Web Application Development, LNCS Vol. 2016, Springer-Verlag, April 2001.

S. Srinivasan and D. Petkovic, "Phonetic Confusion Matrix Based Spoken Document Retrieval. In Proceedings of SIGIR-2000, Greece, July 2000.

B. L. Tseng and C.-Y. Lin, "Personalized Video Summary using Visual Semantic Annotations and Automatic Speech Transcriptions", IEEE Intl. Workshop on Multimedia Signal Processing, US Virgin Islands, Dec. 2002.

H. Valbret, E. Moulines and J.P. Tubach, "Voice transformation using PSOLA technique", Speech Communications, 11:175-187, 1992.

W. Verhelst, M. Roelands, "An overlap-add technique based on waveform similarity (WSOLA) for high quality time-scale modification of speech", IEEE International Conference on Acoustics, Speech, and Signal Processing ICASSP-93, Vol.II, pp. 554-557, 1993.

H. D. Wactlar, M. G. Christel, Y. Gong and A. G. Hauptmann, "Lessons Learned from Building a Terabyte Digital Video Library", IEEE Computer, pp.66-73, Feb 1999.

M. M. Yeung, B. L. Yeo, W. Wolf and B. Liu, "Video Browsing using Clustering and Scene Transitions on Compressed Sequences. In Multimedia Computing and Networking Proc. SPIE, February 1995.

Chapter 2

BEYOND KEY-FRAMES: THE PHYSICAL SETTING AS A VIDEO MINING PRIMITIVE

Aya Aner-Wolf
Department of Computer Science and Applied Math
Weizmann Institute of Science, Israel
aya@wisdom.weizmann.ac.il

John R. Kender
Department of Computer Science
Columbia University
jrk@cs.columbia.edu

Abstract We present an automatic tool for the compact representation, cross-referencing, and exploration of long video sequences, which is based on a novel visual abstraction of semantic content. Our approach is based on building a highly compact hierarchical representation for long sequences. This is achieved by using non-temporal clustering of scene segments into a new conceptual form grounded in the recognition of real-world backgrounds. We represent shots and scenes using mosaics derived from representative shots, and employ a novel method for the comparison of scenes based on these representative mosaics. We then cluster scenes together into a more useful higher level of abstraction – the physical setting. We demonstrate our work using situation comedies, where each half-hour (40,000-frame) episode is well-structured by rules governing background use. Consequently, browsing, indexing, and comparison across videos by physical setting is very fast. Further, we show that the analysis of the frequency of use of these physical settings leads directly to high-level contextual identification of the main plots in each video. We demonstrate these contributions with a browsing tool which allows both temporal and non-temporal browsing of episodes from situation comedies.

Keywords: Video Indexing, mosaic matching, cross-referencing, clustering, main plots, physical settings, situation comedy, sitcoms, sports broadcasts, news broadcasts, temporal browsing, non-temporal browsing, compact representation, visual abstraction, scene segmentation, video summarization, rubber sheet matching, grammar rules, image alignment, image registration, distance measure.

Introduction

Video media has a strong expressive power which should be efficiently managed. There exists a large amount of video material covering a wide variety of needs, e.g. entertainment, education and more. However, storing large amounts of this media has no benefit if there are no efficient tools for searching and browsing it.

A common approach for organizing video data is the structured approach, where the video is segmented into temporal segments, such as shots and scenes. Such temporal segments generate a more efficient representation of video and ease with its classification. They also capture the highly structured organization according to which many video media are constructed. This linear layout of the video data is more effective and meaningful for browsing purposes, since people tend to recall video in a structured view.

However, people do not perceive video media simply as a linear list of shots or scenes. They would remember and classify videos according to interesting events or concepts that the whole video presented. Automatically capturing such concepts or events might be an impossible task, and would serve as an example for the widely known "semantic gap" problem which exists in many image and video indexing and browsing applications [Smeulders et al., 2000].

We present a new approach for the representation of long video sequences, which aims to bridge over the described semantic gap problem, and which is applicable in many video genres. Our approach presents a new concept, the **"physical setting"**. In contrast to the known temporal segmentation into shots and scenes, the "physical setting" segmentation is based on a non-temporal representation of video. This approach provides not only the most compact representation for long videos, but it also enables the extraction of highly semantic concepts regarding the whole video. Our proposed approach relies on the basic temporal segmentation into shots or/and scenes, and applies the non-temporal segmentation on top if this structure.

Our approach results in a complete semantic hierarchical summarization structure, and could be displayed as a tree-like representation in which each level is more compact. As an example, for a tree representing a half-hour long video segment (a single episode from a situation comedy that we explored), the bottom frame level is composed from approximately $50K$ image frames. The next level is the "shot" level, containing approximately 300 shots, and the following level up the tree is the "scene" level, containing approximately 15 scenes. The highest level of our tree is the "physical settings", with 5-6 representative images. We will now explain each level of the tree in detail.

The first temporal segmentation of the video is its segmentation into shots, which composes the second level of the tree. A shot is a sequence of consecutive frames taken from the same camera, and many algorithms have been proposed for shot boundary detection (e.g. [Lienhart, 1999; Gelgon and P.Bouthemy, 1998; Aner and Kender, 2001]).

Segmenting video into its shot structure is a fundamental procedure in analyzing video. Many video indexing or video browsing applications are implemented using only this structure. This is appropriate for some video genres which lack any higher structure. However, many video genres (e.g. television programs) are well structured and for them further compacting of their summarized representation would be beneficial.

The second temporal segmentation, and the third level in the tree, is the segmentation of shots into scenes. We use the following definition for a scene: a collection of consecutive shots, which are related to each other by the same semantic content. In our examples, consecutive shots which were taken at the same location and that describe an event or a story which is related in context to that physical location are grouped together into one scene. Several suggestions for scene boundary detection include [Hanjalic et al., 1999; Kender and Yeo, 1998] or, as it is also referred to, as "story units" detection [Yeung and Yeo, 1996]. We rely on previous work [Kender and Yeo, 1998] for generating this level of the tree.

The last and highest level is the "physical settings" level. These segments are constructed as groups of scenes, each group taking place in the same physical setting. These settings are well defined in many video genres. We chose to demonstrate our work using situation comedies, or in short, sitcoms. Directors of sitcoms are limited in space and time, and therefore each episode of a sitcom takes place in only 5-6 different physical settings. There are often 2-3 settings which re-occur in many episodes of a specific sitcom. These settings are therefore directly related to the main **theme** of the sitcom. In contrast, each episode has 2-3 physical settings which are unique to that episode, and infer the main **plots** of this episode.

By identifying the physical settings in each episode and comparing these settings across several episodes, we are able to **automatically** determine the main story lines or "plots" in each episodes, as will be explained in detail further on. More generally, this well formed structure is typical of sitcoms and by automatically identifying this structure in half-hour long video sequences, we have a strong reliable tool for automatically detecting that a certain video belongs to the "genre class" of sitcoms.

Many existing video indexing and browsing applications rely on frames for visual representation (mainly of shots). We show that mosaics are more reliable for representing shots than key-frames, since they represent the physical settings more clearly. We therefore use mosaics to represent scenes and physical settings, and show their advantage in clustering shots and scenes. A mosaic is an image which is generated from a sequence of frames which we sample from a single shot. We compute the camera-induced motion for these frame which enables the alignment of the image frames and their projection into a single image (either a plane or another surface). In the mosaic creation process, moving objects are detected and masked out so that the constructed mosaic image contains only the static background.

We represent each scene using a subset of all the mosaics associated with the shots of that scene. This group of mosaics is carefully chosen according to cinematography rules and camera characteristics to best describe the scene. In order to represent physical settings, further analysis based on clustering is used.

In our work we use only the background information of mosaics, since this information is the most relevant for the comparison process of shots; it is further used for shot and scene clustering, and for gathering general information about the whole video sequence.

The representation of each shot could be further enriched by using additional foreground information (e.g. characters). Examples of complementary tools suitable for this task are the synopsis mosaic presented in [Irani and Anandan, 1998] or the dynamic mosaic presented in [Irani et al., 1996], as well as motion trajectories shown in [Bouthemy et al., 1999].

We use examples from several sitcoms and demonstrate our contributions with a browsing tool specifically constructed for sitcoms. It allows both non-temporal browsing of episodes and fast indexing into different scenes of specified physical settings. We also show how the semantic information incorporated in physical settings could be retrieved and used even in other video genres which are not well-structured as sitcoms. The use of repeating physical settings appears in many motion picture films, due to either budget constraints or specific styles of film grammar. In sports videos, our approach allows classifying shots for characteristic event detection.

To summarize, our proposed mosaic-comparison method, along with the newly proposed semantic approach for video summary, results in a strong and reliable tool for indexing and browsing. It is both fast and accurate and general enough for the comparison of whole scenes across different videos. It therefore serves as a basis concept for efficient indexing and browsing of very large video libraries.

2.1 Mosaics of Shots and Scenes

2.1.1 Motivation

Since in many video genres the main interest is the interaction between the characters, the physical location of the scene is visible in all of the frames only as background. The characters usually appear in the middle area of the frame, hence, the background information can only be retrieved from the borders of the frame. Oh and Hua noticed this and segmented out the middle area of the frame. They used color information of the borderlines of the frames for shot transition [Oh et al., 2000a], and later scene transition [Oh et al., 2000b] detection. However, when key-frames are used to represent shots and to compare them, the information extracted from a single key-frame is not sufficient. Most of the physical location information is lost since most of it is concealed by the actors and only a fraction of the frame is used.

Consider the example in Figure 2.1(a), which shows a collection of key-frames taken from a panning shot. In that shot an actress was tracked walking across a room. The whole room is visible in that shot, but the actress appears in the middle of all key-frames, blocking significant parts of the background. This effect becomes worse in close-up shots where half of the frame is occluded by the actor. The solution to this problem is to use mosaics to represent shots. A mosaic of the panning shot discussed above is shown in Figure 2.1(b). The whole room is visible and even though the camera changed its zoom, the focal length changes are eliminated in the mosaic.

In order to construct color mosaics using either projective or affine models of camera motion, and in order to use mosaics to represent temporal video segments such as shots, we make several assumptions. We assume that either the 3D physical scene in the video genres that we analyze is relatively planar, or that the camera motion is relatively slow, or that the relative distance between the surface elements in the 3D plane is relatively small compared with their distance to the camera. For the frame sequences out of which we construct the mosaics, we assume that

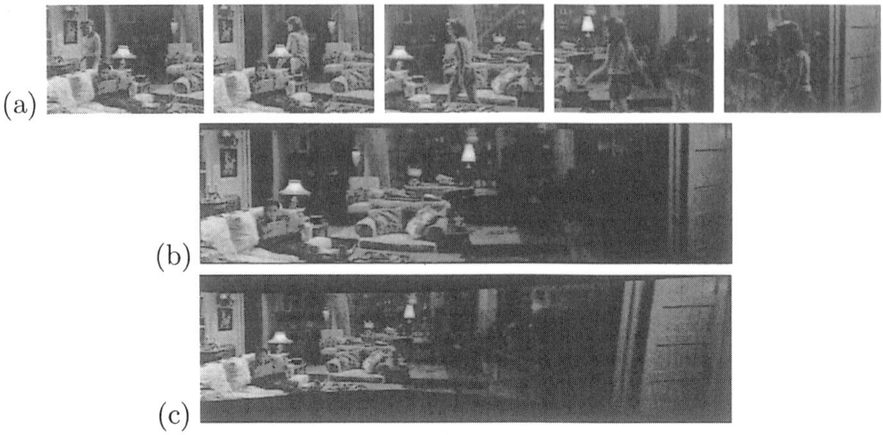

Figure 2.1. (a) Hand-chosen key-frames. (Automatic key-frames generation often does not give complete spatial information). The key-frames are ordered in a temporally reversed order (to match the camera panning direction from right to left). (b) Mosaic representation using affine transformations computed between the frames. Note that the whole background is visible, without occlusion by the foreground objects. (c) Mosaic representation using projective transformations computed between the frames.

the cameras are mostly static, and that most camera motion is either translation and rotation around the main axis, or zoom. Since, in the video genres we analyzed, the physical set is limited in size, the camera's movements and various positions are constrained. In cases where this does not hold for the whole duration of a shot, the shot is divided into several segments such that the mosaic constructed for each segment has no severe global distortions. We also assume that cameras are positioned horizontally, that is, scenes and objects viewed by them will always be situated parallel to the horizon. Since in our analyzed video genres cameras are usually placed indoors, the lighting changes are not as significant as in outdoor scenes.

2.1.2 Construction of Color Mosaics

The use of mosaics for video indexing was proposed in the past by several researchers, and examples include salient stills [Massey and Bender, 1996], video sprites [Lee et al., 1997], video layers [Wang and Adelson, 1994; Vasconcelos, 1998]. However, there aren't any examples of their use for shot comparison and further use for hierarchical representation of video. We base our *color* mosaic construction technique on the *grey level* mosaic construction method proposed in [Irani et al., 1996], and will describe it here only in a brief manner. The first step is the generation of affine transformations between successive frames in the sequence (in this work we sample 6 frames/sec). One of the frames is then chosen as the reference frame, that is, as the basis of the coordinate system (the mosaic plane will be this frame's plane). This frame is "projected" into the mosaic using the identity transformation. The rest of the transformations are mapped to this coordinate system and are used to project the frames into the mosaic plane. The value of each pixel in the mosaic is determined by the median value of all of the pixels that were mapped into it. We note that even though the true transformations between the frames in some shot sequences are projective and not affine, we still only compute affine transformations. This results in some local distortions in the mosaic, but prevents projective distortion, as seen in Figure 2.1(c). This was found more useful for our mosaic construction method which divides the mosaic into equally sized blocks and compares them, as will be described in the following section.

In contrast to the method described in [Irani et al., 1996], here we construct color mosaics, and therefore need to find for every color pixel in the mosaic the color median of all frame pixels projected onto it. Taking the median of each channel will result in colors which might not have existed in the original frames, and we would like to use only true colors. We therefore convert the frames to gray level images while maintaining corresponding pointers from each gray-level pixel to its original color value. For each mosaic-pixel, we form an array of all values from different frames that were mapped onto it, find the median gray value of that array, and use the corresponding color value pointed to by that pixel in the mosaic. We also use outlier rejection, described in [Irani et al., 1996], to detect and segment out moving objects. This both improves the accuracy of the affine transformations constructed between frames as well as results in "clear" mosaics where only the background of the scene is visible, as shown in Figure 2.1(b).

2.1.3 Mosaic Comparison

Comparing mosaics is not a straightforward task. In video genres where a physical setting repeats several times throughout the video, as is the case for video genres we have tested, it is often shot from different view points and at different zoom levels, and sometimes also in different lighting conditions. Therefore, the mosaics generated from these shots are of different size and shape, and the same physical setting will appear different across several mosaics. Moreover, different parts of the same physical scene are visible in different mosaics, since not every shot covers the whole scene location. In order to determine that a group of mosaics all belong to the same physical setting, we often use information from several scenes for which the mosaics show different parts of the setting. Thus, comparing color histograms of whole mosaics is not sufficiently accurate.

A solution to this problem is to divide the mosaic into smaller regions and to look for similarities in consecutive relative regions. Our basic comparison feature is based on the distance between color histograms of small blocks within each mosaic. We generate a 3D color histogram for each block and set the distance between each pair of block using a metric applied on this histogram.

When comparing mosaics generated from certain video genres, we can make assumptions about the camera viewpoint and placement, noting the horizontal nature of the mosaics. Cameras locations and movements are limited due to physical set considerations, causing the topological order of the background to be constant throughout the mosaics. Therefore, the corresponding regions only have horizontal displacements, rather than more complex perspective changes.

This work is, in a sense, similar to that in the area of wide baseline stereo matching, which relies on detecting corresponding features or interest points. Recent work [Schaffalitzky and Zisserman, 2001] has used texture features to match between corresponding segments in images, which are invariant to affine transformations and does not require extracting viewpoint invariant surface regions. Our method handles significant local distortions in the mosaics, which would cause feature detection techniques to be unreliable. For example, lines and corners might appear too blurred or might even appear twice. Moreover, the local region in the mosaic around a feature point will not necessarily produce the same local invariants as the region corresponding the same background from another sequence since it might be the result of several super-imposed images used to construct the mosaic. Our method also has the advantages that it is not sensitive to occluding objects and global non-linear distortions. Similar to [Schaffalitzky and Zisserman, 2001] it is insensitive to some lighting changes.

2.1.4 Rubber Sheet Matching Approach

We follow the idea of rubber-sheet [Gonzalez and Woods, 1993] matching, which takes into account the topological distortions among the mosaics, and the rubber-sheet transformations between two mosaics of the same physical scene. The comparison process is done in a coarse-to-fine manner. Since mosaics of common physical scenes cover different parts of the scene, we first coarsely detect areas in every mosaic pair which correspond to the same spatial area. We require that sufficient portions of the mosaics will match in order to determine them as similar. Since the mosaic is either bigger or has the same size as the original frames in the video, we demand that the width of the corresponding areas detected for a matching mosaic pair should be not less than approximately the original frame width. The height of this area should be at least $\frac{2}{3}$ of the original frame height (we use the upper part of the mosaic), a reason which is motivated by cinematography rules, concerning the focus on active parts [Arijon, 1976].

After smoothing noise in the mosaics with a 5×5 median filter (choosing color median was described in Section 2.1.2), we divide the mosaics into relatively wide vertical strips and compare these strips. We determine an approximate common physical region in the two mosaics by coarsely aligning a sequence of vertical strips in one mosaic with a sequence of vertical strips in the second mosaic. Our method supports coarse alignment of a sequence of k strips with a sequence of up to $2k$ strips, therefore allowing the scale difference between the mosaics to vary between $1 : 1$ and to be as big as $2 : 1$. An example of such mosaics is shown in Figure 2.6(a). This allows us to match mosaics which were generated from shots taken with different focal lengths. In our experiments there was no need to support a scale difference bigger than $2 : 1$, but with some modifications it would be possible to support larger scale differences. The results of this coarse stage detect candidate similar areas in every mosaic pair and are used to crop the mosaics accordingly. If no corresponding regions were found, the mosaics are determined to be dissimilar. In cases where candidate matching regions are found, we use a threshold, determined from sampled mosaic pairs, to discard mosaic pairs with poor match scores. We then apply a more restrictive matching process on the remaining cropped mosaic pairs. We use narrower strips to finely verify similarities and generate final match scores for each mosaic pair. This second, finer step is necessary since global color matches might occur across different settings, but not usually in different relative locations within them.

Finding Best Diagonal. All comparison stages are based on the same method of finding the best diagonal in a distance matrix, which corresponds to finding horizontal or vertical alignment. Example illustrations of such distance matrices can be found in Figures 2.2 (e), (f), and (h). Let $D(i, j)$ be an $N \times M$ matrix where each entry represents a distance measure. We treat this matrix as a four-connected rectangular grid and search for the best diagonal path on that grid. If

$P\{(s,t) \rightarrow (k,l)\}$ is a path from node (s,t) to node (k,l) of length L, then its weight is defined by the average weight of its nodes:

$$WP\{(s,t) \rightarrow (k,l)\} = \frac{1}{L} \sum_{(i,j)\in(s,t)..(k,l)} D(i,j). \tag{2.1}$$

We search for the best diagonal path P with the minimum weight of all diagonals, that also satisfies the following constraints:

1. Length$(P) \geq T_{length}$.

2. Slope$(P) \leq T_{slope}$.

Where T_{length} and T_{slope} specify thresholds for minimum diagonal length and maximum slope value, respectively. The first constraint is determined by the width and height of the original frames in the sequence which determine the minimum mosaics' size (352×240 pixels in our experiments), since we require that sufficient portions of the mosaics will match. For example, the width of the strips in the coarse stage (discussed below) was set to 60 pixels and T_{length} was set to 5 for the horizontal alignment, so that the width of the matched region is at least 300 pixels. The second constraint relates to the different scales and angles of the generated mosaics due to different camera placements. We have found that the scale difference could be as big as $2:1$, resulting in a slope of approximately $26°$. We therefore examine diagonals of slopes that vary between $25°$ - $45°$ in both directions (allowing either the first mosaic to be wider or the second mosaic to be wider). We use intervals of $5°$, a total of 9 possible slopes. However, in order to determine the weight of each diagonal we have to interpolate the values along this straight line. Bilinear interpolation is time consuming and experiments have proved that nearest neighbor interpolation gives satisfactory results. With the straightforward use of lookup tables and index arithmetics, we were able to implement it more efficiently.

Coarse Matching. In this stage we perform coarse horizontal alignment of two consecutive strip-sequences in a mosaic pair in order to detect a common physical area. The width of the strips is set to be 60 pixels each, since we need no more that 5-6 vertical segments and we want a number that further divides into smaller segments (for the finer stage). In order to align two strip-sequences, we generate a strip-to-strip distance matrix $S[i,j]$, where each entry corresponds to

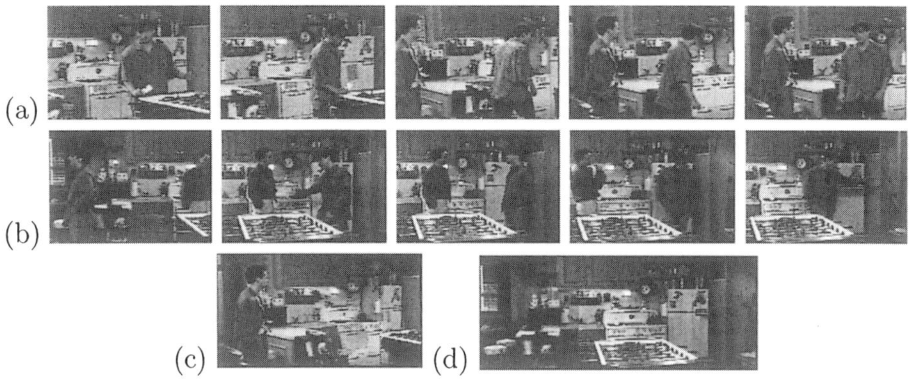

Figure 2.2. (a) Key-frames of the shot from which the mosaic in (c) was generated. (b) Key-frames of the shot from which the mosaic in (d) was generated.

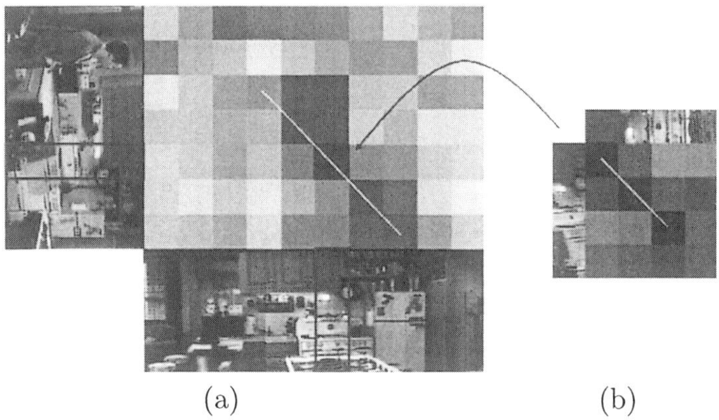

Figure 2.3. (a) Each mosaic is divided into large vertical strips. Each strip-pair (example strips highlighted) is divided into blocks. The values of the B matrix (b) for the marked strips is visualized by grey squares, where each entry represents the difference measure between a single block from the upper strip and a single block from the left strip. Coarse vertical alignment between the strips is determined by the "best" diagonal (thin white line); the average of the values along the diagonal defines the distance between the two strips. The resulting S matrix for the two mosaics is shown in (a), with the strip score from (b) highlighted. Horizontal alignment betwe strip sequences is determined by the "best" diagonal path in S (thin white line).

the distance between a strip s_i from one mosaic to a strip s_j in the second mosaic:

$$S[i,j] = Diff(s_i, s_j). \tag{2.2}$$

where $Diff(s_i, s_j)$ is the difference measure between the strips discussed below. An example of $S[i,j]$ is shown in Figure 2.2(f), where

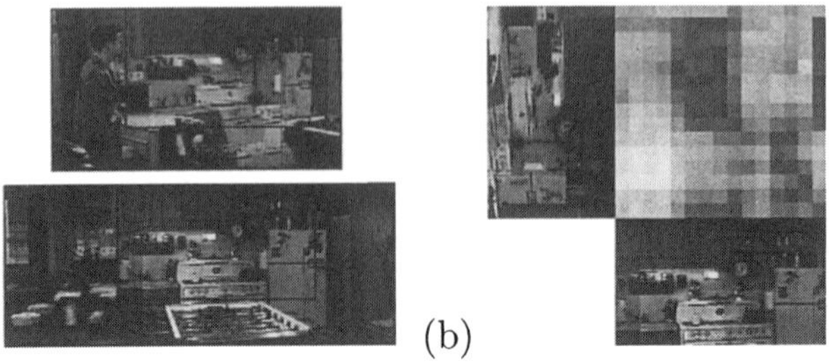

(a) (b)

Figure 2.4. (a) The computed horizontal and vertical alignments (illustrated in Figure 2.3) are used to (a) crop the mosaics to prepare for the verifying finer match in (b), which shows a similar S matrix for the finer stage. Note that the mosaics are matched correctly even though there are scale and viewpoint differences and occluding foreground.

each grey level block corresponds to an entry in the matrix. Finding two strip sequences in the two mosaics that have a "good" alignment and therefore define a common physical area, corresponds to finding a "good" diagonal in the matrix $S[i, j]$.

In comparing two strips that correspond to the same actual physical location, we cannot assume that they cover the same vertical areas, but we can assume that they both have overlapping regions. Therefore, in order to detect their vertical alignment, we further divide each strip into blocks and generate a block-to-block distance matrix $B[k, l]$ for each pair of strips. Each entry in this matrix corresponds to the distance between a block b_k from the first strip and a block b_l from the second strip:

$$B[k][l] = Diff(b_k, b_l). \tag{2.3}$$

where $Diff(b_k, b_l)$ is the distance defined between the color histograms of the two blocks. In this work we used the L_1 norm over HSI color space as our distance metric. [1]. We look for the best diagonal (as explained in Section 2.1.4.0) in the distance matrix $B[k, l]$ and record its start and end points. The value of this diagonal is chosen to set the distance measure between the two strips:

$$S[i, j] = \min_{d \in Diags} \left(\frac{\sum_{(k,l) \in d} B[k][l]}{length(d)} \right). \tag{2.4}$$

[1]a convenient representation was HSI color space in polar coordinates which was easy to invert to and from RGB for quantization verification. We used non-uniform quantized histograms for this space [Aner and Kender, 2002]. Other color spaces and metrics (such as La^*b^* and quadratic form distance) might be more accurate but their added computational cost was not justified.

Where *Diags* is the set of all allowable diagonals in the matrix $B[k, l]$, and each diagonal $d \in Diags$ is given by a set of pairs of indexes (k, l).

The comparison process is shown in Figure 2.3. Two mosaics are shown in Figure 2.2(c) and Figure 2.2(d).he strip comparison process (Eq. 2.3) is graphically displayed in Figure 2.3(b) and the matrix S is aphically displayed in Figure 2.3(a), with the result of the two strip comparison from Figure fig:comp2(b) highlighted. The height of this illustrated grey-level blocks matrix S is the same as width of the mosaic in Figure 2.2(c) and its width is the width of the mosaic in Figure 2.2(d), such that each block represents the strip-to-strip difference between the two mosaics.

We next find the best diagonal in the matrix of strip-to-strip differences $S[i, j]$ in order to find a horizontal alignment between strip sequences, and we record its start and end points. The area of interest is represented by a thin diagonal line across the matrix along which entries have low values (corresponding to good similarity) in Figure 2.3(a). We use these start and end points to crop each mosaic such that only the corresponding matching areas are left. The start and end points of the diagonal in the S matrix set the vertical borders of the cropped mosaics. In order to determine the horizontal borders, we inspect the results of the vertical alignment for every pair of strips along the diagonal path found in S. Since for every strip pair we recorded the start and end points of the best diagonals in its corresponding B matrix, we take the average of these values and use it to set the horizontal border.

Determining Threshold. Once all mosaic pairs are processed, we check the distance values of the diagonal paths found for each pair. We expect that mosaic pairs from different settings or with no common physical area will have high distance values, and we discard them according to a threshold. We determine the threshold by sampling several distance values of mosaic pairs which have common physical areas, which therefore should have the lowest distance values. We choose the highest of these sampled values as our threshold. This sampling method proved correct after inspecting all mosaic pairs distances as shown in Figure 2.5.

To illustrate our results, we manually separated mosaic pairs with common physical areas (two groups on the left) from the rest of the mosaic pairs (group on the right). The sampled pairs are the leftmost group. The threshold chosen from the hand-labeled samples (horizontal line) quite accurately rejects mosaic pairs known to be from different physical areas, although it does permit a few false positive matches. If a diagonal path distance value exceeds this threshold, we determine that the two mosaics do not match. If we find a match, we continue to the

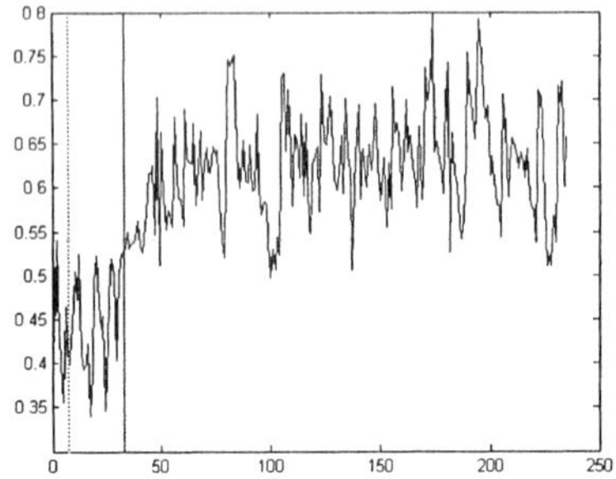

Figure 2.5. Illustration of threshold determination for mosaics distance after the coarse stage. Values in the two leftmost groups belong to mosaic pairs with a known common background, and the rest of the values are on the right. The leftmost group of 10 distance values is the sample group. The horizontal line is the maximum value among this sample group.

finer step where we only use the cropped mosaics: those parts of the mosaics corresponding to the best sub-diagonal found, an example of which is shown in Figure 2.4(a).

Fine Matching. After discarding mosaic pairs which had diagonal paths with large distance values, we refine the measure of their closeness, in order to detect false positives and to more accurately determine physical background similarity. In this finer stage we only compare the cropped parts of the mosaics, applying a more precise and restrictive method than the one used in the coarse stage. The cropped mosaics are displayed in Figure 2.4(b). Note that the mosaics are matched correctly even though there are scale and viewpoint differences, and the occluding foreground is cropped out. We now use thinner strips (20 pixels wide) and also take into account the scale difference between the two mosaics. Assuming that the cropped mosaics cover the same regions of the physical setting, if we divide the narrower of the two cropped mosaics into K thin strips, then the best match will be a one-to-one match with K somewhat wider strips of the wider mosaic, where each strip pair covers the exact physical area. Let $\alpha \geq 1$ be the width ratio between the two mosaics. We re-compute histograms of 20×20 blocks for the narrower cropped mosaic, and histograms of $20\alpha \times 20\alpha$ blocks for the wider cropped mosaic. The best path in the new distance matrix should now have a slope of approximately 45°. Matching in the finer stage is less computationally expensive than the matching in the coarse stage. First, only a few mosaic pairs are re-matched. Second, having adjusted for the mosaic widths, only diagonals parallel to the main diagonal need to be checked. Third, since the cropped mosaics cover the same regions, only complete diagonals rather than all possible sub-diagonals need to be checked. Therefore, we only compute values for the main diagonal and its two adjacent parallel diagonals. These diagonals are automatically known to satisfy the diagonal length and boundary constraints. These operations restrict the total number of strip and block comparisons to be performed and greatly lower computation time, even though there are more strips per mosaic. This final verification of mosaic match values is very fast.

The matching of mosaics could be further enhanced to support a more accurate alignment, by finding a coarse affine transformation between corresponding blocks. However, our coarse-fine approach yields a fast and simple matching method, which performs sufficiently accurately.

Matching Summary. We have presented a method for general mosaic comparison and applied it to mosaics generated from sitcoms. Our method enables matching of mosaics which were generated from shots taken from different locations and with different focal length. This method has only few restrictions, in that it assumes a horizontal nature of mosaics (related to the camera placement, and not necessarily its motion) and was only tested on shots taken indoors, where the lighting changes were generated artificially. However, our mosaic comparison method gave satisfactory results even for distorted mosaics when shots had parallax motion, or for mosaics with global or local distortions due to bad registration. It performs quite well for large scale differences, as shown in Figure 2.6.

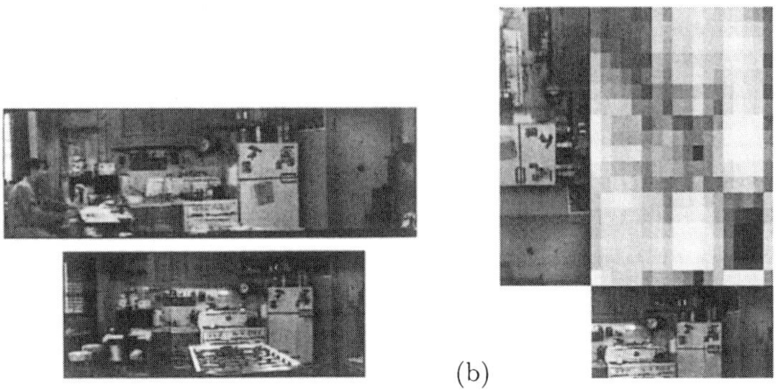

(a) (b)

Figure 2.6. (a) Two mosaics generated from shots in the same episode and (b) their corresponding strip-to-strip difference matrix. Note that the scale difference between them is close to 2 : 1.

By applying a median filter we were able to correct some of the noise which resulted from either inaccurate frame registration or inappropriate projection surface of the mosaic. However, we are aware of methods to improve the appearance of mosaics such as the ones suggested in [Szeliski and Heung-Yeung, 1997] or [Vasconcelos, 1998], if one is able to afford the additional computation time. Nevertheless, these methods are not able to determine the appropriate surface type automatically (whether it is planar, cylindrical or spherical), hence will not guarantee accurate results for our example.

The comparison by alignment method could be made more robust if the strip width is not determined in advance. Instead of two levels of alignment, a more elaborate decreasing sequence of strip widths could be used, starting from relatively wide strips (the whole image width might be used) and converging to smaller strip width (up to pixel-wide strips). This process should converge automatically to the best matching strip width, by computing a parameter that evaluates the 'goodness' of the match at each level.

We believe that even though our comparison by alignment technique will never replace existing narrow baseline stereo matching methods, it could be made more robust to challenge the results of existing wide baseline matching techniques for several applications. In some cases (e.g. distorted images), it may also outperform them.

2.2 Hierarchical Summaries and Video Indexing

We first demonstrate our physical setting based approach using sitcoms, and construct a tree-like representation for each episode in a sitcom. Each episode is first divided into shots using the shot transition detection technique described [Aner and Kender, 2001], and these shots are further divided into scenes using the method described in [Kender and Yeo, 1998]. We represent shots and scenes with mosaics and use them to cluster scenes into the newly proposed "physical settings" segments.

This **new** level of abstraction of video, which concludes the top level of our tree-like representation, does not only form a very compact representation for long video sequences, but also allows us to efficiently compare different videos (different episodes of the same sitcom). By analyzing our comparison results, we can infer the main **theme** of the sitcom as well as the main **plots** of each episode.

2.2.1 Mosaic Scene Representation in Sitcoms

A scene is a collection of consecutive shots, related to each other by some spatial context, which could be an event, a group of characters engaged in some activity, or a physical location. However, a scene in a sitcom occurs in a specific physical location, and these locations are usually repetitive throughout the episode. Some physical locations are characteristic of the specific sitcom, and repeat in almost all of its episodes. Therefore, it is advantageous to describe scenes in sitcoms by their physical location. These physical settings can then be used to generate summaries of sitcoms and to compare different episodes of the same sitcom. This property is not confined to the genre of sitcoms alone. It could be employed for other video data which are constrained by production economics, formal structure, and/or human perceptive limits to re-use their physical settings.

In order to represent a scene, we are only interested in those shots that have the most information about the physical scene. Once these shots are determined, their corresponding mosaics are chosen as the representative mosaics ("R-Mosaics") of the scene. One example of a "good shot" is an extended panning (or tracking) shot, as shown in Figure 2.1. Another example is a zoom shot; the zoomed-out portion of the shot will most likely show a wider view of the background, hence expose more parts of the physical setting. Detecting "good" shots is done by examining the registration transformations computed between consecutive frames in the shot. By analyzing these transformations, we classify shots as panning (or tracking), zoomed-in, zoomed-out, or stationary. We compute affine transformations instead of projective transformations, that describe more accurately the motion within the each shot, but might cause projective distortions in the generated mosaic. The classification of motion shots is done as follows: For each shot, the affine transformations computed are consecutively applied on an initial square, each transformed quadrilateral is then measured for size and distance from the previous quadrilateral. Pan shots are determined by measuring distances between the quadrilaterals, zoom shots are determined by measuring the varying size of the quadrilaterals. We also determine whether there was any parallax motion within the shot, by measuring both size and scale and by measuring the quality of the affine transformation computed (checking accumulated error and comparing to inverse transformations). An illustration is shown in Figure 2.7.

Once the shots are classified, a reference frame for the mosaic image plane is chosen accordingly. For panning and stationary shots, the middle frame is chosen, whereas for zoomed-in and zoomed-out shots the first and last frames are chosen, respectively. We experimentally derived a threshold that allowed us to only choose shots with significant zoom or pan. For shots with significant parallax motion, further processing is needed. We divide such shots into several parts such that each part does not contain much parallax and construct separate mosaics for each part. In our experiments, this was sufficient and the resulting mosaics were accurate enough for comparison purposes.

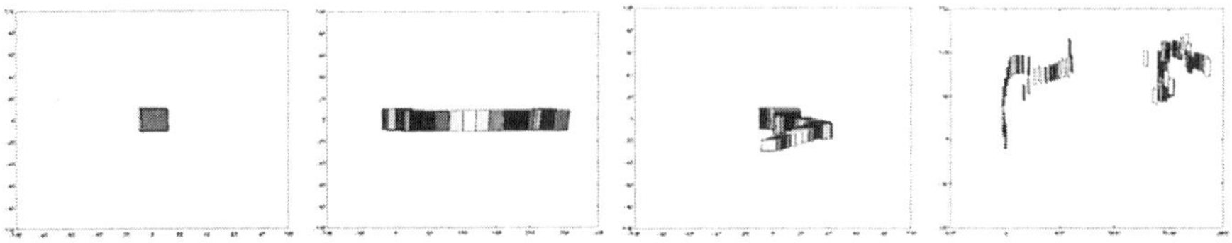

(a) (b) (c) (d)

Figure 2.7. Examples of transformations analysis for different shot types. For each shot, the affine transformation computed were consecutively applied on an initial square, each transformed quadrilateral was colored differently (from a set of 8 colors) for view purposes. (a) Static shot. (b) Panning shot. (c) Zoom-in with pan. (d) Tracking shot with strong parallax in the middle of the shot. This caused the transformation summary display to be divided into two groups.

However, some scenes are mostly static, without pan or zoom shots. Moreover, sometimes the physical setting is not visible in the R-Mosaic because a pan shot was close up. For these scenes, an alternative shot which better represents the scene has to be chosen. We observed that most scenes in sitcoms (and all the static scenes that we processed) begin with an interior "establishing shot", following basic rules of film editing [Arijon, 1976]. The use of an establishing shot appears necessary to permit the human perception of a change of scene; it is a wide-angle shot (a "full-shot" or "long-shot"), taken to allow identification of the location and/or characters participating in that scene. Therefore, we choose the first interior shot of each scene to be an R-mosaic; for a static scene, it is the only R-mosaic. Many indoor scenes also have an exterior shot preceding the interior "establishing shot", which photographs the building from the outside. We can easily detect this shot since it does not cluster with the rest of the following shots into their scene, and also does not cluster with the previous shots into the preceding scene. Instead, it is determined as a unique cluster of one scene (this was also found by [Hanjalic et al., 1999]). We detect those unique clusters and disregard them when constructing the scene representations. In our experiments, using these pan, zoom, and establishing shot rules leads to up to six R-Mosaics per scene.

Once R-mosaics are chosen, we can use them to compare and cluster scenes. The dissimilarity between two scenes is determined by the minimum dissimilarity between any of their R-Mosaics. This is due to the fact that different shots in a scene might contain different parts of the background, and when attempting to match scenes that share the same physical location, we need to find at least one pair of shots (mosaics) from the two scenes that show the same part of the background.

After determining the distance measure between each scene within an episode, we can construct a scene difference matrix where each entry (i, j) corresponds to the difference measure between Scene i and Scene j. An example is shown in Figure 2.8, where the entries in the scene difference matrix were arranged manually so that they correspond to the five physical settings of this episode. More specifically, Scenes $1, 3, 7, 12$ took place in the setting which we marked as "Apartment1", Scenes $2, 6, 9, 11$ took place in "Apartment2", Scenes $5, 10$ took place in "Coffee Shop", Scenes $4, 13$ took place in "Bedroom1" and Scene 8 took place in "Bedroom2".

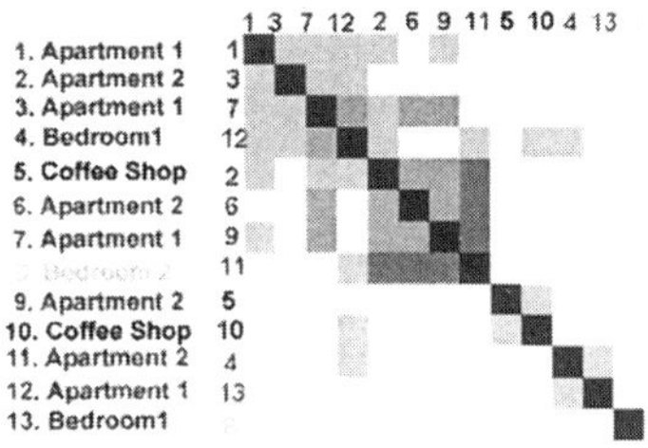

Figure 2.8. Example of clustering scenes for a single episode. On the left is the scene list in temporal order, and on the right is the similarity graph for the scene clustering results, in matrix form (darker = more similar). Scene entries are arranged manually in the order of physical settings. For example, there are 13 scenes in the episode represented in the middle grouping, Scenes $1, 3, 7, 12$ are in "Apartment1", Scenes $2, 6, 9, 11$ are in "Apartment2", Scenes $5, 10$ are in "Coffee Shop", Scenes $4, 13$ are in "Bedroom1" and Scene 8 is in "Bedroom2".

2.2.2 Video Representation by Physical Settings

Although the scene clustering process results typically in 5-6 physical settings, there are 1-6 scenes in each physical setting cluster, and about 1-6 mosaics representing each scene. Ideally, for the purposes of display and user interface, we would like to choose a single mosaic to represent each physical setting. However, this is not always possible. Shots of scenes in the same physical setting, and sometimes even within the same scene, are filmed using cameras in various locations which show different parts of the background. Therefore, two mosaics of the same physical setting might not even have any corresponding regions. We want the representation of a physical setting to include all parts of the background which are relevant to that setting. Therefore, if there does not exist a single mosaic which represents the whole background, we choose several mosaics which together cover the whole background, using the following clustering-based scheme.

We use the results of the mosaic-based comparison algorithm, which recognizes corresponding regions in the mosaics, to determine a "minimal covering set" of mosaics for each physical setting. We approximate this set (since the real solution is an NP-hard problem) by clustering all the representative mosaics of all the scenes of one physical setting and choosing a single mosaic to represent each cluster. This single mosaic is the centroid of the cluster, e.g., it is the mosaic which has the best average match value to the rest of the mosaics in that cluster. This is achieved in two stages: first by identifying separate clusters, where each cluster corresponds to mosaics which show corresponding areas in the background of the particular physical setting; in the second stage a distance matrix between the mosaics of each cluster is computed and the mosaic with the minimal average distance from all mosaics is chosen as the centroid of that cluster.

The result of the mosaic-based scene clustering and the construction of physical settings for three episodes of the same sitcom are shown in Figure 2.9. The order of the entries in the original scene difference matrices were manually arranged to reflect the order of the reappearance of the same physical settings. Dark blocks represent good matches (low difference score). As can be seen in the left column of the middle sitcom, there are 5 main clusters representing the 5 different scene locations. For example, the first cluster along the main diagonal is a 4×4 square representing scenes from "Apartment1". The false similarity values outside this square are due to actual physical similarities shared by "Apartment1" and "Apartment2", for example, light brown kitchen closets. Nevertheless, it did not affect our scene clustering results, as can be seen in the corresponding diagram in the middle column in Figure 2.9(b), and the dendrogram generated by the results of applying a weighted-average clustering algorithm [Jain and Dubes, 1988] to the original scene difference matrix on its left.

2.2.3 Comparing Settings Across Videos

When grouping information from several episodes of the same sitcom, we detect repeating physical settings. It is often the case that each episode has 2-3 settings which are unique to that episode, and 2-3 more settings which are common and recur in other episodes. We summarized the clustering information from three episodes of the same sitcom and computed similarities across the videos based on physical settings. We define the distance between two physical settings to be the minimum

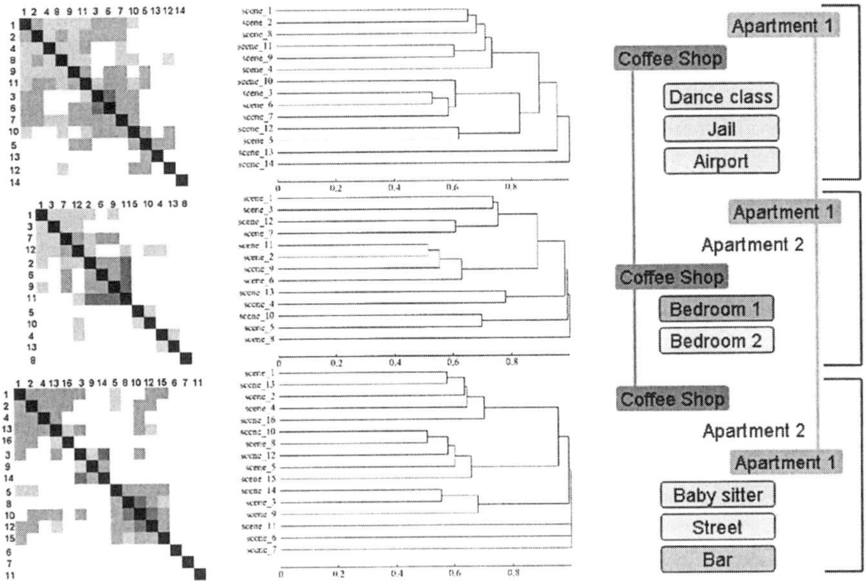

(a) (b) (c)

Figure 2.9. Results from three episodes of the same sitcom. (a) Similarity graphs for scene clustering results, in matrix form as was shown in Figure 2.8. (b) Corresponding dendrograms (generated using the tools available in [Kleiweg,]). Each cluster in each dendrogram represents a different physical setting. (c) Illustration of the results of inter-video comparison of physical settings. Settings appear in the same order they were clustered in (b). Lines join matching physical settings, which are common settings in most episodes of this sitcom: "Apartment1", "Apartment2" and "Coffee Shop". The rest of the settings in each episode identify the main plots of the episode, for example, the main plots in the first episode took place in the settings: jail, airport and dance class.

distance between their R-Mosaics. To compare between three long video sequences, each about 40K frames long, we only need to compare 5 settings of the first episode with 5 settings of the second episode and 6 settings of the third episode.

Comparing settings across episodes leads to a higher-level contextual identification of the **plots** in each episode, characterized by settings which are unique to that episode. For example, in the episode at the top of Figure 2.9 the main plots involve activities in a dance class, jail and airport. We have observed that most sitcoms do in fact involve two or three plots, and anecdotal feedback from human observers suggests that people do relate the plot to the unusual setting in which it occurs. This non-temporal indexing of a video within a library of videos is directly analogous to the Term Frequency/Inverted Document Frequency

(TFIDF [Salton and McGill, 1983]) measure of information retrieval used for documents. That is, what makes a video unique is the use of settings which are unusual with respect to the library. Finally, the descriptive textual labelling of physical settings is done manually. Since there are usually only a few common settings for each sitcom (due to economic constraints on set design and to sitcom writing rules), there are only 2-3 additional settings for every newly added episode.

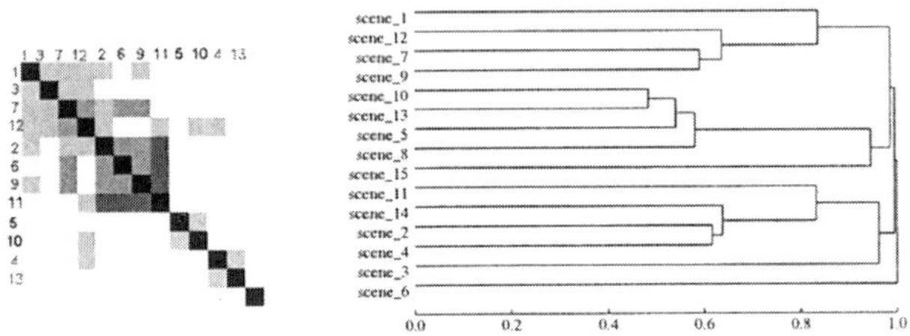

Figure 2.10. Clustering results from a fourth episode: (a) Scene distance matrix with entries manually re-ordered to match physical settings. Here the clustering was different since the first cluster names "Apartment1" is sub-divided into two clusters, one for Scene 1 which took place in the living room, and the second for scenes which took place in the kitchen. (b) Dendrogram generated for the scene distance matrix of this episode.

Another example, of a fourth episode that we have analyzed from the same sitcom, is shown in Figure 2.10. There are 6 true physical settings here: "Apartment 1", "Apartment 2", "Coffee Shop", "Party", "Tennis game" and "Boss's Dinner". In this episode the clustering of scenes together into physical settings was not as straightforward as in the first three episodes. This is due to the fact that the setting of "Apartment1" was not presented in the same manner, since its scenes either took place in the kitchen or in the living room, but not in both. Mosaics which were generated from shots in scenes that took place in the kitchen of "Apartment 1" have no overlap with mosaics which were generated from shots in scenes that took place in the living room of "Apartment 1". This separates scenes of "Apartment1" into two clusters, as shown in the top left part of Figure 2.10(a): Scene 1 belongs to the living room cluster, and Scenes 7, 9 and 12 belong to the kitchen cluster. The clustering also doesn't separate Scene 11 well from the rest of the scenes, since this new setting took place in an apartment with similar colored walls and windows as the colors appearing in the setting "Coffee Shop", into which it is loosely clustered. However, when these settings are compared with settings of different episodes, the splitting of the setting "Apartment 1" into two separate settings is corrected. In the context of all the episodes of this sitcom, the setting of "Apartment 1" includes mosaics of both the living room and the kitchen, causing the two different settings of this episode to be combined together. More specifically, the "Apartment 1" setting cluster already contains mosaics that match both Scene 1 and Scenes 7, 9 and 12 from the new episode.

Table 2.1. Table summary of scenes and physical settings in episodes of the sitcom "Seinfeld". Each column represents a physical setting and each row represent a different episode. Each entry lists the scene numbers of the row's episodes which take place in that column's physical setting. The last column, due to space constraints, represents all unique settings of all episodes.

	Jerry's Monologue	**Jerry's Apartment**	**The Diner**	**Unique Settings**
The Cafe	1, 14, 20	3, 10, 12, 15		Dream Cafe: 4, 6, 8, 11, 17 Street at Dream Cafe: 2, 19 Monica's Apt.: 5, 7, 9, 13, 16, 18
The Big Salad	1, 16	4, 6, 8, 11, 15	5, 9, 12	Stationery Store: 2, 13 Street: 3 Cab: 7 Newman's Apt.: 10 Margaret's Car: 14
The Glasses	1	2, 5, 6, 8, 10		Optical Store: 3, 7, 12, 14 Hospital: 4, 9 Health Club: 11, 15 Jeffrey's Apt: 13
The Junior Mints	1, 15	2, 5, 7, 8, 11, 12, 14	9	Roy's Hospital Room: 3, 6, 13 Hospital Hallway: 4 Hospital Operating Room: 10
The Pie		2, 8	1, 4, 7, 11, 17	Clothing Store: 3, 5, 10, 12 Poppie's Restaurant: 6, 8, 15 Jerry's Car: 13 Other Restaurant: 14, 16

This example demonstrates how the non-temporal level of abstraction of a single video could be verified and corrected by semantic inference to other videos. Scenes and settings that otherwise would not have been grouped together are related by a type of 'learning' from the previously detected "physical setting" structure in other episodes.

To better illustrate our approach of indexing sitcoms according to their physical setting structure, we present an analysis of several episodes from another sitcom, "Seinfeld". A summary table showing the different physical settings used for several episodes is presented in Table 2.1. As could be seen from the table, there are two known settings, "Jerry's Apartment" and "The Diner", which repeat in many episodes, and are therefore directly related to the main theme of the sitcom. In contrast, each episode has several settings which are unique to that episode and are often directly related to the main plot of that episode (the given episodes' names). Examples include the "Dream Cafe" setting and "Laundry room" setting. These settings serve as excellent candidates for recalling the content of these episodes.

2.2.4 A Video Browsing Tool

We have constructed an example video browser specialized for sitcoms, which utilizes our proposed tree-like hierarchical structure. It uses the video data gathered from all levels of representation of the video. At the frame level, the MPEG video format of each episode is used for viewing video segments of the original video. At the shot level, it uses a list of all marked shots for each episode, including the start and end frame of each shot. At the scene level, it uses a list of all marked scenes for each episode, including the start and end shots of each scene. For each scene, a list of representative shots is kept, and their corresponding image mosaics are used for display within the browser. At the physical setting level, it uses a list of all detected physical settings for each episode, with their corresponding hand-labeled descriptions (e.g. "Apartment 1", "Coffee Shop"). Each physical setting has a single representative image mosaic, used for display. The main menu is shown in Figure 2.11.

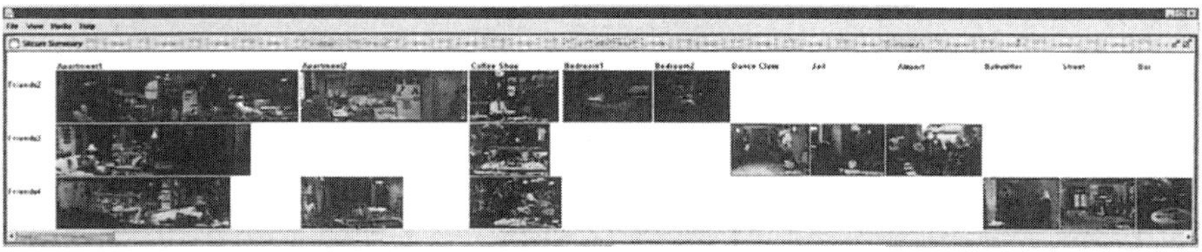

Figure 2.11. The main menu of our proposed browser. It shows all episodes and physical settings, displayed as a table where each episode has entries (mosaic images) only in relevant settings. This is a very compact non-temporal representation of a video.

This browsing tool, described in detail in [Aner et al., 2002], combines the ability to index a library of videos both by compact semantic representations of videos and by temporal representations. The compact visual summary enables cross-referencing of different episodes and fast main plot analysis. The temporal display is used for fast browsing of each episode.

2.2.5 Analysis and Evaluation

Since the concept of segmentation of scenes into physical setting is new and was not discussed in previous work, we evaluate the clustering results by comparing to ground truth. Determining the correct non-temporal segmentation of video into physical settings is a rather subjective matter. However, in sitcoms the physical settings structure is well defined and it is rather straightforward to distinguish between them. This is mainly due to cost considerations which limit the number of different sets used for each sitcom, as well as the very pronounced colors which are often used in different settings, probably to attract viewer attention. As can be seen in Figure 2.9, the scene dissimilarity measure is what determines the accuracy of the physical settings detection. Different clustering methods would result in the same physical settings cluster structure as long as the scenes distance matrix has the correct values. In our experiments, the scene dissimilarity did not result in perfect precision values. Some scene pairs that did not have any corresponding regions had relatively high match values. This, however, did not affect the clustering results for the first three episodes that we analyzed, which were perfect. In the case of the fourth episode, the inter-video comparison of physical settings managed to correct the clustering results for the first setting of "Apartment 1", but the clustering threshold was not as pronounced as in the first three episodes. Depending on this threshold, for large values Scene 11 would be wrongly clustered with Scenes 14, 2 and 4, and for small values Scene 15 would not be clustered with Scenes 5, 8, 10, and 13, as it should.

The complexity of the scene clustering method is very low. Since all mosaic pairs are matched, if there are M mosaics in an episode, then M^2 coarse match stages will be performed, after which only several mosaic pairs will be matched in the finer stage. In our experiments this number of pairs was on the order of $O(2M)$. Once the scene distance matrix is constructed, the physical settings are determined using any clustering algorithm, which we consider as a black box. Since the maximum number of scenes encountered in sitcoms was 15, there are up to 15 elements to cluster, causing every clustering algorithm to run very fast. Finally, when comparing physical settings across episodes, there are only 5 to 6 settings in each episode, each represented by no more than 3 mosaics, which also makes the comparison process very efficient.

To conclude, we show a compact approach for summarizing video, which allows efficient access, storage and indexing of video data. The mosaic-based representation allows direct and immediate access to the physical information of scenes. It therefore allows not only efficient and accurate comparison of shots and scenes, but also the detection and highlighting of common physical regions in mosaics of different sizes and scales.

We also presented a new type of video abstraction. By clustering scenes into an even higher level representation, the **physical setting**, we create a non-temporal organization of video. This structured representation of video enables reliable comparison between different video sequences, therefore allows both intra- and inter-video content-based access. This approach leads to a higher-level contextual identification of **plots** in different episodes of the same sitcom.

2.3 Physical Setting in Different Video Genres

The physical setting has important semantic attributes that could be used to analyze other video genres besides sitcoms. Many produced video genres are confined to few locations or include few distinctive locations that could be used to characterize the video or to extract specific information from the video, depending on the type of genre. This characterizes most television series such as sitcoms and drama series.

Sports Broadcasts. An example for distinctive camera locations is sports broadcasts, where the cameras are often places in confined fixed locations, and are limited in number. Different stages of the game are photographed by different cameras, and characteristic events in the game are often photographed by either a single or a small number of cameras, located in a specific place for that purpose. Examples include foul basket shots in basketball games, fouls in soccer game and more. Sports broadcasting is confined by certain rules governing their editing and the placement of cameras. These rules could be used to automatically detect certain events within each sports genre. For example, in basketball broadcasts, field basket goal throws are usually characterized by a court shot (a shot taken by a camera located high above the basketball court, in zoomed-out position) followed by a close-up shot of the player making the basket goal. By first classifying court shots and close-up shots, such an event could be easily detected by marking all instances of a court shot followed by a close-up shot. More details could be found in [Aner and Kender, 2002]. We also show that mosaics are better for representing shots than key-frames, and result in a more accurate classification.

Motion Pictures. Television programs are not the only media confined to several physical settings. Many low budget films or films of a certain character are confined to small number of settings which reappear throughout the movie. An example of a movie with repeating settings is the movie "Run Lola Run" by Tom Tykwer. Each settings reappears three times throughout the movie, since the main character experiences a particular day in her life three times (Figure 2.12).

Figure 2.12. Sampled frames (in reverse temporal order, from right to left) from a shot, with their corresponding mosaic.

Other examples include movies where the characters pay frequent visits to a certain person or location which is central to the plot of the movie. Example movies are "Silence of the Lambs" by Jonathan Demme, "Prince of Tides" by Barbara Streisand, "Good Will Hunting" by Gus Van Sant, and more.

Home Videos. The use of camera locations for shot classification and comparison could be used in non-edited video data as well. A useful application would be to generate an edited video out of raw unedited video footage. Many home videos do not have an organized shot structure, and often include very long shots taken by a freely moving camera. It would therefore be helpful to segment the shots further into smaller segments, resulting in a more organized sub-shot structure. This segmentation is done by analyzing the motion within each shot, and dividing the shot into segments in which the camera motion is restricted. This would ease the generation of an edited video by sampling a subset of these shots, and grouping them together. An example application for home video was presented in [Aner-Wolf and Wolf, 2002], where an edited wedding video was constructed. The sampling of a group of sub-shots was performed using an existing still image summary of the wedding, generated from the wedding photo album.

2.4 Summary

The scene's physical setting is one of its most prominent properties in many video genres. Each scene takes place at one specific physical setting, which usually changes between two consecutive scenes. Moreover, there is a strong correlation between the scene's location and its respective part in the general plot. If every shot was labeled according to its physical location, the tasks of scene segmentation and video summarization in general would be much easier.

However, since we usually do not have this labeling into physical locations, an alternative is required. In this work we suggest the use of mosaics automatically constructed from the video instead. A mosaic is more powerful than a collection of key-frames since it also holds information about the spatial order of its constructing frames. It also eliminates much of the redundancy caused by overlap in the spatial domain, and is robust to many temporal changes (e.g. moving objects).

Mosaics are not an ideal representation - they are not well defined for complex settings and general camera motion (causing distortions in the mosaics). They are also not invariant to camera viewing point and even for the choice of the reference frame. However, we claim that they are flexible enough to serve as a good representation and indexing of video, when used with the mosaic alignment and comparison method we presented here.

Our alignment method makes use of several assumptions about the video from which the mosaics are constructed. These assumptions are based directly on the underlying grammar rules of common video genres. One assumption is the assumption of lateral camera motion. Another assumption is the assumption of controlled lighting. Both assumptions hold for many video genres including most television programs such as sitcoms, dramas, sports broadcasts, news broadcasts, which are shot indoors. By relying on these assumptions we were able to propose a fast alignment algorithm and an associated distance measure between shots, scenes and physical settings. Although the alignment algorithm is not an accurate registration algorithm (the transformation between mosaics of the same settings is not given by a limited number of parameters), we show that the distance measure is effective.

We believe that our comparison by alignment technique could be extended to more general image retrieval applications. In contrast to many image retrieval techniques that use global color features of single images, our technique incorporates spatial information by applying a coarse alignment between the images. It is robust to occluding objects and will match images for which only partial regions match (for example, top-left region of one image matches bottom-right region of the second image).

Using mosaic comparison as our engine, we have constructed a complete system which takes raw video data and produces video summaries. Our proposed summary has a new semantic meaning based on a hierarchical structure of video and on its physical settings. It enables efficient cross-referencing between whole videos and fast browsing of these videos. We demonstrated its use in cross-referencing different episodes of the same sitcom which further infers their main plots.

References

Aner, A. and Kender, J. R. (2001). A unified memory-based approach to cut, dissolve, key frame and scene analysis. In *Proceedings of IEEE International Conference on Image Processing.*

Aner, A. and Kender, J. R. (2002). Video summaries through mosaic-based shot and scene clustering. In *Proceedings of European Conference on Computer Vision.*

Aner, A., L.Tang, and Kender, J. R. (2002). A method and browser for cross-referenced video summaries. In *Proceedings of International Conference on Multimedia and Expo.*

Aner-Wolf, A. and Wolf, L. (2002). Video de-abstraction or how to save money on your wedding video. In *Workshop on Applications of Computer Vision.*

Arijon, D. (1976). *Grammar of the Film Language.* Silman-James Press.

Bouthemy, P., Dufournaud, Y., Fablet, R., Mohr, R., Peleg, S., and Zomet, A. (1999). Video hyper-links creation for content-based browsing and navigation. In *Workshop on Content-Based Multimedia Indexing*, Touluse, France.

Gelgon, M. and P.Bouthemy (1998). Comparison of automatic shot boundary detection algorithms. In *European Conference on Computer Vision*, volume 1.

Gonzalez, R. C. and Woods, R. E. (1993). *Digital Image Processing.* Addison Wesley.

Hanjalic, A., Lagendijk, R. L., and Biemond, J. (1999). Automated high-level movie segmentation for advanced video retrieval systems. In *IEEE Transactions on Circuits and Systems for Video Technology*, volume 9, pages 580–588.

Irani, M. and Anandan, P. (1998). Video indexing based on mosaic representations. In *IEEE Trans. on Pattern Analysis and Machine Inteligence*, volume 86, pages 905 – 921.

Irani, M., Anandan, P., Kumar, J. B. R., and Hsu, S. (1996). Efficient representation of video sequences and their applications. In *Signal processing: Image Communication*, volume 8, pages 327–351.

Jain, A. K. and Dubes, R. C. (1988). *Algorithms for Clustering Data.* Prentice Hall, Englewood Cliffs, NJ.

Kender, J. R. and Yeo, B.-L. (1998). Video scene segmentation via continuous video coherence. In *Proceedings of the IEEE Conference on Computer Vision and Pattern Recognition.*

Kleiweg, P. Clustering software available at http://odur.let.rug.nl/ kleiweg/clustering/clustering.html.

Lee, M., Chen, W., Lin, C., Gu, C., Markoc, T., Zabinsky, S., and Szeliski, R. (1997). A layered video object coding system using sprite and affine motion model. In *Proceedings IEEE Transactions on Circuits and Systems for Video Technology*, volume 7, pages 130–145.

Lienhart, R. (1999). Comparison of automatic shot boundary detection algorithms. In *SPIE Storage and Retrieval for Still Image and Video Databases VII*, volume 3656, pages 290–301.

Massey, M. and Bender, W. (1996). Salient stills: Process and practice. In *IBM Research Journal*, volume 35.

Oh, J., Hua, K. A., and Liang, N. (2000a). A content-based scene change detection and classification technique using background tracking. In *Proceedings of the IS&T/SPIE Conference on Multimedia Computing and Networking*, pages 254–265.

Oh, J., Hua, K. A., and Liang, N. (2000b). Efficient and cost-effective techniques for browsing and indexing large video databases. In *ACM SIGMOD on Management of Data*, pages 415–426.

Salton, G. and McGill, M. (1983). *Introduction to modern information retrieval.* New York: McGraw-Hill.

Schaffalitzky, F. and Zisserman, A. (2001). Viewpoint invariant texture matching and wide baseline stereo. In *Proceedings of the International Conference on Computer Vision.*

Smeulders, A., Worring, M., Santini, S., and Gupta, A. (2000). Content based image retrieval at the end of the early years. In *International Journal on Pattern Analysis and Machine Intelligence*, volume 22.

Szeliski, R. and Heung-Yeung, S. (1997). Creating full-view panoramic image mosaics and environment maps. In *SIGGRAPH.*

Vasconcelos, N. (1998). A spatiotemporal motion model for video summarization. In *Proceedings of the IEEE Conference on Computer Vision and Pattern Recognition.*

Wang, J. and Adelson, E. (1994). Representing moving images with layers. In *IEEE Transactions on Image Processing*, volume 3, pages 625–638.

Yeung, M. and Yeo, B. (1996). Time-constrained clustering for segmentation of video into story units. In *Proceedings of IEEE International Conference on Pattern Recognition.*

Chapter 3

TEMPORAL VIDEO BOUNDARIES

Nevenka Dimitrova, Lalitha Agnihotri
Philips Research
345 Scarborough Rd.
Briarcliff Manor, NY 10510, USA
nevenka.dimitrova@philips.com

Radu Jasinschi
Philips Research
Prof. Holstlaan 4
5656 AA Eindhoven, The Netherlands
radu.jasinschi@philips.com

Abstract Automatic video content analysis and retrieval is crucial for dealing with large amounts of video data. While most of the work has been done on local video segmentation, object detection, genre classification, and event detection, little attention has been given to a systematic approach to temporal video boundary segmentation taking into account overall structural properties of the video.

In this chapter we first categorize the types of temporal boundaries in video into micro-, macro-, and mega-boundaries. We generalize the concept of a video boundary to include information about video segments, taking into account the combination of different attributes present in the different modalities. For each category we present a mathematical framework, detection method and experimental results.

With this new unified approach we want to have a framework for an autonomous video content analysis system that would operationally analyze continuous video sources over long periods of time. This is very important for consumer video applications where metadata is unavailable, incomplete or inaccurate.

Keywords: Video segmentation, temporal segmentation, scene detection, commercials detection, content analysis, temporal video boundaries, autonomous analysis, micro-boundaries, macro-boundaries, mega-boundaries.

Introduction

There is a great increase in content options for the consumer: multiplex movie houses, hundreds of TV channels, billions of web documents. Users want a balance between feeling in control and coping with the number of options. What we want in the end are scalable interfaces that support "do it for me" users who rely on the technology to make the selection and "let me drive" users who want to be completely in control. For both cases, video content analysis has a crucial role to play in the development of sophisticated technologies required for content-aware applications that include personalized filtering and automatic retrieval of video content. Content aware applications uncover information about the structure and meaning of content and give users access to content at the semantic level. These applications have to automatically and reliably detect the internal structure of the video content [Dimitrova, 2000b].

In consumer domain, the devices and applications need to be very robust, work on a continuous broadcast feed 24 hours a day, seven days a week, work with metadata standards, and deal with situations when metadata is not available. In addition, availability of metadata in many cases brings a whole new set of issues such as inaccurate values and incomplete or not detailed enough information. For example, if the metadata specifies begin time and end time for the "Academy awards" then a digital video recorder that uses this metadata to automatically record the program will stop exactly after three hours, and will miss the most exciting part of the program which includes the main Oscar awards. Now, if the digital video recorder device (which could be either a software or a hardware implementation) was "aware" of the internal timing, structure and meaning of the content, by more detailed metadata which is either inserted manually or extracted automatically using content analysis, then much more information can be available for additional browsing features. In this case, the full Oscars would be recorded with indices to important segments for individual awards, summary information about each award, labels for commercials in a commercial break, over the full length of the program. This level of detailed metadata can enable the user to first of all enjoy the full program, then select the segments by the type of awards, actors, anchors for particular awards, special awards, and music titles. The requirement for the technology then is to segment the full program from the broadcast with correct program boundaries, commercial break boundaries, individual awards segments and other "stories".

Although considerable amount of work has been reported in object and event detection, and summarization, still, the basic problem is to detect the temporal boundaries at different levels [Truong, 2002; Niblack, 1997; Chang, 1997; Vasconcelos, 1998]. We see that in the literature video segmentation has been used to mean shot segmentation, however we propose to examine the temporal boundary problem at different levels of video content structure analysis. To this extent, at the local level, the shortest observable temporal segments are usually bounded within a sequence of contiguously shot video frames, thus defining the micro-boundaries. At the next level, we consider the macro-boundaries, i.e. the temporal boundaries between different parts of the narrative or the segments of a TV program, (e.g. stories in a news program or, the guests in a talk show). A boundary between a program and any non-program material, such as commercial breaks and other promotional material is called a mega-boundary.

In the literature, we have been relying on a well accepted definition of shot segmentation as the first step in analyzing and indexing video. However, with more research in the recent years and also by learning more about the domain of cinematography we see that there are exceptions to these definitions. We believe that we need definitions that are more fundamental, encompass the film definitions and adopted practice, and allow for less exceptions.

We should note here that the MPEG–7 standard has made an effort to define and standardize terms related to temporal video boundaries. The different parts of the MPEG–7 standard offer a comprehensive set of multimedia description tools to create so–called descriptions, which can be used by applications that enable quality access to content [Sanchez, 2002]. In MPEG–7, the Video Segment Description Scheme describes a video or groups of frames, not necessarily connected in space or time, because there are description tools for defining regions, intervals, and their connectivity. MPEG–7 also defines the Analytic Edited Video Description Scheme for the description of an edited video segment (such as shots and transitions) from an analytic point of view – that is, the description is made (either automatically, with supervision, or manually) after video editing. The Analytic Edited Video Description Scheme adds, among others, description tools for the spatio-temporal decomposition from the viewpoint of video editing (shots and gradual transitions), and editing level location and reliability. This definition comes from the realization that transitions can be a result of the editing effects as well as compositional effects within a shot: for example, only one half of the screen has a shot transition while the other half stays constant ("splitscreen").

In this chapter, we introduce definitions for temporal boundaries and propose methods to detect these boundaries. In Section 3.2, we introduce definitions and methods for micro-boundary detection. In Section 3.3

we define macro-boundaries; we present story detection as an example of macro-boundary detection. In Section 3.4 we present commercial detection as an example of mega-boundary detection.

3.1 Temporal Video Boundaries: Definitions

First, we discuss the three types of temporal video boundaries. This is followed by a formal definition of video boundary.

3.1.1 Three Types of Boundaries

- *Micro-boundaries.* These boundaries are associated to the smallest video units – video micro-units – for which a given attribute is constant or slowly varying. The attribute can be any feature in the visual, audio or text domain. For example, if we consider shots as collections of video frames for which luminance and/or motion have approximate constant values, then their boundaries, which can be sharp cuts or slowly varying cuts, e.g., fades, are micro-boundaries.

- *Macro-boundaries.* These are boundaries between collections of video micro-segments that are clearly identifiable organic parts of an event defining a structural (action) or thematic (story) unit.

- *Mega-boundaries.* These are boundaries between collections of macro-segments which typically exhibit a structural and feature (e.g. audio-visual) consistency.

3.1.2 Formal Boundary Definition

Before giving a general formal definition of temporal video boundaries, we discuss some basic notions. First, a video contains three types of modalities: (i) visual, (ii) audio, and (iii) textual. Within each modality we have, in overall, three levels: (i) low-, (ii) mid-, and high-level. These levels describe the amount of details described in each modality in terms of granularity and abstraction [Jasinschi, 2001b; Jasinschi, 2002]. For each modality and for each level there is a set of attributes. These can be formalized as vectors: $A_{\alpha}^{(m)}[t_k] \equiv (A_{\alpha,1}^{(m)}[t_k], \ldots, A_{\alpha,N_{\alpha}^{(m)}}^{(m)}[t_k])$, where m denotes the modality ($m = 1$ is visual, $m = 2$ is audio, and $m = 3$ is textual), α is the index for the attribute, such as, for $m = 1$, $\alpha = 1$ indexes color, $N_{\alpha}^{(m)}$ is the total number of vector components, and t_k is the time instant. For example, if we have a 16-bin color histogram, $N_{\alpha}^{(m)}$ is 16. The time instant can be expressed in frame numbers or milliseconds. For each video frame, either consecutive or subsampled, or other temporal unit, we define an attribute vector. This vector can change from one instant to another – chosen instants that repeat periodically after a fixed interval Δt – or can remain about the same. The detection of temporal video boundaries is based on the amount of change, and its associated statistics change at a given instant, compared to a given threshold.

Given a set of vectors $\{A_\alpha^{(m)}[1], \ldots, A_\alpha^{(m)}[t_k - \Delta t], A_\alpha^{(m)}[t_k]\}$, defined up to time t_k, their average value is the vector $\mu_{\alpha,t_k}^{(m)}$. The deviation of this set of vectors w.r.t. to $\mu_{\alpha,t_k}^{(m)}$ is given by the vector $\{\Delta A_\alpha^{(m)}[1], \ldots, \Delta A_\alpha^{(m)}[t_k - \Delta t], \Delta A_\alpha^{(m)}[t_k]\}$, where $\Delta A_\alpha^{(m)}[t] \equiv A_\alpha^{(m)}[t] - \mu_{\alpha,t_k}^{(m)}$. We could also have a second order deviation of $\{\Delta A_\alpha^{(m)}[1], \ldots, \Delta A_\alpha^{(m)}[t_k - \Delta t], \Delta A_\alpha^{(m)}[t_k]\}$ w.r.t. its average $\delta\mu_{\alpha,t_k}^{(m)}$ thus giving $\{\Delta^2 A_\alpha^{(m)}[1], \ldots, \Delta^2 A_\alpha^{(m)}[t_k - \Delta t], \Delta^2 A_\alpha^{(m)}[t_k]\}$, where $\Delta^2 A_\alpha^{(m)}[t] \equiv \Delta A_\alpha^{(m)}[t] - \delta\mu_{\alpha,t_k}^{(m)}$. Next, we define different methods to estimate temporal video boundaries.

1 *Local method.* This method does not have memory, i.e. the difference is computed between consecutive frames. Given a threshold λ_{local}, and a distance metric $\mathcal{M}(\cdot)$, determine if $\mathcal{M}\ (A_\alpha^{(m)}[t_k] - A_\alpha^{(m)}[t_k - \Delta t]) > \lambda_{local}$. If this is true, then there exists a boundary at instant t_k.

2 *Global method.* This method has memory, i.e. the difference is computed over a series of frames. Take a sliding temporal window W, centered at instant t, with symmetric or asymmetric support. Also, take threshold λ_{global} and a distance metric $\mathcal{M}(\cdot)$. Determine if $\mathcal{M}\ (A_\alpha^{(m)}[t_k] - \mu_{\alpha,t_k}^{(m)}) > \lambda_{global}$. If this is true, a boundary exists at time t_k.

3.2 Micro-Boundaries

The concepts of "shot" and "scene" are used in different ways in multimedia. We discuss some of these definitions next, and after that present our operational definition. In the movie industry a shot is a single, continuously exposed piece of film - however long or short - without any edits or cuts. "Shot" is the term used in the completed film - whilst during filming it is called a "take". On the other hand, in multimedia literature a shot can be a short video segment for which the luminance and/or color is approximately constant; a representative frame of a shot is called keyframe. In the film literature, a scene is a place or setting where the action takes place. A scene can be made up of one shot or many shots that depict a continuous action or event. The notion of a computable scene has been introduced to define a segment of audio-visual data that is consistent with respect to certain low-level properties which preserve the syntax of the original video [Sundaram, 2002].

We should note here that, although in the literature it is well accepted that scenes comprise one or more shots, there are complete movies or long sections of movies that defy this rule. Instead, a single shot consists of multiple scenes. Examples of movies that are seemingly comprised of a single shot containing multiple scenes include "Running Time" [1997], directed by Josh Becker and "Rope" [1948], directed by Alfred Hitchcock. Also, consumer videos are often made in a single shot or relatively few shots (e.g. a birthday party) while there could be a significant change in scenery from indoor to outdoor, including conversations of different groups of people at a party, etc.

We introduced the notion of *family histograms* and *superhistograms* [Dimitrova, 1999]. A family histogram is a data structure that represents the color information of a family of frames. A superhistogram is a data structure that contains the information about non-contiguous family histograms and their respective temporal locations within a larger video segment (e.g. a whole TV program). The extraction process consists of keyframe detection, color quantization, histogram computation, histogram family comparison and merging, ranking of the histogram families to extract a superhistogram. We use several histogram similarity measures in this research to see the effect of different measures on the final result. Also, there are several dimensions to think about when making comparisons between frames based on (i) the amount of memory (time) (ii) contiguity of compared families, (iii) the representation for a family. We can think of memory as one dimension, that is how much history of the computed family histograms we keep: a process that has "zero memory" compares the current frame only with a previous frame histogram, while a process with long memory compares the current frame with a number of previous family histograms. As we show later in the chapter, the results are quite different. Another dimension is time step, which describes whether we make contiguous or non–contiguous comparisons. The third dimension is the choice of a representative histogram of the family. This can be the average histogram, or the histogram of a selected frame (e.g. the last frame). There are different merging strategies with different results as we explore further in this chapter.

3.2.1 Family of Frames

A family of frames is a set of frames (not necessarily contiguous) that exhibits uniform features (e.g. color, texture, composition). Here we use color histogram representations for images for our exploration of family of frames for scene boundary detection. An image histogram is a vector representing the color values and the frequency of their occurrence in the image. Family histograms and superhistograms are color representations of video segments and full length videos respectively. Our concept of family histograms and superhistograms is based on computing frame histograms, finding the differences between consecutive histograms and merging similar histograms.

In order to find the similarity between frames we experimented with multiple histogram difference measures. In the family histogram computation step, for each selected frame (all frames or using linear temporal subsampling) we compute the histogram and then search the previously computed family histograms to find the closest family histogram match. This uses the local method for boundary detection. The comparison between the current histogram, H_C, and the previous family histograms, H_P, can be computed using one of the following methods for calculating histogram difference.

We experimented with histogram difference using L1 and L2 metrics, bin-wise histogram intersection, and histogram intersection. In our experiments, the L1 and bin-wise histogram intersection give better results so we just review the formulae for these two measures.

Histogram difference using an $L1$ metric between the histogram of the current frame $H_C(1), \ldots, H_C(N)$ and the histogram of the previous frame $H_P(1), \ldots, H_P(N)$ is computed by using the following formula:

$$D_{L1} = \sum_{i=1}^{N} \mid H_C(i) - H_P(i) \mid$$

Here, N is the total number of color bins used. We normalize the value by dividing with the total number of pixels. The normalized difference values are between 0 and 1, where values close to 0 mean that the images are similar, and those close to 1 mean that the images are dissimilar.

Another method to compute the distance is by using histogram intersection. The formula for bin-wise histogram intersection is given by:

$$D_B = \sum_{i=1}^{N} \frac{\min(H_C(i), H_P(i))}{\max(H_C(i), H_P(i))}$$

Here, lower values mean that images are dissimilar and higher values mean that images are similar. In order to be consistent, we compute the difference D_B using: $D_B = 1 - B/N$.

The boundary computation consists of frame selection (contiguous or subsampled), choice of color space and color quantization, histogram computation, histogram family comparison and merging, extracting the micro-boundaries and storing the family histogram for further clustering and retrieval. In the following sections we present details of the micro-boundary detection and evaluation.

3.2.2 Micro-Boundary Detection

At the start of a video sequence, the first family consists of the first frame and its histogram. For subsequent frames, given the difference between two family histograms a decision on whether to start a new family of frames is made. If the difference is less than a certain threshold, then the current histogram is merged into the family histogram. The family histogram is a data structure consisting of; (i) pointers to each of the constituent histograms and frame numbers; (ii) a merged family histogram (or original family histogram if it has not been merged yet), and a variable representing total duration which is initialized to the duration of the micro-segment represented by the current histogram. Merging of family histograms is performed according to the following formula:

$$H_{fam}(l) = \sum_{i}^{I} \left(\frac{dur_i}{total\_dur_{fam}} \times H_i(l) \right)$$

In this formula l represents the bin number, fam is an index for the family and H_{fam} is a vector representing the family histogram, i is an index representing the segment number in the family histogram which ranges from 1 to the number of segments in a family I, dur_i is the duration of the family i. $H_i(l)$ is the numerical value indicating the number of pixels in bin l for key frame number i; and $total\_dur_{fam}$ is a variable representing the total duration of all segments already in the family.

There are multiple ways to compare and merge families based on the choice of contiguity and memory: contiguous vs. non-contiguous comparison, and comparison with a short term and longer term memory:

1 Contiguous with zero memory: A new frame histogram is compared with previous frame histogram. This uses the local boundary method introduced in Section 3.1.2.

2 Contiguous with limited memory: A new frame histogram is compared with previous family histogram. This uses the global boundary method, where the temporal window extends over past frames and family histograms.

3 Non contiguous with unlimited memory: A new frame histogram is compared with all previous family histograms within the same video. This also uses the global boundary method with very long temporal window.

4 Hybrid: First a new frame histogram is compared using the contiguous frames and then the generated family histograms are merged using the non contiguous case.

For example, for browsing applications in which we need *scene boundary* detection, we can use family histograms and vary the memory – temporal window in the boundary detection methods – and the contiguity of comparison. Figure 3.1 shows an example for each of the above comparison and merging methods for family histogram computation. As the figure shows, the non contiguous case is very useful because gives the clustering of frame families over the entire video, and a way to characterize the entire video program. The contiguous case is useful for scene boundary detection.

Figure 3.2 presents an example of family boundary computation for the first 15 minutes from Robert Altman's movie "The Player". We should note here that this video segment is about 15 minutes long and comprises a seemingly continuous video sequence without any editing effects while the camera is flying over the grounds of a film studio following the conversations of multiple groups of people.

3.2.3 Experiments

We performed several experiments with boundary detection where scene boundary is defined as separating a visually continuous sequence that exhibits uniform color. We have benchmarked the scene boundary detection on a frame by frame basis for a data set of 15 minutes of CNN headline news (corresponding to 27000 frames). There were 266 "families" in the ground truth. The ground truth was arrived at by collaboratively watching the video frames and deciding where a boundary is by two authors. It was sometimes quite hard to come to a unanimous decision that a boundary existed in the video. The number of detected families varies for different color spaces and quantization schemes. For example, there were 274 families detected using HSB space with 30 bins. We experimented with color quantization 9, 30, 90, 900 bins in HSB, 512

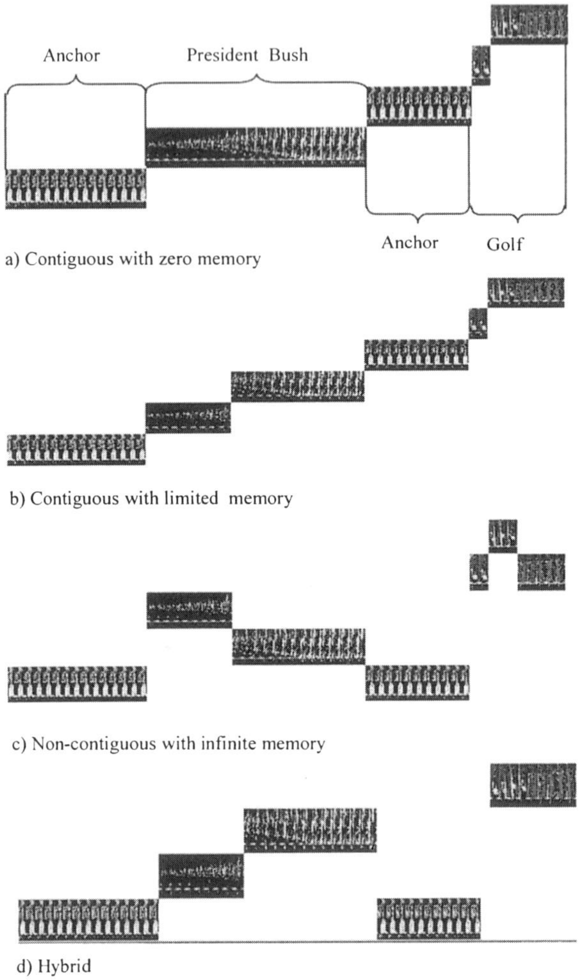

Figure 3.1. Four different approaches to micro-boundary computation for CNN test data.

bins in RGB. We applied multiple histogram comparison measures: L1, L2, bin-wise histogram intersection, and histogram intersection.

The number of correctly identified micro-segments varies based on the chosen threshold and method of comparison. We present in figure 3.3 the precision vs. recall for scene boundary detection for CNN news test video. In total, we tried 100 threshold values for 5 different difference measures and for two boundary detection methods. The curve shows that for threshold values from 0 to 100 both methods, contiguous with limited memory and contiguous with zero memory, show comparable

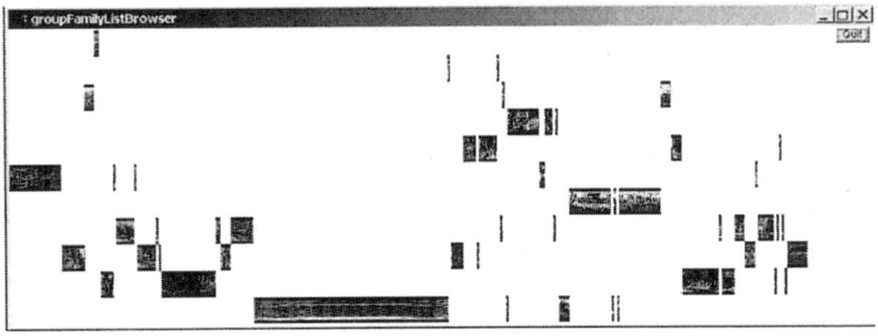

Figure 3.2. Micro-boundaries for the beginning of the movie "The Player".

performance. The best values are obtained for threshold 10 with the L1 comparison method for contiguous with limited memory boundary method. The recall and precision obtained are 94% and 85% respectively. Figure 3.4 presents the results for HSB space quantized to 9 bins, using L1 distance measure. The recall-precision curves for these two methods show that both methods: contiguous with memory and contiguous with zero memory have comparable performance. The histogram intersection method shows better recall overall for the values between 30 and 60, while the L1 method shows consistent precision for the same values. Overall, all the investigated methods have similar curves and show comparable performance – the trick is to use the correct range of thresholds for each difference measure.

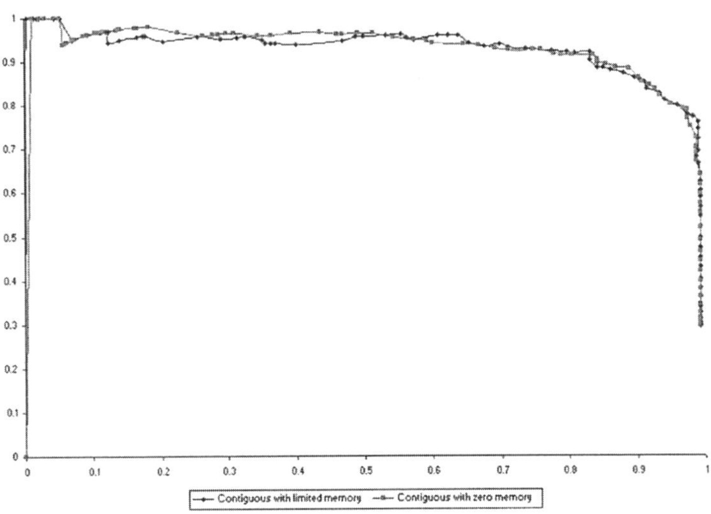

Figure 3.3. Precision vs. recall for scene boundary using histogram intersection method.

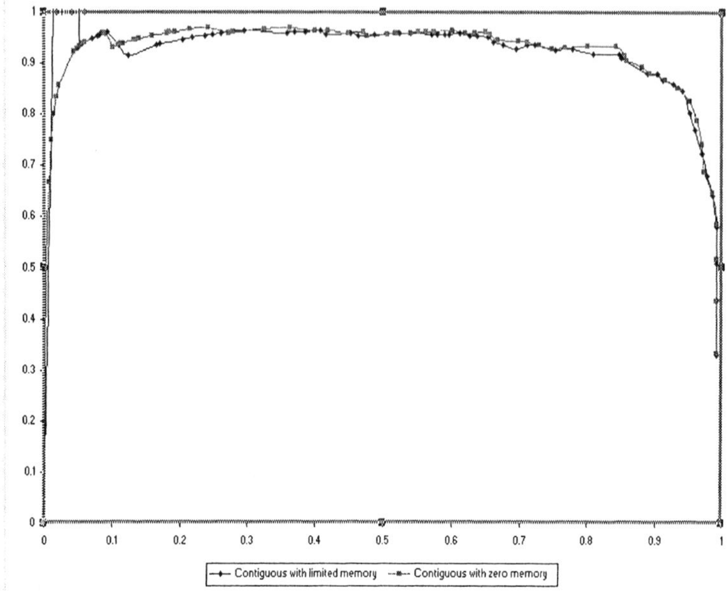

Figure 3.4. Precision vs. recall for scene boundary using L1 distance measure.

3.3 Macro-Boundaries

A story, from the semantic point of view, is a complete narrative structure, conveying a continuous thought or event. This fundamental property is used in text based documents to segment and index stories. From a purely textual point of view, a story can be a complete independent unit, or it can be interrelated with other stories. We import these notions into the domain of multimedia databases by saying that within a video (linear temporal structure) there exists a collection of stories that may or may not be interconnected. In multimedia databases, we have, in addition to textual cues (transcript), visual and audio cues [Hauptmann, 1998; Merlino, 1997]. These should be combined with textual (semantic) cues. Here, we describe a method for segmenting stories in a news program. What differentiates our method from other methods is that ours is able to segment stories even in the absence of textual cues, that is, based solely on audio and visual cues. Basic to the method proposed here is the observation that multimedia stories are characterized by multiple intervals of constant or slowly varying multimedia attributes, such as, color family histograms, mid-level audio categories, and text categories. The constancy is measured with respect to a numerical description of the attribute, such as a set of probabilities.

We distinguish two types of uniform segment detection: unimodal and multimodal. Unimodal – within the same modality – segment detection means that a video segment exhibits "same" characteristic over a period of time using a single type of modality such as camera motion, presence of certain objects such as videotext or faces or color palette in the visual domain. Multimodal segment detection means that a video segment exhibits a certain characteristic taking into account attributes from different modalities. We first describe uniform segment detection using a single modality: color super-histogram, audio segmentation and classification and topic segmentation, and then give an example of cross-modal segment detection. Next, we describe multimodal pooling as a method to determine boundaries based on applying votes from multiple modalities. Finally we present the multimodal story segmentation as an embodiment of the macro boundary detection.

3.3.1 Single Modality Segmentation

In order to delineate visually uniform video segments, we use a terse representation of the family histograms as described in Section 3.2.1. The visual segmentation in our case relies on the non-continuous micro-boundary segmentation method with unlimited memory as described in Section 3.2.2.

The purpose of audio segmentation and classification is to divide the audio signal into portions of different classes (e.g. speech, music, noise, or combinations thereof) [Li, 2000]. The first step is to partition a continuous bitstream of audio data into non–overlapping segments, such that each segment is homogenous in terms of its class. Each audio segment is then classified, using low–level audio features such as bandwidth, energy and pitch, MFCC, into seven mid-level audio categories. The seven audio categories used in our experiments include silence, single speaker speech, music, environmental noise, multiple speakers' speech, simultaneous speech and music, and simultaneous speech and noise.

A text transcript is extracted from either the closed captions (available in the US) or speech-to-text conversion (automated speech recognition). The timing information for the segmentation comes from the time stamp for each new line of closed captions or the time stamps of the speech to text conversion program. We have to note here that both the time stamps and the transcript are imperfect. The transcript is segmented and categorized with respect to a predefined topic list, for example, war, politics, golf, etc. A frequency-of-word-occurrence metric is used to compare incoming news stories against profiles of manually pre–categorized stories, and to determine the category that best fits it and the degree of fit.

3.3.2 Multimodal Segment Pooling

As discussed in the previous section, the video is divided into homogenous micro-segments. For each media modality, we can have multiple attributes according to which we perform micro-segmentation. For example, one type of micro-segment may correspond to the overall color palette having a fixed set of values or another type of micro segment being characterized by a given type of camera motion, such as a constant zoom in or out. This constitutes the horizontal structure. The segments can be pooled within each modality for the different attributes – the intra-modality case. They can also be pooled across different modalities and different attributes - the inter modality case. Together, these pooling processes form the vertical structure. Pooling can be performed deterministically or probabilistically. It uses domain knowledge. It results in super-segments of video, that is, video segments that have a longer duration than the individual blocks and that are indexed according to their story content and the various multimedia attributes.

3.3.3 Multimodal Segmentation

Once the unimodal segmentation is performed, we can combine the segments from different modalities using multimodal segmentation. We use the descent method through the stack produced by the different unimodal micro-segmentation approaches. The descent method is defined by the process of combining uniform segments of different multimedia attributes, given that one of these attributes is dominant. Next, we describe elements of this method. First, we describe the pre-merging steps, and then the descent steps.

Pre-merging Steps. The role of the pre-merging steps is to detect micro-segments that exhibit uniform properties, remove the potential small gaps and determine attribute templates for further segmentation. The following steps are used to perform uniform segment detection, merging and story segmentation.

Uniform segment detection. Determine time intervals (start and end times) in which a given attribute in a given modality is “uniform”. If an attribute is characterized by a given probability distribution, as in the case of color superhistograms and audio categories, then the segments of uniformity are described by similar probability values; this is illustrated in figure 3.1. A given attribute A_{α}^{m} for a given modality m, such that for $m = 1, 2$, and 3 we associate visual, audio, and text, respectively, and type is uniform in the time interval. For example, if m is 1, then is 1 for color, 2 for motion, and 3 for shape, etc.

Intra-modal segment clustering. In this step, we collect all time intervals related to uniform attribute values for each attribute and domain. If $\Delta T^k_{A^m_\alpha}$ denotes the kth time interval for a given attribute A^m_α, then $\bigcup_k \Delta T^k_{A^m_\alpha}$ represents all the time intervals in a given video associated with a given TV program or movie which share uniform attribute values for A^m_α. This is called the intra-modal segment clustering within the same modality and attribute. This method is deterministic: each interval has the same "weight". Now, within each attribute, we can perform a more specific clustering. For example, time intervals can have uniform color values, but of different hues. Therefore, we can take the union of all the time intervals that have, not only uniform color values, but also colors with the same hue range, e.g., dark blue for news program backgrounds.

Attribute template determination. The attribute template is a combination of numbers that characterize the attribute. This template is either of fixed structure (static), or it changes over a predefined period of time (dynamic). A template is descriptive of a single attribute of a modality. For example, a color template can be a color histogram for an anchor scene, it is incrementally built out of a set of color histograms, and an audio template [Jasinschi, 2001a] is given by a combination of probabilities or votes per each of the seven categories. The attribute template is associated with a given story characteristic.

Dominant attribute determination. In this step, we select the dominant attribute that guides the segmentation and indexing of video stories. For example, text, when present, has had in most cases the dominant role to guide this process because it relates the semantic content. In this case, textual information from the transcript is used to generate temporal units of "uniform" text. Here we determine the dominant attribute by examining large test sets. In the case of audio template we examined over nine news programs.

Template application. Select, for each attribute within the other modalities, a set of video segments, by matching that attribute with its corresponding template. We associate a reliability factor to each matching, i.e., a number on a given scale that tells us how "good" the matching was per video segment. An overall reliability factor is computed by combining these factors for all video segments that are matched for that attribute. This final number will tell us how good the matching was.

Descent Methods. The descent method for story segmentation is realized via multimedia segment combination across multiple modalities. The attributes are temporarily aligned and each attribute with its segments of uniform values is associated with a line, as shown in figure 3.5. The first line on top shows segments of uniform text, the second line down shows uniform audio segments, and the third line shows uniform visual micro-segments. The descent method next performs the union and/or intersection of the micro-segments from selected modalities.

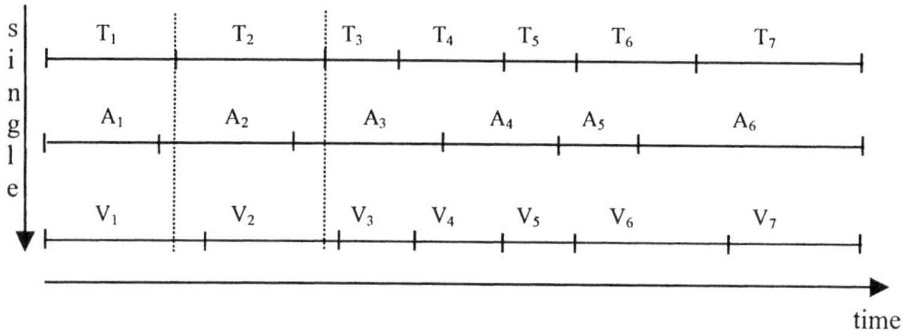

Figure 3.5. Uniform segments for text, audio and visual attributes.

The single descent method describes the process of generating story segments by combining these segments. In the single descent, the idea is to start at the top of the stacked uniform segments and perform a single pass through the layers of unimodal segments that in the end yields the merged segments. In this process of descending down through the different attribute lines, we perform simple set operations such as union and intersection. One approach is to start with the textual segments as the dominant attributes and then combine them with the audio and visual segments. The multiple descent method on the other hand describes the process of multiple passes through stacked uniform segments.

In general, the start and end (duration) times for the different attributes do not coincide, and different criteria for stopping the descent process have to be used. The single descent method corresponds to a single pass through the segments w.r.t. figure 3.5. Starting from a segment for a dominant attribute, we perform the following operations.

- *Single descent with intersecting union:* we descend down through the lines of different attributes, and we include in the union all intersecting segments with this (dominant) attribute. In figure 3.5, given T_2 as dominant text, we include A_2 and A_3 and V_1 and V_2.

- *Single descent with intersection:* we descend down through the lines of different attributes, and we include all intersecting regions with this attribute. In figure 3.5 given T_2 as dominant text, only those parts of A_2 and A_3 and V_1 and V_2 that fall inside the marked vertical boundary of T_2 are being used.

- *Single descent with secondary voting attributes:* we descend down through the lines of different attributes, and we include all intersecting regions with this attribute. The onset and offset border is determined by at least two votes of n other attributes. There could be voting attributes for onset and offset of segments, separately. This means that audio speech can be a voting attribute for onset but not offset of the story segment.
- *Single descent with conditional union:* we descend down through the lines of different attributes, and we include in the union all intersecting regions that exceed a certain length.

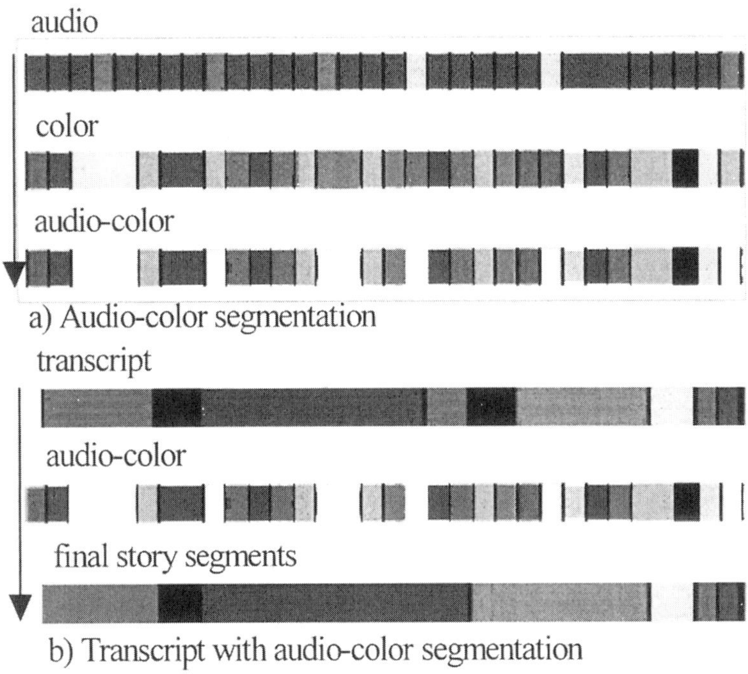

Figure 3.6. The Single Descent merging method using CNN test data.

3.3.4 Experiments

We experimented with a single descent process with conditional union, using text transcript as the dominant attribute, uniform visual micro-segments as described in 3.2.1 and uniform audio segments [Li, 2000]. The steps of the process are as follows:

1 Extract uniform audio segments and apply news audio template. This template [Jasinschi, 2001a] is characterized by the dominance of the speech and speech with simultaneous music audio categories.

2 Extract uniform color segments and create longer segments using a mask with color template (a histogram with specific bins to fit a news studio scene).

3 Merge audio segments from step 1 and visual segments from step 2 by taking the intersection of visual and audio segments. In this manner we eliminate all the extraneous segments that are not speech or speech with music and produce AV segments.

4 Descent with conditional union: Merge segments from the previous step with transcript-based segments. Given the dominant modality, compute the intersection of the time intervals defined by $\sum_{k,\alpha,\beta} S(t_k, \Delta T^k_{A^3_{\alpha,\beta}})$, that is, $\bigcup_{k,\alpha,\beta} \Delta T^k_{A^3_{\alpha,\beta}}$, with all the time intervals of all the attributes within the other two domains; this is realized by: (i) taking the sum of all the time intervals for all (relevant) attributes per modality: $\bigcup_{m=1,2} \bigcup_{k,\alpha,\beta} \Delta T^k_{A^m_{\alpha,\beta}}$, (ii) take the intersection of the total number of time intervals associated with the dominant attribute which results in: $((\bigcup_{k,\alpha,\beta} \Delta T^k_{A^3_{\alpha,\beta}}) \bigcap (\bigcup_{m=1,2} \bigcup_{k,\alpha,\beta} \Delta T^k_{A^m_{\alpha,\beta}})$. In our case, we take the transcript as a dominant attribute, and expand boundaries if there is intersection with AV segments. If a transcript segment is denoted as $T = (x_1, x_2)$ and the audiovisual segment is $AV = (y_1, y_2)$, then the resulting merged segment is $R = (r_1, r_2)$ where $r_1 = \min\{x_1, y_1\}$ and $r_2 = \max\{x_2, y_2\}$ if the intersecting region is at least a certain percentage of the length of the smaller segment.

The process is illustrated using inputs from CNN Headline News programs. In Figure 3.6a) we show the intermediate results after the step 3. The first line represents "audio" segments where brown means speech, red means silence and cyan means noise. The second line represents the dominant colors of different families of frames and the third line represents the merged audiovisual segments that have speech and inherent color uniformity. The audio color segments represent an intersection of color segments that have speech or speech and background noise. In case of transcript absence, the non-white segments in the third line would represent story segments and the process will stop. Figure 3.6b) shows the results after the descent merging process with conditional union. The first line represents transcript segments (which can be only a few seconds small), each block representing a different story segment and each color representing a different topic. The second line represents the audio-color segments. The third line has the final merged story segments. As can be seen, the final story segments are now longer due to the merging, based on the audio-color segments. From transcript line down to the final story segment line we see that story numbers 3 (brown) and 4 (red) are merged; similarly, story numbers 5 (navy) and 6 (pink) are merged. In this process, the current implementation does not take into account the closeness between topics when merging story segments, so story 3 (golf) is merged with story 4 (crime) and story 5 (economy) is merged with 6 (finance). We are working on an improved version where the topic relatedness is also taken into account.

The goal of the presented multimodal segmentation is to create macro-boundaries that are more accurate than the boundaries produced by individual modalities. In the news example, there is a lag between the story beginning and the production of transcript. The correct boundary starts with the anchor's verbal introduction of the story. While this is true in news, in narrative content, such as movies and documentaries, there is a problem of continuous music that helps in the continuity of story telling and passing time. We believe that the proposed framework can help us explore the boundary issues in narrative content.

3.4 Mega-Boundaries

Producing abstract video representations of a TV broadcast requires reliable methods for detecting mega-boundaries by isolating program segments out of the full broadcast material. The mega-boundaries are defined between macro-segments that exhibit different structural and feature consistency (e.g. different genres.) Here we present a commercial detection method as an example of mega-boundary detection. A mega-boundary is located between a commercial break with several commercials and a news program with one or more stories.

In the literature there are many methods that have been proposed for detecting commercials extending back more than 20 years [Blum, 1992] [Bonner, 1982] [Agnihotri, 2003] [Iggulden, 1997] [Y. Li, 2000] [Lienhart, 1997] [McGee, 1999] [Nafeh, 1994] [Novak, 1988]. A common method is detection of high activity rate and black frame detection coupled with silence detection usually associated with the beginning of a commercial break. These methods show partially promising results [Lienhart, 1997]. Use of monochrome images, scene breaks, and action (the amount of edge pixels changing over consecutive frames and motion vector length) as indicative features have also been reported [Lienhart, 1997]. Blum et al. used black frame and "activity" detectors [Blum, 1992]. Activity is the rate of change in luminance level between two different sets of frames. Commercials are generally rich in activity. When a low amount of activity is detected, the commercial is deemed to have ended. Unfortunately, it is difficult to determine what is "activity' and what is the duration of the activity. In addition, black frames are also found in dissolves. Any sequence of black frames followed by a high action sequence can be misjudged and skipped as a commercial. Another technique by Iggulden is using the distance between black frame sequences to determine the presence of a commercial [Iggulden, 1997]. Dimitrova et al. introduced a method that automatically spots repetitive patterns such as commercials, stores and identifies them in future programs [Dimitrova, 2002]. However, the commercial has to be identified to the system before it can recognize it. Nafeh proposed a method for classifying patterns of television programs and commercials based on learning and discerning of broadcast audio and video signals using a neural network [Nafeh, 1994].

3.4.1 Features for Commercial Detection

There are different families of algorithms that can be used for mega-boundary detection based on low and mid-level features. Also, the features can be extracted in either spatial or compressed domain. We have considered the following features:

- black frame, or more generically unicolor frame. In this respect our video testing material is from countries where unicolor frames play an important role in commercial break editing.
- high visual activity, computed using cut distance (i.e. distance between consecutive abrupt shot changes),
- letterbox format, i.e. if the video material is presented in a 4:3 vs. 16:9 aspect ratio.

We performed an extensive analysis of the available features for a data set of eight hours of TV programs from various genres. The feature data was extracted using temporal subsampling, totaling about 600000 frames. Then we divided the set of features into two classes: (i) features that can aid in determining the location of the commercial break (triggers) and (ii) features that can determine the boundaries of the commercial break (verifiers). In the first step, the algorithm checks for triggers. The algorithm, then verifies if the detected segment is a commercial break. In our model we use the two types of features in a Bayesian Network, where the first layer comprises the features that act as triggers, and the second layer comprises the features that act as verifiers.

3.4.2 Triggers and Verifiers

Our approach to commercial detection relies on the framework of triggers and verifiers. The presence of certain features can act as a trigger i.e. a probability that the video segment actually represents a commercial break.

In many countries, black frames (or unicolor frames) are used as delineators of commercials within a commercial break and as markers for the beginning and ending of a whole commercial break. This leads to the assumption that we should expect to encounter unicolor frames within a predetermined threshold (e.g. 50 seconds). After n black frames sequences ($n = 3$ in our experiments) the probability of a commercial being present is increased and a potential commercial end is searched for.

Furthermore, we can apply constraints on the duration of commercials. Usually commercial breaks are longer than one minute and shorter than a predefined number of minutes (e.g. seven minutes.) Also commercial breaks have to be temporally distant for a prespecified length of time. (e.g. one minute.) This latter constraint is important for connecting the segments that potentially represent commercials. This is to prevent from falsely detecting long commercial breaks that actually cover a commercial break and an action scene from a movie.

We use the time interval between detected black or unicolor frames as triggers. Once a potential commercial is detected, then other features are tested to increase or decrease the probability of presence of a commercial break. Presence of a letterbox change or high cut rate expressed in terms of low cut distance is used as a verifier. All the features listed above could be used as commercial triggers, however a much more complex algorithm should then be used for verification. We performed feature analysis over an extensive data set to see how much each feature adds to the probability of a commercial break. Based on the experiments so far, we do not have a positive result on the reliability of other features as commercial triggers.

In the case of a letterbox change, the probability that the given area is a commercial break is increased. In the case of low cut distance (or high cut rate), the probability of a commercial being present is again increased. If the cut rate is below a certain threshold then the probability is decreased. Average keyframe distance is defined as the average shots duration between the last n cuts (experimentally, we derived that $n = 5$). The threshold used for the cut distance can be varied from 6 to 10 seconds for good results. Again, segments which are close by are connected to infer the whole commercial break. There are artful commercials which are very slow and therefore increase the average cut distance temporarily. We allow for the cut distance to be high for a certain amount of time (e.g. 30 seconds) before decreasing the probability of being in a commercial break. As with the black frame indicators, we have placed constraints on the duration of the commercials. In the experiments we have stated that commercial breaks cannot be shorter than one minute and cannot be longer than six minutes. An additional constraint is that commercial breaks are spaced at a certain distance (e.g. one and a half minute apart).

3.4.3 Bayesian Belief Network Model

This trigger-verifier model can be represented using Bayesian Belief networks [Pearl, 1988]. They are directed acyclical graphs (DAG) in which: (i) the nodes correspond to (stochastic) variables, (ii) the arcs describe a direct causal relationship between the linked variables, and (iii) the strength of these links is given by conditional probability distributions (cpds). First we review the notation for the general case and then present our model for the trigger-verifier approach.

Let the set $\Omega(x_1, \ldots x_N)$ of N variables define a DAG. For each variable there exists a sub-set of variables of Ω, Π_{x_i}, the parent set of x_i, i.e., the predecessors of x_i in the DAG, such that $P(x_i \mid \Pi_{x_i}) = P(x_i \mid x_1, \ldots, x_{i-1})$, where $P(\cdot|\cdot)$ is a cpd, strictly positive. Now, given the joint probability density function (pdf) $P(x_1, \ldots, x_N)$, using the chain rule, we obtain

$$P(x_1, \ldots, x_N) = P(x_N \mid x_{N-1}, \ldots, x_1) \times \cdots \times P(x_2 \mid x_1)P(x_1).$$

According to this equation, the parent set Π_{x_i} has the property that x_i and $\{x_1, \ldots, x_N\} \setminus \Pi_{x_i}$ are conditionally independent.

In Figure 3.7 we present the Bayesian Belief Network for our specific trigger-verifier model. The model shows that the root nodes are: black frame detection, B, cut detection, C, and letterbox detection L. We start with a black (unicolor) frame detection, B, which is then used for detection of a sequence of black frames, Q, to serve as a separator detector S. Separator detection is further used as a trigger for potential commercial detector, T. The verification node takes the keyframe distance, D, and the letterbox presence, L, in order to verify the onset or offset of a commercial break.

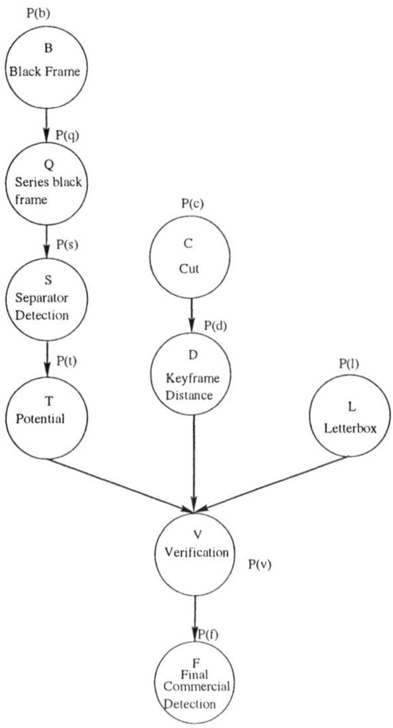

Figure 3.7. Bayesian Belief Network Model for Commercial Detection

Now, given the joint probability density function $P(v_1, \ldots, v_N)$, using the chain rule, we obtain the probability for the verification node:

$$P(v_i) = \sum_{j=1}^{J} \sum_{k=1}^{K} \sum_{m=1}^{M} P(v_i \mid t_j, d_k, l_m) P(t_j) P(d_k) P(l_m)$$

where, probability for potential commercial is given as: $P(t_j) = \sum_{w=1}^{W} P(t_j \mid s_w) P(s_w)$, the probability for separator is given as $P(s_w) = \sum_{o=1}^{O} P(s_j \mid q_o) P(q_o)$, the probability for sequence of black frames is given as $P(q_o) = \sum_{r=1}^{R} P(q_o \mid b_r) P(b_r)$, and the probability for keyframe distance is given as $P(d_k) = \sum_{u=1}^{U} P(d_k \mid c_u) P(c_u)$.

3.4.4 Evolved Algorithm

It is a challenge to have robust commercial detection methodology for various platforms, content formats, and broadcast styles that are used all over the world. Wide deployment of such an algorithm not only requires the development of new algorithms but also updating and tuning of parameters for existing algorithms. Due to the intermittent nature of the features used for commercial detection, and platform restrictions, the commercial detection relies on a set of thresholds to keep the implementation as simple as discussed above. We evolved these thresholds using genetic algorithms (GAs) to optimize the performance.

Genetic algorithms implement a form of Darwinian evolution and have been found to be very effective at discovering high performance solutions to some very complex problems. For this work we have chosen to use Eshelman's CHC algorithm [Eshelman, 1991]. CHC has proven itself very robust for a range of difficult problems with little or no tuning of its control parameters. CHC is a generational style GA with three distinguishing features. First, selection in CHC is monotonic: only the best M individuals, where M is the population size, survive from the pool of both the offspring and parents. Second, CHC prevents parents from mating if their genetic material is too similar (i.e., incest prevention). Controlling the production of offsprings in this way maintains genetic diversity and slows population convergence. Finally, CHC uses a "soft-restart" mechanism. When convergence has been detected, or the search stops making progress, the best individual found so far in the search is preserved. The rest of the population is reinitialized, using the best string as a template and flipping some percentage (i.e., the divergence rate) of the template's bits. This is known as a divergence and introduces new diversity into the population to continue search (details given in [Agnihotri, 2003].)

All experiments were run with CHC's default parameters: population size of 50 chromosomes, and divergence rate of 35% (meaning that 35% of the bits in the best chromosome are randomly flipped in a soft restart).

The set of free parameters to the commercial detector algorithm constitutes the chromosome. For example, experiments 1–4 used a 27–bit chromosome. Each parameter was coded as a binary string where the range and precision needed determine the number of bits required. The first parameter used in experiment 1 has a range from 100–3250 with a precision of 50. This means that we need 6 bits to encode the 63 values $((3250 - 100)/50 = 63)$. This approach gives the needed control over range and precision and permits the binary crossover operator described below.

Each chromosome was decoded into a set of parameters for the commercial detector and this detector was given a test video stream. Every frame of that stream was labeled by the detector as either commercial or program. The correct label for these frames had previously been determined by a human labeler, so each detector-assigned label could be classified as a true positive (TP), a false positive (FP), a true negative

(TN) or a false negative (FN). From these counts, precision and recall were computed as described in Section 3.4.5.

Ideally we should have used a GA that takes two numbers (precision and recall in our case) as fitness measure and tries to find a chromosome that has high values for both these measures. Since CHC requires a scalar performance, we tried various combinations of recall and precision to yield a single value that gives good fitness of a chromosome. In the end, the highest precision and recall performance was achieved with the simple sum of precision and recall.

3.4.5 Results

We obtained two data sets of broadcast video. One contains about 8 hours of European TV broadcast (data set 1) and comprised 13 different TV programs of various genres including movies, news, sports programs, talk shows, and sitcoms. This broadcast sample contains 28 different commercial breaks, for a total duration of about 1.5 hours of commercials. The second data set contains about 4 hours of US content from 11 different television programs including sports, movies, games, talk shows, MTV music videos, and news. The second set includes 35 different commercial breaks, for a total duration of more than one hour.

To find the accuracy of commercial identification we need to determine the number of commercials correctly identified (correct), the number of commercials missed (false negatives: FN), and the number of program segments identified as commercials (false positives: FP). Following the classical definition from information retrieval, the recall is the number of correctly identified (TP) commercials divided by the actual number of commercials present which is given by sum of TP and false negatives (FN). The precision is defined as the number of correctly identified commercials divided by the total number of segments identified as the commercials which is given by the sum of TP and false positives (FP). In formulas, they are expressed as follows:

$$Recall = \frac{TP}{TP + FN}$$

$$Precision = \frac{TP}{TP + FP}$$

The recall is the fraction of commercials that the algorithm successfully detects, in terms of program duration. For the user, losses in recall mean that some commercials will not be skipped. The precision is the percentage of actual commercials within what the algorithm detects as commercials. For the user, losses in precision mean that pieces of non-commercials content will be skipped together with the commercials. We prefer configurations offering high precision, as users would certainly prefer a system that lets a part of the commercial break go falsely detected to a system that skips culmination scenes in an action movie.

The original algorithm was developed and tested on the first data set for which we had a benchmark set by the expert engineer. We needed to determine what fitness measure would be best to use in order to find the best set of parameter values. We performed five different experiments in which we changed the number of parameters and also the fitness metrics.

The performance achieved on validation is the expected performance for commercial detection from this algorithm. The evolved detectors are performing in same region as the best performance achieved by an expert after many months of experimentation and fine tuning. The performances yielded by the experiments in recall and precision for the validation set are 80.8% and 92.6%, 80.8% and 92.6%, 79.7% and 87.4%, 81.3% and 94.3%, for the first four experiments. Experiment 4 had the best results with a fitness measure of $R+P$ and was used in Experiment 5 to produce a commercial detector with recall and precision of 88% and 90.0%. The reason that the validation set did not perform as well as the training set could be due to over-fitting by the GA on the training set.

The next two experiments (experiments 6 and 7) were performed with the second data set, and are shown in Figure 3.8. This data set was acquired after the above work had already been done and serves as a test of our claim that a robust GA is a valuable tool for adjusting a commercial detector to new data. Independently, the expert searched for (using the entire validation set) and reported a detector whose performance on the validation data is shown as the leftmost asterisk in the Figure (58.0% recall and 94.5% precision). We see that the GA located points that look superior to the expert's, but these points may represent overfitting to the training data. When we run two detectors obtained from applying the GAs on the validation set, we see that performance deviates and no longer dominates the expert's point (64.8% and 85.8%, and 79.5% and 92.3%) At this point, we experimented with additional features that assess the likelihood that the video frame is in letterbox format. Believing this may be valuable additional information for commercial detection, the detector algorithm was augmented with logic that considered this new feature. The expert, being in possession of the experiment 6 results, used them as the starting point in an effort to discover good parameter settings for the new letterbox thresholds. The result of this effort (again using the entire validation set) is the rightmost asterisk in Figure 3.8 (77.0% and 95.5%). In experiment 7 the GA searched the augmented threshold set. Three validation tests from this run are also shown at 65.0% and 96.0%, 82.0% and 97.0%, and 68.9% and 94.0%. We see that

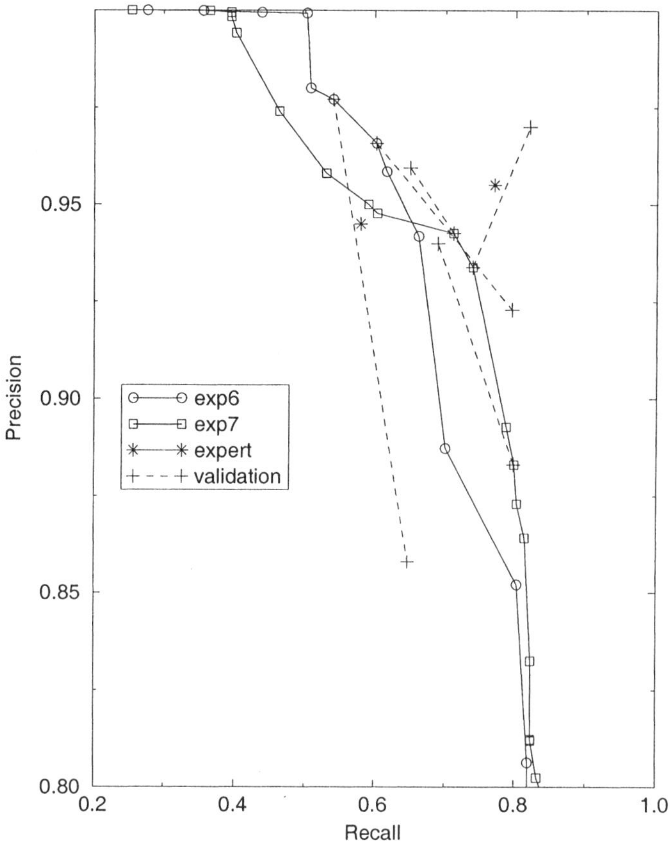

Figure 3.8. Experimental results using GAs

the GA located the best performance seen to date on 82% and 97%. We believe that this demonstrates the potential of the GAs for this emerging application. The speed at which we converge to result as opposed to manual search is very short. GAs are well known for search in massive spaces and not getting stuck at local minimas, unlike gradient search methods.

3.5 Conclusions

In this chapter we presented methods for temporal boundary segmentation in video sequences. This is an attempt to both clarify the usage of the terms in the literature and to formalize the detection of these boundaries. We presented visual scene segmentation as an example of micro-boundary detection, multimodal story segmentation as an example of macro-boundary detection, and commercial detection as an example of mega boundary detection.

Acknowledgements

We would like to acknowledge the technical discussions and contributions from our colleagues: Thomas McGee, Sylvie Jeannin, David Schaffer, Dongge Li, Gang Wei, Norman Haas, Ruud Bolle, Jan Nesvadba, Herman Elenbaas, Jacquelyn Martino and John Zimmerman on topics related to temporal video boundaries. We would like to thank Wouter Leibbrandt, Wim Verhaegh, Angel Janevski and Jan Korst for many constructive comments on this manuscript.

References

N. Dimitrova, I. Sethi and Y. Rui, Media Content Management, in Design and Management of Multimedia Information Systems: Opportunities and Challenges edited by Mahbubur Rahman Syed, Idea Publishing Group, 2000.

B. T. Truong, S. Venkatesh, C. Dorai, "Application of Computational Media Aesthetics Methodology to Extracting Color Semantics in Film", ACM Multimedia Juan Les Pin, December 1–6, 2002.

W. Niblack, J.L. Hafner, T. Breuel, D. Ponceleon, "Updates to the QBIC System," SPIE, vol. 3312, pp. 150-161, 1997.

S.-F. Chang, W. Chen, H. J. Meng, H. Sundaram, and D. Zhong, "VideoQ: An Automated Content Based Video Search System Using Visual Cues," ACM Multimedia, 1997.

N. Vasconcelos, A. Lippman, "Bayesian Modeling Of Video Editing And Structure: Semantic Features For Video Summarization And Browsing," IEEE ICIP 98(153-157).

Jos Mara Martnez Sanchez, Rob Koenen, Fernando Pereira: MPEG-7: The Generic Multimedia Content Description Standard, Part 1 and 2. IEEE MultiMedia 9(2,3): 78-87 (2002).

R. S. Jasinschi and J. Louie, "Automatic TV Genre Classification based on Audio Patterns", Proc. of IEEE 27th EUROMICRO Conference, 370-375, Warsaw, Poland, September 2001.

R. S. Jasinschi, N. Dimitrova, T. McGee, L. Agnihotri, J. Zimmerman, and D. Li, "Integrated Multimedia Processing for Topic Segmentation and Classification", Proc. of IEEE ICIP 2001, Thessaloniki, Greece, October 2001.

R. S. Jasinschi, N. Dimitrova, T. McGee, L. Agnihotri, J. Zimmerman, D. Li, and J. Louie, "A Probabilistic Layered Framework for Integrating Multimedia Content and Context Information," Proc. of IEEE ICASSP 2002, Florida 2002.

N. Dimitrova, J. Martino, L. Agnihotri, H. Elenbaas, Superhistograms for video representation, IEEE ICIP 1999, Kobe, Japan.

Alexander G. Hauptmann, Michael J. Witbrock, Story Segmentation and Detection of Commercials In Broadcast News Video, ADL-98 Advances in Digital Libraries Conference, Santa Barbara, CA, April 22-24, 1998.

Andrew Merlino, Daryl Morey, Mark Maybury, Broadcast News Navigation using Story Segmentation, ACM Multimedia Conference, 1997.

D. Li, I. K. Sethi, N. Dimitrova, and T. McGee, "Classification of General Audio Data for Content-Based Retrieval," Pattern Recognition Letters 2000

D. W. Blum, "Method and Apparatus for Identifying and Eliminating Specific Material from Video Signals, " US patent 5,151,788, September 1992.

E. L. Bonner and N. A. Faerber, "Editing system for video apparatus," US patent 4,314,285, February 1982.

L. Agnihotri, N. Dimitrova, T. McGee, S. Jeannin, D. Schaffer, J. Nesvadba "Evolvable Visual Commercial Detectors", IEEE Conference on Vision and Pattern Recognition, Madison, Wisconsin June 16-22, 2003.

L.J. Eshelman, "The CHC Adaptive Search Algorithm: How to Have Safe Search When Engaging in Nontradi-tional Genetic Recombination," Foundations of Genetic Al-gorithms, Gregory Rawlins (ed.), Morgan Kaufmann, 1991.

J. Iggulden, K. Fields, A. McFarland, J. Wu, "Method and Apparatus for Eliminating Television Commercial Messages, " US Patent 5,696,866, Dec. 7, 1997.

Y. Li and C.C.J. Kuo, "Detecting commercial breaks in real TV programs based on audiovisual information," Proc. Of SPIE Proc. on Internet Multimedia Management System (USA), vol.4210, p.225-236, Boston, 2000.

R. Lienhart, C. Kuhmunch and W. Effelsberg, "On the Detection and Recognition of Television Commercials," in Proc. Of IEEE International Conference on Multimedia Computing and Systems, pp. 509-516, 1997.

T. McGee and N. Dimitrova, Parsing TV Program Structures for Identification and Removal of Non-story Segments, SPIE Conference on Storage and Retrieval for Image and Video Databases VII (ei24) 1999.

N. Dimitrova, T. McGee, L. Agnihotri, "Automatic signature-based spotting, learning and extracting of commercials and other video content", US patent US6469749, Serial No. 417288, issued on 10/22/2002.

J. Nafeh, "Method and Apparatus for Classifying patterns of Television Programs and Commercials Based on Discerning of Broadcast Audio and Video Signals," US patent 5,343,251, Aug. 30, 1994.

A. P. Novak, "Method and System for Editing Unwanted Program Material from Broadcast Signals," US patent 4,750,213, Jun. 7,1988.

J. Pearl, "Probabilistic Reasoning in Intelligent Systems: Networks of Plausible Inference," Morgan Kaufmann Publishers, Inc., San Mateo, California, 1988.

H. Sundaram and S.-F. Chang, "Determining Computable Scenes in Films and their Structures using Audio-Visual Memory Models," presented at ACM Multimedia, Marina Del Rey, 2000.

Chapter 4

VIDEO SUMMARIZATION USING MPEG-7 MOTION ACTIVITY AND AUDIO DESCRIPTORS

A Compressed Domain Approach to Video Browsing

Ajay Divakaran, Kadir A. Peker, Regunathan Radhakrishnan, Ziyou Xiong and Romain Cabasson
Mitsubishi Electric Research Laboratories,
Cambridge, MA 02139
{ajayd,peker,regu,zxiong,romain}@merl.com

Abstract We present video summarization and indexing techniques using the MPEG-7 motion activity descriptor. The descriptor can be extracted in the compressed domain and is compact, and hence is easy to extract and match. We establish that the intensity of motion activity of a video shot is a direct indication of its summarizability. We describe video summarization techniques based on sampling in the cumulative motion activity space. We then describe combinations of the motion activity based techniques with generalized sound recognition that enable completely automatic generation of news and sports video summaries. Our summarization is computationally simple and flexible, which allows rapid generation of a summary of any desired length.

Keywords: MPEG-7, motion activity, video summarization, audio-visual analysis, sports highlights, news video browsing, descriptors, compressed domain, summarizability, fidelity of summary, motion activity space, activity descriptors, key-frame extraction, non-uniform sampling, activity normalized playback, sound recognition, sound clustering, speaker change detection, hidden Markov model (HMM), Gaussian mixture model (GMM), cast identification.

Introduction

Past work on video summarization has mostly employed color descriptors, with some work on video abstraction based on motion features. In this chapter we present a novel approach to video summarization using the MPEG-7 motion activity descriptor [Jeannin and Divakaran, 2001]. Since our motivation is computational simplicity and easy incorporation into consumer system hardware, we focus on feature extraction in the compressed domain. We first address the problem of summarizing a video sequence by abstracting each of its constituent shots. We verify our hypothesis that the intensity of motion activity indicates the difficulty of summarization of a video shot. We do so by studying the variation of the fidelity of a single key-frame with change in the intensity of motion activity as defined by the MPEG-7 video standard. Our hypothesis motivates our proposed key-frame extraction technique that relies on sampling of the video shot in the cumulative intensity of motion activity space. It also motivates our adaptive playback frame rate approach to summarization. We then develop a two-step summarization technique by first finding the semantic boundaries of the video sequence using MPEG-7 generalized sound recognition and then applying the key-frame extraction based summarization to each of the semantic segments.

The above approach works well for video content such as news video in which every video shot needs to be somehow represented in the final summary. In sports video however, all shots are not equally important since key events occur only periodically. This motivates us to develop a set of sports highlights generation techniques that rely on characteristic temporal patterns of combinations of the motion activity and other audio-visual features.

4.1 Background and Motivation

4.1.1 Motion Activity Descriptor

The MPEG-7 [Jeannin and Divakaran, 2001] motion activity descriptor attempts to capture human perception of the "intensity of action" or the "pace" of a video segment. For instance, a goal scoring moment in a soccer game would be perceived as a "high action" sequence by most human viewers. On the other hand, a "head and shoulders" sequence of a talking person would certainly be considered a "low action" sequence by most. The MPEG-7 motion activity descriptor has been found to accurately capture the entire range of intensity of action in natural video. It uses quantized standard deviation of motion vectors to classify video segments into five classes ranging from very low to very high intensity.

4.1.2 Key-frame Extraction from Shots

An initial approach to key-frame extraction was to choose the first frame of a shot as the key-frame. It is a reasonable approach and works well for low-motion shots. However, as the motion becomes higher, the first frame is increasingly unacceptable as a key-frame. Many other subsequent approaches (see [Hanjalic and Zhang, 1999] for a survey) have built upon the first frame by using additional frames that significantly depart from the first frame in addition to the first frame. Another category consists of approaches that rely on clustering and other computationally intensive analysis. Neither category makes use of motion features and are computationally intensive. The reason for using color is that it enables a reliable measure of change from frame to frame. However, motion-compensated video also relies on measurement of change from frame to frame, which motivates us to investigate schemes that use motion vectors to sense the change from frame to frame in a video sequence. Furthermore, motion vectors are readily available in the compressed domain hence offering a computationally attractive avenue. Our approach is similar to Wolf's (see [Hanjalic and Zhang, 1999]) approach in that we also make use of a simple motion metric and in that we do not make use of fixed thresholds to decide which frames will be key-frames. However, unlike Wolf, instead of following the variation of the measure from frame to frame, we propose that the simple shot-wide motion metric, the MPEG-7 intensity of motion activity descriptor, is a measure of the summarizability of the video sequence

4.1.3 The Fidelity of a Set of Key-Frames

The fidelity measure [Chang *et al.*, 1999] is defined as the Semi-Hausdorff distance between the set of key-frames S and the set of frames R in the video sequences. A practical definition of the Semi-Hausdorff distance d_{sh} is as follows: Let the key frame set consist of m frames $S_i, i = 1..m$, and let the set of frames R contain n frames $R_i, i = 1..n$. Let the distance between two frames S_i and R_i be $d(S_i, R_i)$. Define d_i for each frame R_i as

$$d_i = min(d(S_j, R_i)), j = 1, m$$

Then the Semi-Hausdorff distance between S and R is given by

$$d_{sh}(S, R) = max(d_i), i = 1, n$$

Most existing dissimilarity measures satisfy the properties required for the distance over a metric space used in the above definition. In this chapter, we use the color histogram intersection metric proposed by Swain and Ballard (See [Chang *et al.*, 1999]).

4.2 Motion Activity as a Measure of Summarizability

We hypothesize that since high or low action is in fact a measure of how much a video scene is changing, it is a measure of the "summarizability" of the video scene. For instance, a high speed car chase will certainly have many more "changes" in it compared to say a news anchor shot, and thus the high speed car chase will require more resources for a visual summary than would a news anchor shot. Unfortunately, there are no simple objective measures to test such a hypothesis. However, since change in a scene often involves change in the color characteristics as well, we first try to investigate the relationship between color-based fidelity as defined in Section 4.2.2, and intensity of motion activity. Let the key-frame set for shot A be S_A and that for shot B be S_B. If S_A and S_B both contain the same number of key-frames, then our hypothesis is that if the intensity of motion activity of shot A is greater than the intensity of motion activity of shot B, then the fidelity of S_A is less than the fidelity of S_B.

4.2.1 Establishing the Hypothesis

We extract the color and motion features of news video programs from the MPEG-7 test-set, which is in the MPEG-1 format. We first segment the programs into shots. For each shot, we then extract the motion activity features from all the P-frames by computing the standard deviation of motion vector magnitudes of the macro-blocks of each P frame, and a 64 bin RGB Histogram from all the I-frames, both in the compressed domain. Note that intra-coded blocks are considered to have zero motion vector magnitude. We then compute the motion activity descriptor for each I-Frame by averaging those of the previous P-frames in the Group of Pictures (GOP). The I-Frames thus all have a histogram and a motion activity value associated with them. The motion activity of the entire shot is got by averaging the individual motion activity values computed above. From now on, we treat the set of I-frames in the shot as the set of frames R as defined earlier. The simplest strategy for generating a single key-frame for a shot is to use the first frame, as mentioned earlier. We thus use the first I-frame as the key-frame and compute its fidelity as described in Section 4.2.2. We find empirically that a key-frame with Semi-Hausdorff distance at most 0.2 is of satisfactory quality, by analyzing examples of "talking head" sequences. We

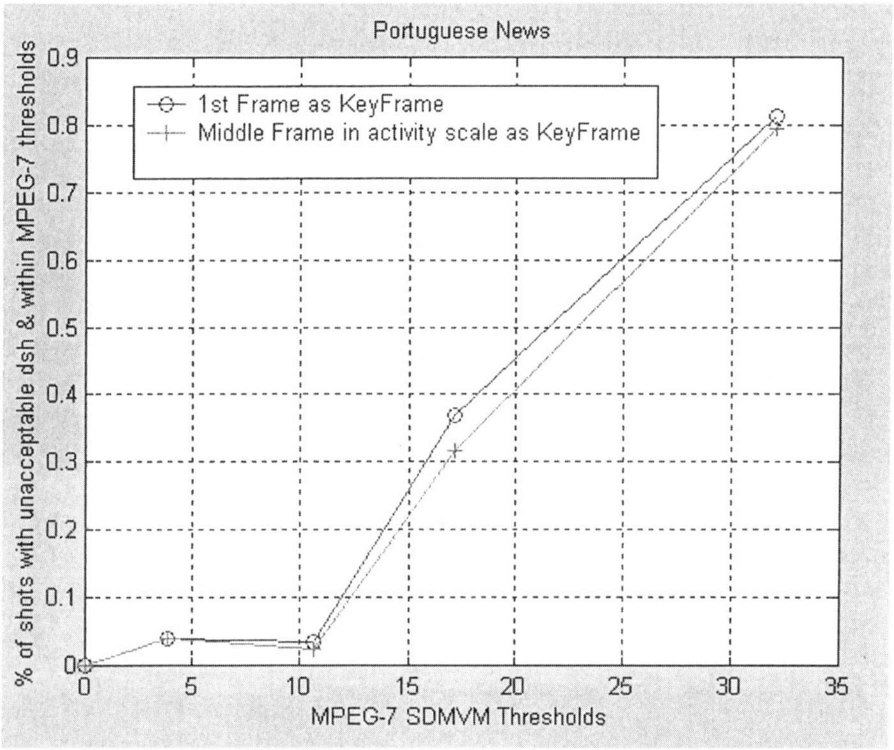

Figure 4.1. Verification of Hypothesis and choice of single key-frame Motion activity (Standard Deviation of Motion Vector Magnitude) vs. percentage duration of unacceptable Shots (Portuguese News from MPEG-7 Test Set jornaldanoite1.mpg)

can therefore classify the shots into two categories, those with key-frames with d_{sh} less than or equal to 0.2 i.e. of acceptable fidelity and those with key-frames with d_{sh} greater than 0.2, i.e. unacceptable fidelity. Using the MPEG-7 motion activity descriptor, we can also classify the shots into five categories ranging from very low to very high activity. We then find the percentage duration of shots with d_{sh} greater than 0.2 in each of these categories for the news program News1 (Spanish News) and plot the results in Figure 4.1. We can see that as the motion activity goes up from very low to very high, the percentage of unacceptably summarized shots also increases consistently. In other words, the summarizability of the shots goes down as their motion activity goes up. Furthermore, the fidelity of the single key frame is acceptable for 90 percent of the shots in the very low intensity of motion activity category. We find the same pattern with other news programs. We thus find experimental evidence that with news program content, our hypothesis is valid. Since news programs are diverse in content, we would expect this result to apply to a wide variety of content. Since we use the MPEG-7 thresholds for motion activity, our result is not content dependent.

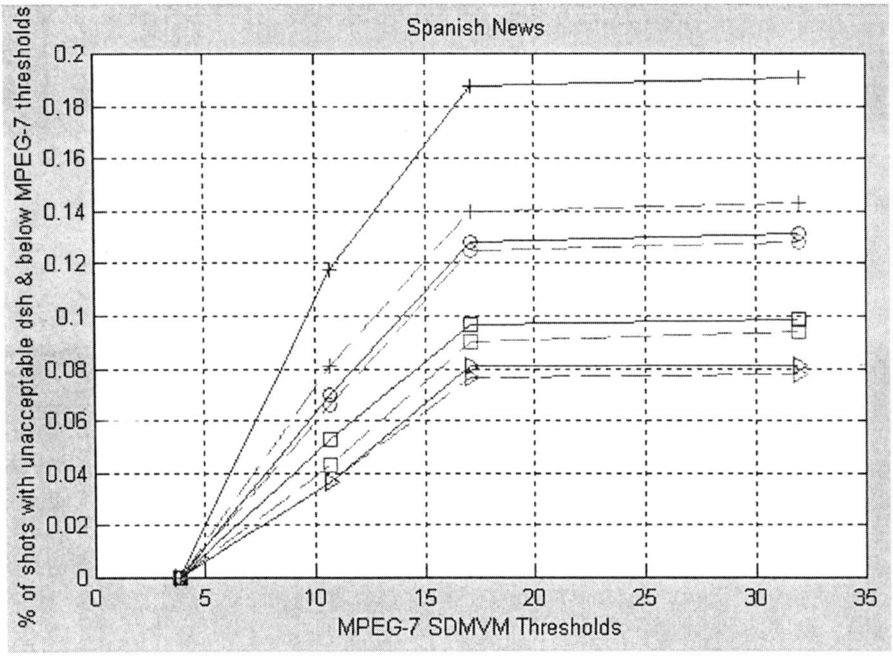

Figure 4.2. Motion activity (Standard Deviation of Motion Vector Magnitude) vs. percentage duration of unacceptable Shots (Spanish News from MPEG-7 Test Set) The firm line represents the "optimal" key-frame strategy while the dotted line represents the progressive key-frame extraction strategy. Each shape represents a certain number of key-frames, the + represents a single frame, the circle two frames, the square three frames and the triangle five frames.

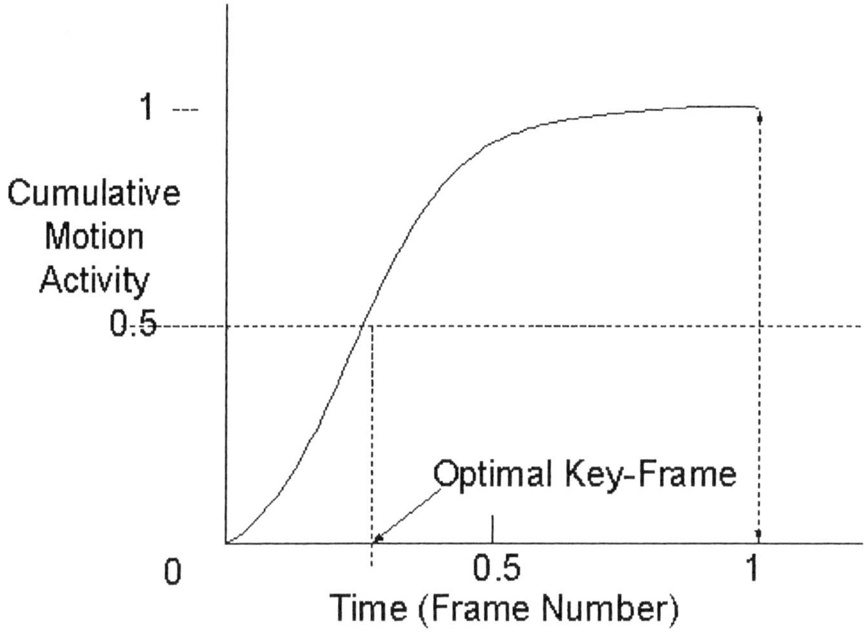

Figure 4.3. Illustration of single key-frame extraction strategy. Note that there is a simple generalization for n key-frames

Table 4.1. Comparison with Optimal Fidelity Key-Frame

Motion Activity	Δd_{sh} First Frame	Δd_{sh} Proposed KF	Number of Shots
Very Low	0.0116	0.0080	25
Low	0.0197	0.0110	133
Medium	0.0406	0.0316	73
High	0.095	0.0576	28
Very High overall avg.	0.0430	0.022	16

4.2.2 A Motion Activity based Non-Uniform Sampling Approach to Key-Frame Extraction

If as per Section 4.2.1 intensity of motion activity is indeed a measure of the change from frame to frame, then over time, the cumulative intensity of motion activity must be a good indication of the cumulative change in the content. Recall that in our review of previous work we stated that being forced to pick the first frame as a key-frame is disadvantageous. If the first frame is not the best choice for the best first key-frame, schemes that use it as the first key-frame such as those surveyed in [Hanjalic and Zhang, 1999] start off at a disadvantage. This motivates us to find a better single key-frame based on motion activity. If each frame represents an increment in information then the last frame is at the maximum distance from the first. That would imply that the frame at which the cumulative motion activity is half the maximum value is the best choice for the first key-frame. We test this hypothesis by using the frame at which the cumulative motion activity is half its value for the entire shot as the single key-frame instead of the first key-frame for the Spanish News sequence and repeating the experiment in the previous section. We find that the new key-frame choice out-performs the first frame, as illustrated in Figure 4.1. Since previous schemes have also improved upon using the first frame as a key-frame, we need to compare our single key-frame extraction strategy with them. For each shot, we compute the optimal single key-frame as per the fidelity criterion mentioned in Section 4.2.2. We compute it by finding the fidelity of each of the frames of the video, and then finding the frame with the best fidelity. We use the fidelity of the aforementioned optimal key-frame as a benchmark for our key-frame extraction strategy by measuring the difference in d_{sh} between the optimal key-frame obtained through the exhaustive computation mentioned earlier and the key-frame obtained through our proposed motion-activity based strategy. We carry out a similar comparison for the first-frame based strategy as well. We illustrate our results in Table 4.1. Note that our strategy produces key-frames that are nearly optimal in fidelity. Furthermore, the quality of the approximation degrades as the intensity of motion activity increases. In other words, we find that our strategy closely approximates the optimal key-frame extraction in terms of fidelity while using much less computation. This motivates us to propose a new nearly optimal strategy, which is very similar to the activity-based sampling proposed in the next section [Peker *et al.*, 2001] as follows. To get n key-frames, divide the video sequence into n equal parts on the cumulative motion activity scale. Then use the frame at the middle of the cumulative motion activity scale of each of the segments as a key-frame, thus getting n key-frames. Note that our n key-frame extraction strategy scales linearly with n unlike the exhaustive computation described earlier, which grows exponentially in complexity because of the growth in the number of candidate key-frame combinations. It is for this reason that we do not compare our n frame strategy with the exhaustive benchmark. We illustrate our strategy in Figure 4.3.

Note that our criterion of acceptable fidelity combined with the n frame strategy enables a simple and effective solution to the two basic problems of key-frame extraction:

- How many key-frames does a shot require for acceptable fidelity?
- How do we generate the required number of key-frames?

4.2.3 A Simple Progressive Modification

Progressive key-frame extraction is important for interactive browsing since the user may want further elaboration of summaries that he has already received. Since our key-frame extraction is not progressive, we propose a progressive modification of our technique. We start with the first frame, and then choose the last frame as the next key-frame because it is at the greatest distance from the first frame. We carry this logic forward as we compute further key-frames by choosing the middle key-frame in cumulative motion activity space as the third key-frame, and so on recursively. The modified version is slightly inferior to our original technique but has the advantage of being progressive. In figure 4.2, we illustrate a typical result. We have tried our approach with several news programs from different sources [Divakaran *et al.*, 2002; Divakaran *et al.*, 2001].

4.3 Constant Pace Skimming Using Motion Activity

4.3.1 Introduction

In the previous section, we showed that the intensity of motion activity (or pace) of a video sequence is a good indication of its "summarizability." Here we build on this notion by adjusting the playback frame-rate or the temporal sub-sampling rate. The pace of the summary serves as a parameter that enables production of a video summary of any desired length. Either the less active parts of the sequence are played back at a faster frame rate or the less active parts of the sequence are sub-sampled more heavily than are the more active parts, so as to produce a summary with constant pace. The basic idea is to skip over the less interesting parts of the video.

4.3.2 Constant Activity Sub-Sampling or Activity Normalized Playback

A brute force way of summarizing video is to play it back at a faster-than-normal rate. Note that this can also be viewed as uniform sub-sampling. Such a fast playback has the undesirable effect of speeding up all the portions equally thus making the high motion parts difficult to view, while not speeding up the low motion parts sufficiently. This suggests that a more useful approach to fast playback would be to play back the video at a speed that provides a viewable and constant level of motion activity. Thus, the low activity segments would have to be speeded up considerably to meet the required level of motion activity, while the high activity segments would need significantly less speeding up if at all. In other words, we would speed up the slow parts more than we would the fast parts. This can be viewed as adaptive playback speed variation based on motion activity, or activity normalized playback. Another interpretation could be in terms of viewability or "perceptual bandwidth." The most efficient way to play the video is to make full use of the instantaneous perceptual bandwidth, which is what the constant activity playback achieves. We speculate that the motion activity is a measure of the perceptual bandwidth as a logical extension of the notion of motion activity as a measure of summarizability. Let us make the preceding qualitative description more precise. To achieve a specified activity level while playing back video, we need to modify the activity level of the constituent video segments. We first make the assumption that the intensity of motion activity is proportional to the motion vector magnitudes. Hence, we need to modify the motion vectors so as to modify the activity level. There are two ways that we can achieve this:

- Increasing/Decreasing the Playback Frame-Rate - As per our earlier assumption, the intensity of motion activity increases linearly with frame rate. We can therefore achieve a desired level of motion activity for a video segment as follows:

  Playback frame rate =(Original Frame rate)*(Desired level of motion activity/original level of motion activity)

- Sub-Sampling the Video Sequence - Another interpretation of such playback is that it is adaptive sub-sampling of the frames of the segment with the low activity parts being sub-sampled more heavily. This interpretation is especially useful if we need to summarize remotely located video, since we often cannot afford the bandwidth required to actually play the video back at a faster rate.

In both cases above, the summary length would then be given by

Summary length=(Sum of frame activities)/desired activity

Note that we have not yet specified the measure of motion activity. The most obvious choices are the average motion vector magnitude and the variance of the motion vector magnitude [Jeannin and Divakaran, 2001; Peker and Divakaran, 2001]. However, there are many variations possible, depending on the application. For instance, we could use the average motion vector magnitude as a measure of motion activity, so as to favor segments with moving regions of significant size and activity. As another example, we could use the magnitude of the shortest motion vector as a measure of motion activity, so as to favor segments with significant global motion.

The average motion vector magnitude provides a convenient linear measure of motion activity. Decreasing the allocated playing time by a factor of two, for example, doubles the average motion vector magnitude. The average motion vector magnitude $\hat{r}$ of the input video of N frames can be expressed as:

$$\hat{r} = (\frac{1}{N}) \sum_{i=1}^{N} r_i$$

where the average motion vector magnitude of frame i is r_i. For a target level of motion activity r_{target} in the output video, the relationship between the length L_{output} of the output video and the length L_{input} of the input video can be expressed as:

$$L_{output} = \frac{\hat{r}}{r_{target}} L_{input}$$

While playing back at a desired constant activity is possible in theory, in practice it would require interpolation of frames or slowing down the playback frame rate whenever there are segments that are higher in activity than the desired level. Such interpolation of frames would be computationally intensive and difficult. Furthermore, such an approach does not lend itself to generation of a continuum of summary lengths that extends from the shortest possible summary to the original sequence itself.

The preceding discussion motivates us to change the sampling strategy to achieve a guaranteed minimum level of activity as opposed to a constant level of activity, so we are able to get a continuum of summaries ranging from the sequence being its own summary to a single frame summary. With the guaranteed minimum activity method, we speed up all portions of the input video that are lower than the targeted minimum motion activity r_{target} so that they attain the targeted motion activity using the above formulations. The portions of the input video that exceed the targeted motion activity can remain unchanged.

At one extreme, where the guaranteed minimum activity is equal to the minimum motion activity in the input video, the entire input video becomes the output video. When the guaranteed minimum activity exceeds the maximum motion activity of the input video, the problem reduces to the above constant activity case. At the other extreme, where the targeted level of activity is extremely high, the output video includes only one frame of the input video as a result of down-sampling or fast play.

The length of the output video using the guaranteed minimum activity approach can be determined as follows. First, classify all of the frames of the input video into two sets. A first set S_{higher} includes all frames j where the motion activity is equal to or higher than the targeted minimum activity. The second set S_{lower} includes all frames k where the motion activity is lower than the targeted motion activity. Then, the length of the input video is expressed by:

$$L_{input} = L_{higher} + L_{lower}.$$

The average motion activity of frames j that belong to the set S_{lower} is

$$\hat{r}_{lower} = \frac{1}{N_{lower}} \sum_{j}^{N_{lower}} r_k$$

and the length of the output converted is

$$L_{output} = \frac{\hat{r}_{lower}}{r_{target}} L_{lower} + L_{higher}$$

It is now apparent that the guaranteed minimum activity approach reduces to the constant activity approach because when L_{higher} becomes zero, the entire input video needs to be processed.

4.3.3 How Fast Can You Play the Video?

While in theory it is possible to play the video back at infinite speed, the temporal Nyquist rate limits how fast it can be played without becoming imperceptible by a human observer. A simple way of visualizing this is to imagine a video sequence that captures the revolution of a stroboscope. At the point where the frame rate is equal to the rate of revolution, the stroboscope will seem to be stationary. Thus, the maximum motion activity level in the video segment determines how fast it can be played. Furthermore, as the sub-sampling increases, the video segment reduces to a set of still frames or a "slide show." Our key-frame extraction technique of Section 4.2 can therefore also be seen as an efficient way to generate a slide show since it uses the least feasible number of frames. It is obvious that there is a cross-over point where it is more efficient to summarize the video segment using a slide show instead of with a video or "moving" summary. How to locate the cross-over point is an open problem. We hope to address this problem in our ongoing research.

Figure 4.4. Illustration of adaptive sub-sampling approach to video summarization. The top row shows a uniform sub-sampling of the surveillance video, while the bottom row shows an adaptive sub-sampling of the surveillance video. Note that the adaptive approach captures the interesting events while the uniform sub-sampling mostly captures the highway when it is empty. The measure of motion activity is the average motion vector magnitude in this case.

4.3.4 Experimental Procedure, Results and Discussion

We have tried our adaptive speeding up of playback using diverse content and got satisfactory results. We find that with surveillance footage of a highway (see Figure 4.4), using the average motion vector magnitude as the measure of motion activity, we are able to produce summaries that successfully skip across the parts where there is insignificant traffic, and focus on the parts with significant traffic. We get good results with the variance of motion vector magnitude as well. We have been able to focus on parts with large vehicles, as well as on parts with heavy traffic. Note that our approach is computationally simple since it relies on simple descriptors of motion activity.

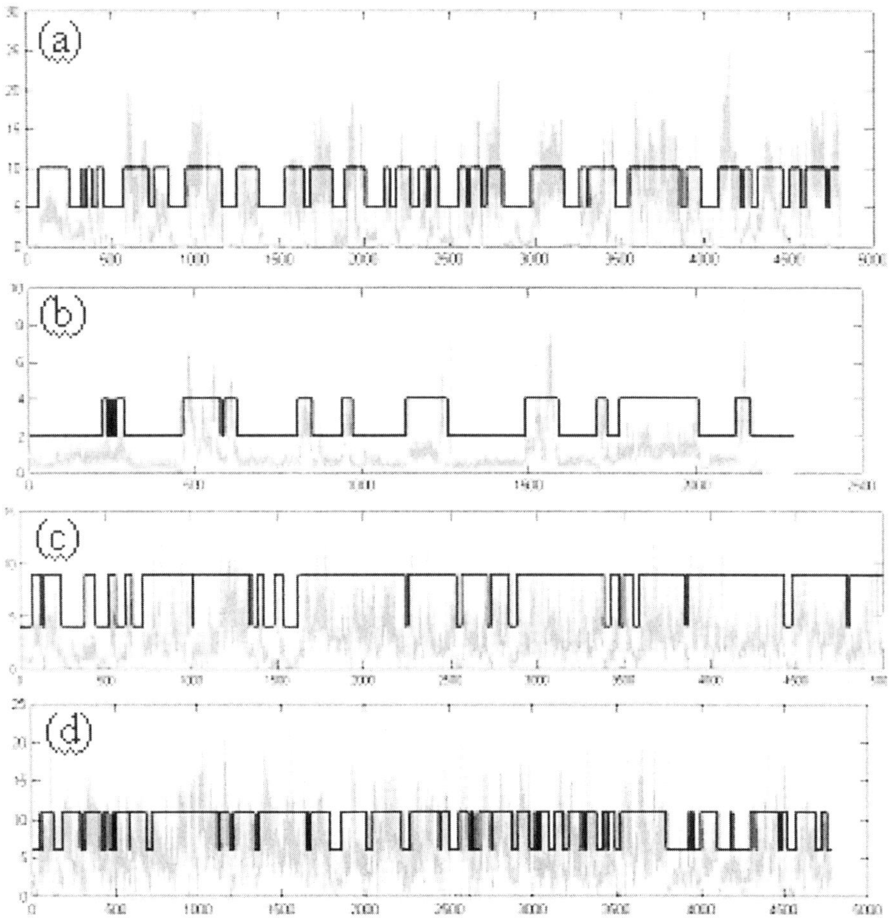

Figure 4.5. Motion activity vs. frame number for four different types of video content, with the smoothed and quantized versions superimposed: a) Golf. b) News segment. c) Soccer. d) Basketball.

As illustrated by our results, the constant pace skimming is especially useful in surveillance and similar applications in which the shots are long and the background is fixed. Note that in such applications, color-based techniques are at a disadvantage since the semantics of the interesting events are much more strongly related to motion characteristics than to changes in color.

We have also tried this approach with sports video and with news content with mixed success. When viewing consumer video such as news or sports, skipping some parts at fast forward and viewing others at normal speed may be preferable to continuously changing the playback speed. For this purpose, we smooth the activity curve using a moving average and quantize the smoothed values (Figure 4.5). In our implementations, we used two-level quantization with the average activity as the threshold. For news and basketball (Figure 4.5 b and d), we used manually selected thresholds.

For the golf video, the low activity parts are where the player prepares for his hit, followed by the high activity part where the camera follows the ball, or closes up on the player. For the news segment, we are able to separate the interview parts from the outdoor footage. For the soccer video, we see low activity segments before the game really starts, and also during the game where the game is interrupted. The basketball game, in contrast with soccer, has a high frequency of low and high activity segments. Furthermore, the low activity parts are when the ball is in one side of the court and the game is proceeding, and the high activity occurs mostly during close-ups or fast court changes. Hence, those low activity parts should be played at normal speed, while some high activity parts can be skipped. In short, we can achieve a semantic segmentation of various content types using motion activity, and use domain knowledge to determine where to skip or playback at normal speed. Then, we can accordingly adapt our basic strategy described in Section 4.3, to different kinds of content.

The foregoing discussion on sports video indicates that the key to summarizing sports video is in fact in identifying interesting events. This motivates us to investigate temporal patterns of motion activity that are associated with interesting events in Section 4.5.

For news video, it is perhaps better to use a slide show based on our key-frame extraction technique since the semantics of the content are not directly coupled with the motion characteristics of the content. However, it works best when the semantic boundaries of the content are known. In that case, a semantic segment can be segmented into shots and key-frames extracted for each shot so as to produce a set of key-frames for the entire semantic segment. This motivates us to investigate automatic news video semantic, or topic, boundary detection using audio features in Section 4.4.

4.4 Audio-Assisted News Video Browsing

4.4.1 Motivation

The key-frame based video summarization techniques of Section 4.2 are evidently restricted to summarization of video shots. Video in general, and news video in particular, consists of several distinct semantic units, each of which in turn consists of shots. Therefore it would be much more convenient to somehow choose the semantic unit of interest and then view its key-frame based summary in real-time, than to generate a key-frame based summary of the entire video sequence and then look for the semantic unit of interest in the summary. If a topic list is available in the content meta-data, then the problem of finding the boundaries of the semantic units is already solved, so that the user can first browse the list of topics and then generate and view a summary of the desired topic. However, if the topic list is unavailable, as is the case more often than not, the semantic boundaries are no longer readily available. We then need to extract the semantic/topic boundaries automatically. Past work on news video browsing systems has emphasized news anchor detection and topic detection, since news video is typically arranged topic-wise and the news anchor introduces each topic at the beginning. Thus knowing the topic boundaries enables the user to skim through the news video from topic to topic until he has found the desired topic, which he can then watch using a normal video player.

Topic detection has been mostly carried out using closed caption information, embedded captions and text obtained through speech recognition, by themselves or in combination with each other (see [Hanjalic *et al.*, 2001; Jasinschi *et al.*, 2001] for example). In such approach, text is extracted from the video using some or all of the aforementioned sources and then processed using various heuristics to extract the topic(s).

News anchor detection has been carried out using color, motion, texture and audio features. For example, in [Wang *et al.*, 2000] Wang et al carry out a speaker separation on the audio track and then use the visual or video track to locate the faces of the most frequent speakers or the "principal cast." The speaker separation is carried out by first classifying the audio segments into the categories of speech and non-speech. The speech segments are then used to train Gaussian Mixture Models (GMM's) for each speaker that enable speaker separation through fitting of each speech segment with the different GMM's.

Speaker separation has itself been a topic of active research. The techniques mostly rely on extraction of low level audio features followed by a clustering/classification procedure.

Speaker separation and principal cast identification provide a solution to the problem of topic boundary detection. Unfortunately, the proposed methods in the literature are highly complex in computation and hence do not lend themselves well to consumer video browsing systems. Furthermore, in a video browsing system, in addition to principal cast identification we would also like to identify further semantic characteristics based on the audio track such as speaker gender, as well as carry out searches for similar scenes based on audio.

4.4.2 MPEG-7 Generalized Sound Recognition

The above discussion motivates us to try the sound recognition framework proposed by Casey [Casey, 2001] and accepted by the MPEG-7 standard. In this framework, reduced rank spectra and entropic priors are used to train Hidden Markov Models for various sounds such as speech, male speech, female speech, barking dogs, breaking glass etc. The training is done off-line with training data so that the category identification is done by using the Viterbi algorithm on the various HMM's, which is computationally inexpensive. For each sound segment, in addition to the sound category identification, a histogram of percentage duration spent in each state of the HMM is also generated. This histogram serves as a compact feature vector that enables similarity matching.

4.4.3 Proposed Principal Cast Identification Technique

Our procedure [Divakaran *et al.*, 2003] is illustrated in Figure 4.6. It consists of the following steps:

1 Extract motion activity, color and audio features from the News video.

2 Use the sound recognition and clustering framework as shown in figure 4.6 to find speaker changes.

3 Use motion and color to merge speaker clusters and to identify principal speakers. The locations of the principal speakers provide the topic boundaries.

4 Apply the motion based browsing described in Sections 4.2 and 4.3 to each topic.

In the following section, we describe the speaker change detection component of the whole system.

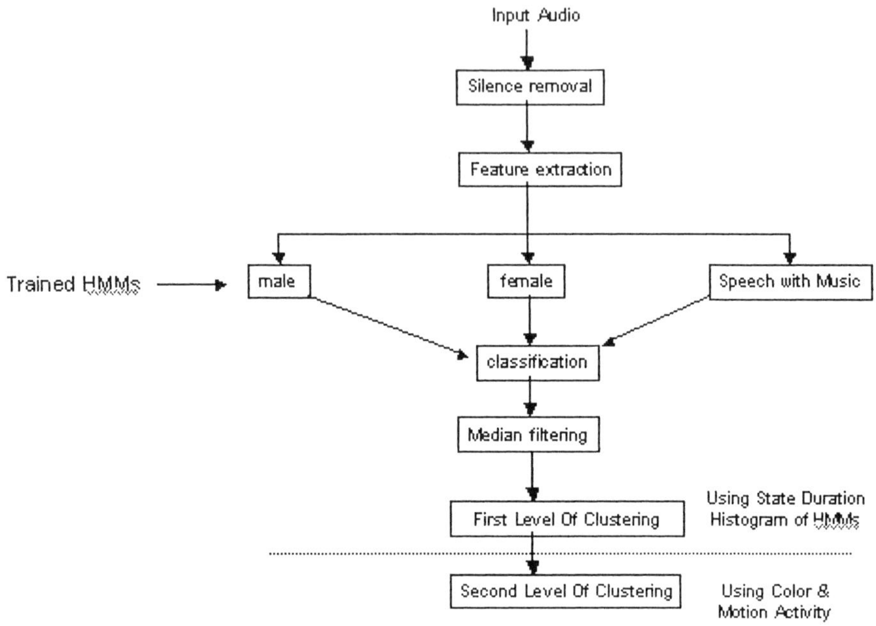

Figure 4.6. Audio Feature Extraction, Classification and Segmentation for Speaker Change Detection

Speaker Change Detection Using Sound Recognition and Clustering. The input audio from broadcast news is broken down into sub-clips of smaller durations such that they are homogenous. The energy of each sub-clip is calculated so as to detect and remove silent sub-clips. MPEG-7 features are extracted from non-silent sub-clips and are classified into one of the three sound classes namely male, female and speech with music.

At this point, all male and female speakers are separated. Median filtering is performed to eliminate spurious changes in speakers. In order to identify individual speakers within male and female sound class, an unsupervised clustering step is performed based on the MPEG-7 state duration histogram descriptor. This clustering step is essential to identify individual male and female speakers after classification of all sub-clips into one of three sound classes. Each classified sub-clip is then associated with a state duration histogram descriptor.

The state duration histogram can also be interpreted as a modified representation of GMM. Each state in the trained HMM can be thought of as a cluster in feature space, which can be modeled by a Gaussian.

Note that the state duration histogram represents the probability of occurrence of a particular state. This probability can be interpreted as the probability of a mixture component in a GMM. Thus, the state duration histogram descriptor can be considered as a reduced representation of GMM, which in its unsimplified form is known to model a speaker's utterance well. Note, since the histogram is derived from the HMM, it also captures some temporal dynamics which a GMM cannot. We are thus motivated to use this descriptor to identify clusters belonging to different speakers in each sound class.

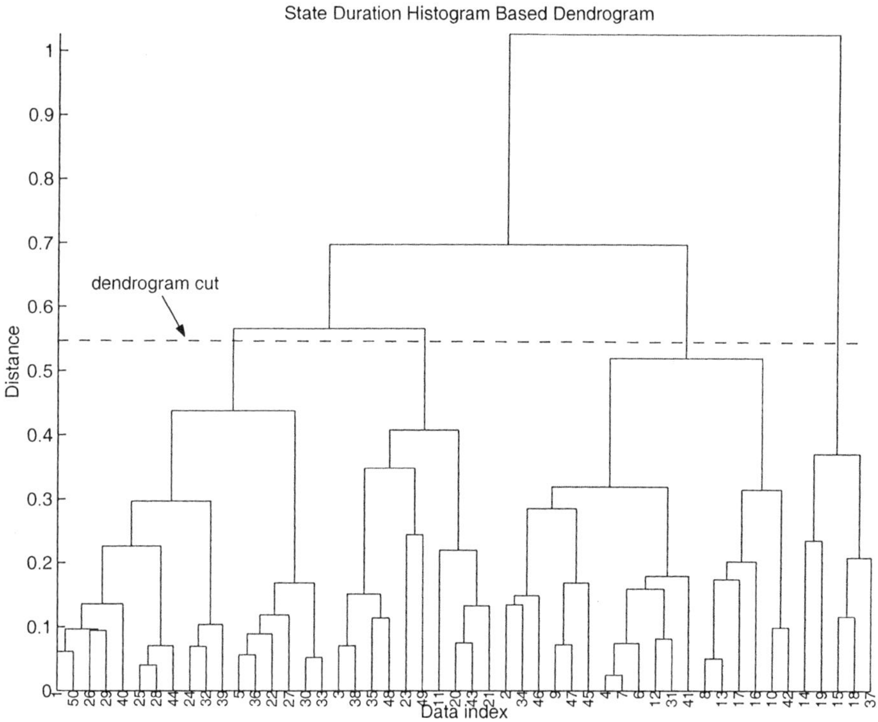

Figure 4.7. Example Dendrogram Construction and Cluster generation for a contiguous set of female speech segments

The clustering approach adopted was bottom-up agglomerative dendrogram construction based. In this approach, a distance matrix is first obtained by computing pairwise distance between all utterances to be clustered. The distance metric used is a modification of Kullback-Leibler distance to compare two probability density functions (pdf). The modified Kullback-Leibler distance between two pdfs H and K is defined as below:

$$D(H,K) = \Sigma h_i log(\frac{h_i}{m_i}) + m_i log(\frac{k_i}{m_i})$$

where $m_i = \frac{h_i + k_i}{2}$ and $1 \leq i \leq Number\ of\ bins\ in\ the\ histogram$

Then a dendrogram is constructed by merging two closest clusters according to the distance matrix until there is only one cluster. Then, the dendrogram is cut to obtain the clusters of individual speakers (See 4.7).

Second level of clustering using motion and color features. Since clustering is done only on contiguous male/female speech segments, we achieve speaker segmentation only in that portion of the whole audio record of the news program. A second level of clustering is required to establish correspondences between clusters from two distinct portions. Motion and color cues extracted from the video can be used for the second level of clustering. Once the clusters have been merged, it is easy to identify principal cast and hence semantic boundaries. Then, the combination of the principal cast identification and motion-based summary of each semantic segment enables quick and effective browsing of the news video content.

4.4.4 Experimental Procedure and Results

Data-Set. Since broadcast news contains mainly three sound classes viz., male speech, female speech and speech with music, we collected training examples for each of the sound classes from three and a half hours of news video from four different TV channels manually. The audio signals are all mono-channel, 16 bits per sample with a sampling rate of 16 kHz. The database for training HMMs is partitioned into 90%-10% training/testing set for cross-validation. The test sequences for speaker change detection were two audio tracks from TV broadcast news: *News1* with duration 34 minutes and *News2* with duration 59 minutes.

Feature Extraction. The input audio signal from the news program is cut into segments of length three seconds and silent segments are removed. For each non-silent three second segment, MPEG-7 features are extracted as follows. Each segment is divided into overlapping frames of duration 30 ms with 10 ms overlapping for consecutive frames. Each frame is then multiplied by a Hamming window function: $w_i = (0.5 - 0.46\cos(2\pi i/N)), i = 1 \ldots N$, where N is the number of samples in the window. After performing an FFT on each windowed frame, the energy in each of the sub-bands is computed and the resulting vector is projected onto the first 10 principal components of each sound class.

We also extract from compressed domain, the MPEG-7 intensity of motion activity for each P-frame and a 64 bin color histogram for each I-frame from the video stream of the news program.

Table 4.2a. Classification Results on News1 with Average Recognition Rate = 80.384%

	Female	Male	Speech and Music
Female	116	52	13
Male	6	184	2
Speech and Music	15	46	264

Table 4.2b. Classification Results on News2 with Average Recognition Rate = 84.78%

	Female	Male	Speech and Music
Female	370	50	8
Male	34	248	1
Speech and Music	47	49	391

Classification and Clustering. The number of states in each of the HMMs was chosen to be 10 and each state is modeled by a single multi-variate Gaussian. Note that the state duration histogram descriptor relates to GMM only if the HMM states are represented by a single Gaussian. Viterbi decoding is performed to classify the input audio segment by picking the model for which the likelihood value is maximum which is followed by median filtering on the labels obtained for each three second segment so as to impose time continuity. For each contiguous set of labels, agglomerative clustering is performed using the state duration histogram descriptor to obtain a dendrogram as shown in figure 4.7. The dendrogram is then cut at a particular level relative to the maximum height of the dendrogram to obtain individual speaker clusters.

The accuracy of the proposed approach for speaker change detection depends on the following two aspects: the classification accuracy of the trained HMMs for segmenting the input audio into male and female speech classes, and the accuracy of the clustering approach to identify individual speakers in a contiguous set of male/female speech utterances. Tables 4.2a and 4.2b show the classification performance of the HMM on each of the test broadcast news sequences *without* any post-processing on the labels.

Tables 4.2a and 4.2b indicate that many male and female speech segments are classified as speech with music. These segments actually cor-

Table 4.2c. Speaker Change Detection Accuracy on two test news sequences **A** Number of speaker change time stamps in ground truth; **B** Number of speaker change time stamps obtained after clustering step; **C** Number of 'TRUE' speaker change time stamps; **D** Precision = [C]/[A] in %; **E** Recall = [C]/[B] in %

	A	B	C	D	E
News1	68	90	46	67.64	51.11
News2	173	156	87	50.28	55.77

respond to outdoor speech segments in the broadcast news and have been misclassified due to the background noise.

Since clustering was done only on contiguous male or female speech segments instead of the whole audio record, the performance of the system is evaluated as a speaker change detection system even though we achieve segmentation in smaller portions. We compare speaker change positions output by the system against ground truth speaker changes, and count the number of correct speaker change positions. Table 4.2c summarizes the performance of the clustering approach on both the test sequences.

The accuracy of the proposed algorithm for speaker change detection is only moderate for both the news programs, for the following reasons. The dendrogram pruning procedure adopted to generate clusters was the simplest one and hence results would improve if multi-level dendrogram pruning procedure was adopted. Some of the speaker changes were missed by the system because of misclassification of the outdoor speech segments into speech with music class. Moreover, there was no post-processing on the clustering labels to incorporate some domain knowledge. For example, a cluster label sequence such as s_1, s_2, s_1, s_2 in which speakers alternate frequently, is highly unlikely in a news program and would simply mean that s_1 and s_2 belong to the same speaker cluster. However, even with such a moderate accuracy in audio analysis it is shown below that by combining motion and color cues from video, the principal cast from news program can be obtained.

In order to obtain correspondence between speaker clusters from distinct portions of the news program, we associate each speaker cluster with a color histogram, obtained from a frame with motion activity less than a threshold. Obtaining a frame from a low-motion sequence increases the confidence of its being one from a head-shoulder sequence. A second clustering step is then performed based on color histogram to

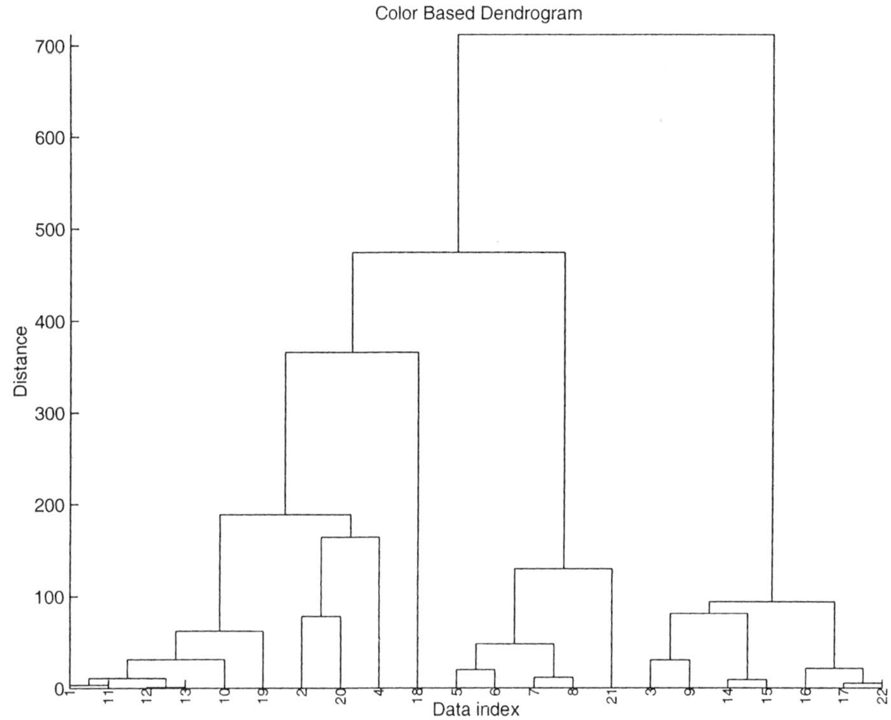

Figure 4.8. Second level of clustering based on color histograms of frames corresponding to male speaker clusters

merge clusters obtained from pure audio analysis. Figure 4.8 shows the second level clustering results. After this step, principal cast clusters can be identified as either the clusters that occupy significant periods of time or clusters that appear at different times, throughout the news program. Due to copyright issues, we unfortunately cannot display any of the images corresponding to the clusters. In future work, we hope to use public domain data such as news video from the MPEG-7 video test-set.

4.4.5 Future Work

Our future work will focus on first improving the audio classification using more extensive training. Second, we will further improve clustering of audio by allowing multi-level dendrogram cutting. Third, we will further refine the combination of motion activity and color to increase the reliability of principal cast identification.

4.5 Sports Highlights Detection

Most sports highlights extraction techniques depend on the camera motion, and thus require accurate motion estimation for their success. In the compressed domain, however, since the motion vectors are noisy, such accuracy is difficult to achieve. Our discussion in Section 4.3.4 motivates us to investigate temporal patterns of motion activity as a means for event detection since they are simple to compute. We begin by devising strategies for specific sports based on domain knowledge. We find that using motion activity alone gives rise to too many false positives for certain sports. We then resort to combining simple audio and video cues to eliminate false positives. Our results with the combination as well as the results from Section 4.4 motivate us to apply the generalized sound recognition framework to a unified highlights extraction framework for soccer, golf and baseball. Our current challenge is therefore to combine visual features with the sound recognition framework. We hope to build upon our experience with combining low-level cues.

Note that the problem has two parts viz. detecting an interesting event and then capturing its entire duration. In this chapter we focus on the first part since in our target applications, we are able to solve the second part by merely using an interactive interface that allows conventional fast-forward and rewind.

4.5.1 Highlights Extraction for Golf

In [Peker *et al.*, 2002] we describe a simple technique to detect Golf highlights. We first carry out a smoothing of the motion activity of the video sequence. In the golf video, we look for long stretches of very low activity followed by high activity. These usually correspond to the player working on his shot, then hitting, followed by the camera following the ball or zooming on the player. We mark the interesting points in the golf video in this way. We generate a highlight sequence by merely concatenating ten-second sections that begin at the interesting points marked. We get interesting results but miss some events, most notably the putts since they are often not associated with rapid camera motion.

4.5.2 Extraction of Soccer Video Highlights

For soccer games, we capitalize on domain specific constraints that help us locate the highlights. Our basic intuition is that an interesting event always has the following associated effects. The game stops and stays stopped for a non-trivial duration, and the crowd noise goes up either in anticipation of the event, or after the event has taken place

This suggests the following straightforward strategy to locate interesting events or highlights. Locate all audio volume peaks. The peaks correspond to increase in crowd noise in response to the interesting event. At every peak, find out if the game stopped before it and stayed stopped for a non-trivial duration. Similarly find out if the game stopped after the peak and stayed stopped. The concatenation of the stops before and after the audio peak, if valid, forms the highlight associated with that audio peak. We describe the details of the computation of the audio peaks and the start and stop patterns of motion activity in [Cabasson and Divakaran, 2003]

Experimental Results. We have tried our strategy [Cabasson and Divakaran, 2003] with seven soccer games from Korea, Europe and the United States of America including a women's soccer game. We find that we miss only one goal and capture all the other goals in all the games. We also capture several other interesting parts of the game that do not lead to a goal being scored such as attempts at goals, major injuries etc. Despite its success with diverse content, this technique has a significant drawback which is its reliance on a low-level feature like audio volume, which may not always be a good indicator of the content semantics. We are thus motivated once again to resort to generalized sound recognition.

4.5.3 Audio Events Detection based Highlights Extraction from Baseball, Golf and Soccer Games in a Unified Framework

We describe an audio-classification based approach in which we explicitly identify applause/cheering segments, and use those to identify highlights. We also use our audio classification framework to set up a future investigation of fusion of audio and video cues for sports highlights extraction.

Audio Classification Framework. The system constraints of our target platform rule out having a completely distinct algorithm for each sport and motivate us to investigate a common unified highlights framework for our three sports of interest, golf, soccer and baseball. Since audio lends itself better to extraction of content semantics, we start with audio classification.

We illustrate the audio classification based framework in Figure 4.9. In the audio domain, there are common events relating to highlights across different sports. After an interesting golf or baseball hit or an exciting soccer attack, the audience shows appreciation by applauding or

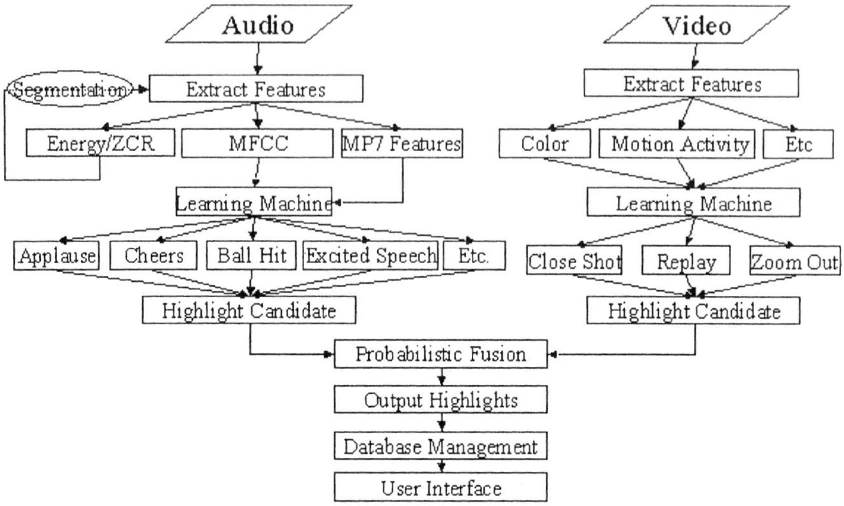

Figure 4.9. Highlights Extraction Framework: We have partially realized the video and the probabilistic fusion.

cheering. The duration of the applause/cheering is an indication of the "significance" of the moment. Furthermore, in sports broadcast video, there are also common events related to commercial messages that often consist of speech or speech and music. Our observation is that the audience's cheering and applause are more general across different sports than is the announcer's excited speech. We hence look for robust audio features and classifiers to classify and recognize the following audio signals: applause, cheering, ball hits, music, speech and speech with music. The former two are used for highlights extraction and the latter three are used to filter out the uninteresting segments. We employ a general sound recognition framework based on HMMs trained for each of the classes. The HMMs operate on Mel Frequency Cepstral Coefficients (MFCC). Each segment is 0.5 seconds long while each frame is 30 ms long. We show that our classification accuracy is high and thus we are motivated to extract the highlights based on the results of the classification.

We collect the continuous or uninterrupted stretches of applause/cheering. We retain all the segments that are a certain percentage of the maximum duration of the applause and cheering. Our default choice is 33%. Note that this gives us a simple threshold with which to tune the highlights

	[A]	[B]	[C]	[D]	[E]	[F]	[G]
[1]	58	47	35	60.3%	74.5%	151	23.1%
[2]	42	94	24	57.1%	25.5%	512	4.7%
[3]	82	290	72	87.8%	24.8%	1392	5.2%
[4]	54	145	22	40.7%	15.1%	1393	1.6%

Table 4.3. Classification results for the four games. [1]: golf game 1; [2]: golf game 2; [3] baseball game; [4] soccer game. [A]: Number of Applause and Cheering Portions(NACP) in Ground Truth Set; [B]: NACP by Classifiers WITH Post-processing; [C]: Number of TRUE ACP by Classifiers; [D]: Precision $\frac{[C]}{[A]}$; [E]: Recall $\frac{[C]}{[B]}$ WITH Post-processing; [F]: NACP by Classifiers WITHOUT Post-processing; [G]: Recall $\frac{[C]}{[F]}$ WITHOUT Post-processing.

extraction for interactive browsing. Finally, we add a preset time cushion to both ends of each selected segment to get the final presentation time stamps. The presentation then consists of playing the video normally through a time-stamp pair corresponding to a highlight and then skipping to the next pair.

Note that the duration of the applause/cheering also enables generation of sports highlights of a desired length as follows: We can sort all the applause/cheering segments in a descending order of duration. Then given a time budget, we can spend it by playing each segment down the list until the budget is exhausted. While the above technique is promising, we find that it still has room for improvement as can be seen in Table 4.3. First, the classification accuracy needs to be improved. Second, using applause duration alone is probably simplistic. Its chief strength is that it uses the same technique for three different sports. Since we do not expect a high gain from increased classification accuracy alone, we are motivated to combine visual cues with the audio classification with the hope that we may get a bigger gain in highlight extraction efficacy.

4.5.4 Future Work

In ongoing research, we propose to combine the semantic strength of the audio classification with the computational simplicity of the techniques described in Section 4.5.

We are thus motivated to investigate combination of audio classification with the motion activity pattern matching. We illustrate our general framework in Figure 4.9. Note that the audio classification and the video feature extraction both produce candidates for sports highlights. We then propose to use probabilistic fusion to choose the right candidates. Note also that the proposed video feature extraction goes well beyond the motion activity patterns that we described earlier.

Our proposed techniques have the advantage of simplicity and fair accuracy. In ongoing work, we are examining more sophisticated methods for audio-visual feature fusion.

4.6 Efficacy of Summarization

Using the capture of goals in soccer as a measure of the accuracy of highlights has the big advantage of zero ambiguity but also has the disadvantage of incompleteness since it ignores all other interesting events that could arguably be even more interesting. We are currently working on a framework to assess the accuracy of a sports highlight in terms of user satisfaction, so as to get a more complete assessment. Such a framework would require a carefully set up psycho-visual experiment that creates a ground truth for the "interesting" and "uninteresting" parts of a sports video.

More structured content such as news lends itself to easier assessment of the success of the summarization. However, note that our fidelity based computations for example, did not address semantic issues. The assessment of the semantic success of a summary is still an open problem although techniques such as ours provide part of the solution.

4.7 Data Mining vs. Video Mining: A Discussion

Finally we consider the video mining problem in the light of existing data mining techniques. The fundamental aim of data mining is to discover patterns. In the results we have presented here, we have attempted to discover patterns in audio-visual content through principal cast detection, sports highlights detection, and location of "significant" parts of video sequences. Note that while our techniques do attempt to satisfy the aim of pattern discovery, they do not directly employ common data mining techniques such as time-series mining or discovery of association rules. Furthermore, in our work, the boundary between detection of a known pattern and pattern discovery is not always clear. For instance, looking for audio peaks and then motion activity patterns around them could be thought of as merely locating a known pattern, or on the other hand, could be thought of as an association rule formed between the audio peak event and the temporal pattern of motion event, through statistical analysis of training data.

Our approach to video mining is to think of it as content adaptive or blind processing. For instance, using temporal association rules over multiple-cue labels could throw up recurring patterns that would help locate the semantic boundaries of the content. Similarly, we could mine the time series stemming from the motion activity values of video frames. Our experience so far indicates that techniques that take advantage of the spatio-temporal properties of the multi-media content are more likely to succeed than methods that treat feature data as if it were generic statistical data. The challenge however is to minimize the content dependence of the techniques by making them as content adaptive as possible. We believe that this is where the challenge of video mining lies.

4.8 Conclusions

We presented video summarization techniques based on sampling in the cumulative intensity of motion activity space. The key-frame extraction works well with news video and is computationally very simple. It thus provides a baseline technique for summarization. It is best used to summarize distinct semantic units, which motivates us to identify such units by using MPEG-7 generalized sound recognition. We also addressed the related but distinct problem of generation of sports highlights by developing techniques based on the MPEG-7 motion activity descriptor. These techniques make use of domain knowledge to identify characteristic temporal patterns of high and low motion activity along with audio patterns that are typically associated with interesting moments in sports video. We get promising results with low computational complexity. There are a few important avenues for further improvement of our techniques. First, the audio-assisted video browsing can be made more robust and further use made of the semantic information provided by the audio classification. Second, we should develop content-adaptive techniques that adapt to variations in the content, from genre to genre or within a genre. Third, we should investigate incorporation of visual semantics such as the play-break detection proposed in [Xie *et al.*, 2002]. The main challenge then is to maintain and enhance our ability to rapidly generate summaries of any desired length.

Acknowledgements

The authors would like to thank Padma Akella and Pradubkiat Bouklee for carrying out the basic design and implementation of the video browsing demonstration system. We would also like to thank Anthony Vetro for numerous discussions and suggestions. We would like to thank Dr. Huifang Sun for his guidance and encouragement. We would like to thank Dr. Tommy Poon for his enthusiastic support and comments. We would also like to thank Shih-Fu Chang for many helpful discussions and suggestions. We would like to thank Michael Casey for providing his audio classification expertise and software. We would like to thank our colleagues Dr. Tokumichi Murakami, Mr. Takashi Kan and Mr. Kohtaro Asai for their steady support and encouragement over the years. We would like to thank our colleagues Dr. Masaharu Ogawa, Mr. Kazuhiko Nakane, Mr. Isao Otsuka and Mr. Kenji Esumi, for their valuable application oriented comments and suggestions.

References

Jeannin, S., and A. Divakaran. *MPEG-7 Visual Motion Descriptors,* IEEE Transactions on Circuits and Systems for Video Technology, Vol 11, No. 6, pp. 720-724, June 2001.

Peker K.A. and A. Divakaran *Automatic Measurement of Intensity of Motion Activity of Video Segments* , Proc. SPIE Conference on Storage and Retrieval for Media Databases, January 2001.

Chang, H.S., S. Sull and S.U. Lee, *Efficient video indexing scheme for content-based retrieval*, IEEE Transactions on Circuits and Systems for Video Technology, Vol. 9, No. 8, pp. 1269-1279, December 1999.

Hanjalic A., and H. Zhang, *An Integrated Scheme for Automated Video Abstraction Based on Unsupervised Cluster-Validity Analysis*, IEEE Transactions on Circuits and Systems for Video Technology, Vol. 9, No. 8, December 1999.

Peker K. A., A. Divakaran and H. Sun, *Constant pace skimming and temporal sub-sampling of video using motion activity*, Proc. IEEE International Conference on Image Processing (ICIP), Thessaloniki, Greece, October 2001.

Divakaran A., K.A. Peker and R. Radhakrishnan, *Video Summarization with Motion Descriptors*,Journal of Electronic Imaging, October 2001.

Divakaran A., K.A. Peker and R. Radhakrishnan, *Motion Activity-based Extraction of Key-Frames from Video Shots*, Proc. IEEE International Conference on Image Processing (ICIP), Rochester, NY, USA, October 2002.

A. Hanjalic, G. Kakes, R.L. Lagendijk, and J. Biemond, *Dancers: Delft advanced news retrieval system,*" in SPIE Electronic Imaging 2001: Storage and retrieval for Media Databases, San Jose, USA., 2001.

R.S. Jasinschi, N. Dimitrova, T. McGee, L. Agnihotri, J. Zimmerman, and D. Li, *Integrated multimedia processing for topic segmentation and classification,* in ICIP-2001, Thessaloniki, Greece, 2001, pp. 366-369

Divakaran A., R. Radhakrishnan, Z. Xiong and M. Casey *A Procedure for Audio-Assisted Browsing of News Video using Generalized Sound Recognition*, Proc. SPIE Conference on Storage and Retrieval for Media Databases, January 2003.

Peker K.A., R. Cabasson and A. Divakaran *Rapid Generation of Sports Highlights using the MPEG-7 Motion Activity Descriptor,* Proc. SPIE Conference on Storage and Retrieval for Media Databases, January 2002.

Cabasson R. and A. Divakaran *Automatic Extraction of Soccer Video Highlights using a combination of motion and audio features*, Proc. SPIE Conference on Storage and Retrieval for Media Databases, January 2003.

Casey M. *MPEG-7 Sound Recognition Tools,* IEEE Transactions on Circuits and Systems for Video Technology, Vol 11, No. 6, June 2001.

Wang Y., Z. Liu and J-C. Huang, *Multimedia Content Analysis*, IEEE Signal Processing Magazine, November 2000.

Y. Rui, A. Gupta, and A. Acero, *Automatically extracting highlights for TV baseball programs,"* Eighth ACM International Conference on Multimedia, pp. 105–115, 2000.

W. Hsu, *Speech audio project report,* Class Project Report, 2000, www.ee.columbia.edu/~winston

L. Xie, S.F. Chang, A. Divakaran, and H. Sun, *Structure analysis of soccer video with hidden markov models,* Proc. Interational Conference on Acoustic, Speech and Signal Processing, (ICASSP-2002), May 2002, Orlando, FL, USA.

P. Xu, L. Xie, S.F. Chang, A. Divakaran, A. Vetro, and H. Sun, *Algorithms and system for segmentation and structure analysis in soccer video,* Proceedings of IEEE Conference on Multimedia and Expo, pp. 928–931, 2001.

L. Rabiner and B.-H. Juang, *Fundamentals of Speech Recognition,* Prentice Hall, 1993.

Z. Xiong, R. Radhakrishnan, A. Divakaran, and T.S. Huang, *Audio Events Detection based Highlights Extraction from Baseball, Golf and Soccer Games in A Unified Framework,* ICASSP 2003, April 6-10, 2003.

Chapter 5

MOVIE CONTENT ANALYSIS, INDEXING AND SKIMMING VIA MULTIMODAL INFORMATION

Ying Li
IBM T.J. Watson Research Center
19 Skyline Dr., Hawthorne, NY 10532
yingli@us.ibm.com

Shrikanth Narayanan and C.-C. Jay Kuo
Integrated Media Systems Center and Department of Electrical Engineering
University of Southern California, Los Angeles, CA 90089
shri,cckuo@sipi.usc.edu

Abstract A content-based movie analysis, indexing and skimming system is developed in this research. Specifically, it includes the following three major modules: 1) an event detection module, where three types of movie events, namely, two-speaker dialogs, multiple-speaker dialogs, and hybrid events are extracted from the content. Multiple media cues such as audio, speech, visual and face information are integrated to achieve this goal; 2) a speaker identification module, where an adaptive speaker identification scheme is proposed to recognize target movie cast members for content indexing purposes. Both audio and visual sources are exploited in the identification process, where the audio source is analyzed to recognize speakers using a likelihood-based approach, and the visual source is examined to locate talking faces with face detection/recognition and mouth tracking techniques; 3) a movie skimming module, where an event-based skimming system is developed to abstract movie content in the form of a short video clip for content browsing purposes. Extensive experiments on integrating multiple media cues for movie content analysis, indexing and skimming have yielded encouraging results.

Keywords: Movie content analysis, event detection, speaker identification, speaker modeling, talking face detection, skim generation, multimodal analysis, two-speaker dialogs, multiple-speaker dialogs, face detection and recognition, audio analysis, mouth tracking, browsing, integrating multiple media cues, story structure, Viterbi model adaptation.

Introduction

With the fast growth of multimedia information, content-based video analysis, indexing and representation have attracted increasing attention in recent years. Many applications have emerged in areas such as video-on-demand, distributed multimedia systems, digital video libraries, distance education and entertainment. The need for content-based video indexing and retrieval was also recognized by ISO/MPEG, and a new international standard called "Multimedia Content Description Interface" (or in short, MPEG-7) has been initialized since 1998 [MPEG-7, 1999], and finalized in September 2001.

Content-based video analysis aims at obtaining a structured organization of the original video content and understanding its embedded semantics like humans do. Content-based video indexing is the task of tagging semantic video units obtained from content analysis to enable convenient and efficient content retrieval. Although content understanding is an easy task for a human, it remains a very complicated process for a computer because of the limitations of machine perception in unconstrained environments and the unstructured nature of video data. Robust techniques are still lacking today despite a large amount of effort in this area [Yeung *et al.*, 1997; Sundaram *et al.*, 2000; Tsekeridou *et al.*, 2001].

So far, the predominant approach to this problem is to first extract some low- to mid-level audiovisual features such as color, texture, shape, motion, shots, keyframes, object trajectories, human faces, and classified audio classes; then partially derive or understand the video semantics by analyzing and integrating these features. Although encouraging results have been reported in previous work, a semantic gap still exists between the real video content and the video contexts derived from these features.

Content-based video skimming aims at abstracting the video content and presenting its essence to users in form of a short video clip. Video skimming is mainly adopted for video browsing purposes, which forms an inseparable part of a video indexing and retrieval system. Currently, most video skimming work is based on pre-extracted keyframes, which may result in undesired discontinuities in the skim's audio and visual content.

This work first extracts semantic video events from the content. Feature film is the major focus of this work. The extracted event information can be utilized to facilitate movie content browsing, abstraction and indexing. In the second stage, target movie cast members or anchors are identified from the content, based on speech and face information. Finally, a movie skimming system is proposed which summarizes movie content based on the pre-obtained event structure.

The rest of this chapter is organized as follows. We first review some related previous work and give an overview of our approach in Section 5.1. Low-level audiovisual content analysis is briefly reviewed in Section 5.2. Section 5.3 elaborates on the work of movie event extraction and characterization. In Section 5.4, the proposed adaptive speaker identification scheme is detailed. Section 5.5 describes our work on movie skimming generation. Experimental results are reported and discussed in Section 5.6, and finally, concluding remarks are drawn in Section 5.7.

5.1 Approach Overview

5.1.1 Event Detection and Extraction

By *event*, we mean a video paragraph which contains a meaningful theme and progresses under a consistent environment. Because event is a rather subjectively defined concept, below we will review some previous work that addresses similar concepts or has similar research goals.

Shot detection, where a *shot* is defined as a set of contiguously recorded image frames, is usually the first step towards video content understanding. However, while the shot forms the building block of the video content, this low-level structure does not correspond to the underlying video semantics in a direct and convenient way. Thus most recent work tends to understand the video semantics by extracting the underlying video *scenes*, where a scene is defined as a collection of semantically related shots that depicts and conveys a high-level concept or story. For instance, [Rui *et al.*, 1998; Yeung *et al.*, 1996] proposed to extract video scenes by grouping visually similar and temporally adjacent shots. [Huang *et al.*, 1998] developed a scene detection scheme based on the integration of audio, visual and motion information. Similar ideas were also explored in the Informedia project [Hauptmann *et al.*, 1995].

However, while a scene does provide a higher-level video context, not every scene contains a meaningful thematic topic, especially for movies where progressive scenes, which are frequently inserted to establish the story situation, are actually unimportant for content understanding. Therefore, we propose to analyze movies and extract events which operate at a higher semantic level so as to better reveal, represent and abstract the movie content. The reason we choose movie applications is that a movie has a clear story structure which can be well exploited by our approach. Moreover, a movie has many specific characteristics created by complex film editing techniques [Reisz *et al.*, 1968].

Because a movie plot is usually developed through either dialogs or actions, we focus on identifying the following three types of events in this research: *two-speaker dialogs*, *multiple-speaker dialogs*, and *hybrid events* which accommodate events with less speech and more visual action. The detection of dialogs has been explored by some previous work. For instance, [Yeung *et al.*, 1997] characterized a temporal event into either dialog, action or others. In particular, it proposed to detect a dialog by searching a shot sequence with a repetitive nature of two dominant shots such as "A B A B A B". A similar periodic analysis was also employed in [Sundaram *et al.*, 2000] for dialog detection. However, since the arrangement of shot sequences in a dialog basically varies with the film genre and also heavily depends on the directorial style, strict periodic analysis appears to be too restrictive for a general scenario. In addition, the problem becomes more complex when multiple speakers are present. Finally, speech information, which is an important indicator for dialogs, was not considered in either work.

In this work, we aim to fulfill this task by analyzing the movie content structure and exploiting films' special editing features. Moreover, visual cues such as human faces, will be effectively integrated with speech cues to obtain robust results.

5.1.2 Speaker Identification

Automatic speaker identification has been an active research topic for many years with the bulk of the progress facilitated by work on standard speech databases such as YOHO, HUB4, and SWITCHBOARD [Wan *et al.*, 2000; Johnson, 1999]. Recently, with the increase of accessibility to other media sources, researchers have attempted to improve system performance by integrating knowledge from all available media cues. For instance, [Tsekeridou *et al.*, 2001] proposed to identify speakers by integrating cues from both speaker recognition and facial analysis schemes. This system is, however, impractical for generic video types since it assumes there is only one human face in each video frame. Similar work was also reported by [Li and Wei, 2001], where TV sitcoms were used as test sequences. In [Li *et al.*, 2002], a speaker identification system was proposed for movie content indexing, where both speech and visual cues were employed. This system, however, has certain limitations since it only identifies speakers in movie dialogs.

Most existing work in this field deals with supervised identification, where speaker models are not allowed to change once they are trained.

Two drawbacks arise when this approach is applied to feature films: 1) we may not have sufficient training data; because a speaker's voice can have distinct variations along time, especially in feature films, a model built with limited training data cannot model a speaker well for the entire sequence; 2) since we have to go through the movie at least once to collect and transcribe the training data before the actual identification process can be started, time is wasted and system efficiency is decreased.

An adaptive speaker identification system is proposed in this work which offers a better solution for identifying speakers in movies. Specifically, after building coarse models for target speakers during system initialization, we continuously update them on the fly by adapting to speakers' newly contributed data. It is our claim that models adapting to incoming speech data can achieve higher identification accuracy because they can better capture speakers' voice variations along time. Both audio and visual sources will be exploited in the identification process, where the audio source is analyzed to recognize speakers using a likelihood-based approach, and the visual source is parsed to find talking faces using face detection/recognition and mouth tracking techniques.

5.1.3 Video Skimming

Video skimming is used to summarize video content and present users with its essence in the form of a moving storyboard. So far, many research efforts have been reported on video skimming such as the VAbstract system [Pfeiffer *et al.*, 1996], which generated trailers for feature films, and the Informedia project developed at Carnegie-Mellon University [Smith *et al.*, 1997] which generated short video synopses by exploiting text keywords and image information. Various techniques such as textual content analysis [Toklu *et al.*, 2000], speech transcription analysis [Taskiran *et al.*, 2002], and dynamic sampling schemes [Tseng *et al.*, 2002] have been adopted to obtain meaningful skims.

However, while acceptable results were reported, most research efforts developed their skimming schemes based on pre-developed summarization schemes and, as a result, the generated video skims are actually the by-products of those summarization systems. For instance, a general approach is to first locate all keyframes using certain summarization techniques, then either shots or other video segments that contain these keyframes are assembled to form the skim. Two major drawbacks exist in these systems: 1) the skim's semantic flow is discontinuous. This is because that keyframes are usually visually different and temporally apart, thus when we generate a skim by expanding these keyframes, all visual, audio and motion content continuities might be lost; 2) the embedded audio cue is ignored in the skim generation process. The accompanying audio track usually contains very important information, especially for movies, but unfortunately, skimming of the audio source has not been well exploited yet.

This work proposes an event-based video skimming system for feature films which aims to produce better skims by avoiding the above drawbacks. Specifically, given the set of extracted movie events, we first compute six types of low- to high-level features for each event; then we use these features to evaluate the importance of an event when integrated with user preference. Finally, important events are assembled to generate the final skim.

5.2 Audio and Visual Content Pre-analysis

The first step towards visual content analysis is shot detection. A color histogram-based approach is employed to perform this task. Specifically, once a distinct peak is detected in the frame-to-frame histogram difference, we declare it as a shot cut [Li and Kuo, 2003]. An average of 92.5% precision and 99% recall rates were achieved in the current work. In the second step, we proceed to extract one or more keyframes from each shot to represent its underlying content. For simplicity, currently we assign the first and last frames of each shot as its keyframes.

The audio content analysis mainly deals with audio content classification, where each shot is classified into one of the following four classes: *silence*, *speech*, *music*, and *environmental sounds* [Zhang *et al.*, 2001]. Five audio features are extracted for classification purposes, which include the short-time energy function, the short-time average zero-crossing rate, the short-time fundamental frequency, the energy band ratio and the silence ratio. An average of 88% classification accuracy was achieved in the current work.

Facial analysis is mainly performed to detect human faces in the frontal view or faces rotated by plus or minus 10 degrees from the vertical direction. Currently, we use the face detection and recognition library provided by the HP Labs [HP Labs, 1998], which reports 85% detection accuracy and 80% recognition accuracy.

5.3 Movie Event Extraction

Film, a recording art, is practical, environmental, pictorial, dramatic, narrative and musical [Monaco, 1977]. Since a film operates in a limited period of time, all movie shots are efficiently organized by a film maker in such a way that audiences will follow his or her own way of story-telling.

Specifically, this goal is achieved by presenting audiences a sequence of cascaded events that gradually develop the movie plot.

There are basically two ways to develop a thematic topic in an event: through actions or through dialogs. Although they differ a lot in the way of conveying the story, both of them present repetitive visual structures at certain points. This is due to the so-called *montage* effect as described in [Tarkovsky, 1986], "*In order to present two or more processes as simultaneous or parallel, you have to show them one after the other, they have to be in sequential montage.*" This means that, in order to convey conversations, innuendoes or reactions, film makers have to repeat important shots to express the content and motion continuity.

Our objective in this work is to extract three types of events, *i.e.* the two-speaker dialogs, the multiple-speaker dialogs and the hybrid events. Figure 5.1 gives two movie dialog models which are constructed by an analysis of movie editing styles [Reisz *et al.*, 1968]. In particular, Figure 5.1 (a) models a two-speaker dialog, and (b) models a multiple-speaker dialog (here we use three speakers as an example). Each node in the figure represents a shot that contains the indicated speaker(s), and arrows are used to denote the switches between two shots. As we can see from these models that there are certain shot repeating patterns in both cases, although the former one presents more periodic patterns than the latter one. Based on this observation, we propose to extract movie events in the following four steps: 1) shot sink computation, where temporally close and visually similar shots are pooled into a sink; 2) sink clustering and characterization, where each sink is recognized to be either periodic, partly-period, or non-periodic; 3) event extraction and classification; and 4) post-processing with integrated speech and face cues. Each of these steps is detailed in the following subsections.

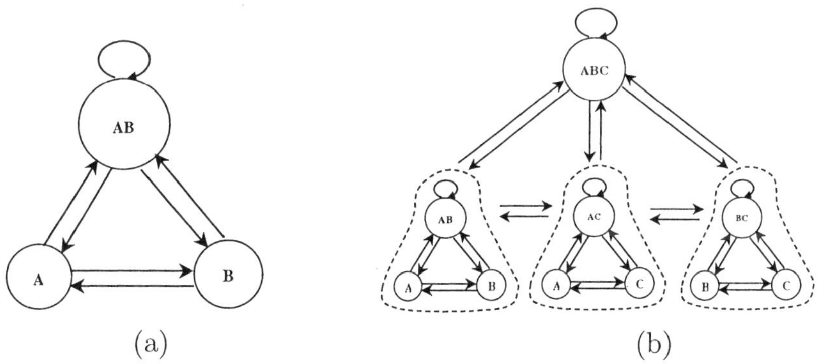

Figure 5.1. Typical movie dialog models for: (a) a two-speaker dialog (speakers A and B), and (b) a multiple-speaker dialog (speakers A, B and C).

5.3.1 Computing Shot Sinks Using Visual Information

Since an event is generally characterized by a repetitive visual structure, our first step is to extract all video paragraphs that possess this feature. A new concept called *shot sink* is defined for this purpose. Particularly, a shot sink contains a pool of shots which are temporally close and visually similar. Shot sinks are generated using the proposed window-based sweep algorithm described below.

Window-based Sweep Algorithm. Given shot i, this algorithm will find all shots that are visually similar to i. Considering that an event practically occurs within a certain temporal locality, we naturally restrict this search range to a window of length $winL$. Denoting shots i and j's keyframes by b_i, e_i, and b_j, e_j $(i < j)$, extracted as described in Section 5.2, we compute shots i and j's similarity as

$$\begin{aligned} Dist_{i,j} &= \frac{1}{4}(w_1 \times dist(b_i, b_j) + w_2 \times dist(b_i, e_j) \\ &+ w_3 \times dist(e_i, b_j) + w_4 \times dist(e_i, e_j)) \end{aligned}$$

where $dist(b_i, b_j)$ could be either the Euclidean distance or the histogram intersection between b_i and b_j's color histograms. w_1, w_2, w_3 and w_4 are four weighting coefficients computed as

$$\begin{array}{ll} w_1 = 1 - \frac{L_i}{winL}, & w_2 = 1 - \frac{L_i + L_j}{winL}, \\ w_3 = 1, & w_4 = 1 - \frac{L_j}{winL}, \end{array} \tag{5.1}$$

where L_i and L_j are shot lengths in the unit of frames. The four coefficients are derived as follows. First, since we want to find all similar shots within the window (hence the name "*sweep*"), we shall not lower their visual similarity because of their physical separation. Thus, we set w_3 to be 1 since e_i and b_j form the closest frame pair. Second, due to motion continuity, the similarity between b_i and b_j becomes smaller as shot i gets longer so that we set w_1 to be $1 - \frac{L_i}{winL}$ where $winL$ is introduced for normalization. We can derive the formulas for w_2 and w_4 in similar ways.

Now, if $Dist_{i,j}$ is less than a predefined threshold $shotT$, we consider shots i and j to be similar, and put shot j into shot i's sink. Basically we will run this algorithm for every shot. However, if one shot has already been included in a sink, we will skip this shot and continue with the next.

5.3.2 Clustering Shot Sinks Using K-means Algorithm

In this stage, we cluster and characterize each sink into one of the following three predefined classes: *periodic*, *partly-periodic* and *non-periodic*, based on the evaluated degree of shot repetition. For instance, if shot i's sink contains shots i, $i+2$, $i+4$ and $i+6$, we will classify it into the first class since a very strict shot repetition pattern is observed. However, if shot i's sink only contains itself, this sink will be discarded and excluded from further consideration. To quantitatively determine the sink periodicity, we apply the following three steps.

1. For each sink, calculate the relative temporal distance between each pair of neighboring shots. For example, if shot i's sink contains shots i, $i+2$, $i+4$, $i+7$ and $i+10$, then the distance sequence would be 2, 2, 3, 3.

2. Compute mean μ and standard deviation σ for each sink's distance sequence and set them as its features. Thus for the sink in the above example, it will have mean 2.5 and standard deviation 0.5. Intuitively, a sink belonging to the periodic class will have a smaller standard deviation than the one belonging to the non-periodic class.

3. Group all sinks into the three desired classes using K-means algorithm in terms of their features. The K-means algorithm performs unsupervised clustering and can circumvent the trouble of determining thresholds. Furthermore, the K-means algorithm is a least-squares partitioning method that naturally divides a collection of objects into K groups. Hence, it is more tolerant to "noisy" data as compared to other approaches.

Figure 5.2 shows the clustering results for two test movies. As we can see, all shot sinks have been well categorized into three groups, where the leftmost group belongs to the periodic class and the rightmost belongs to the non-periodic class.

5.3.3 Extracting and Classifying Events

At this step, we proceed to extract events by grouping all temporally overlapped sinks into one event. This is because no shots which are semantically inter-related with each other will belong to different events, since different events have different thematic topics. Moreover, shots that do not belong to these sinks but are physically covered by their temporal ranges will be included into the same event as well. After extracting all events, we proceed to classify them into the three desired classes based on the following three heuristically derived rules.

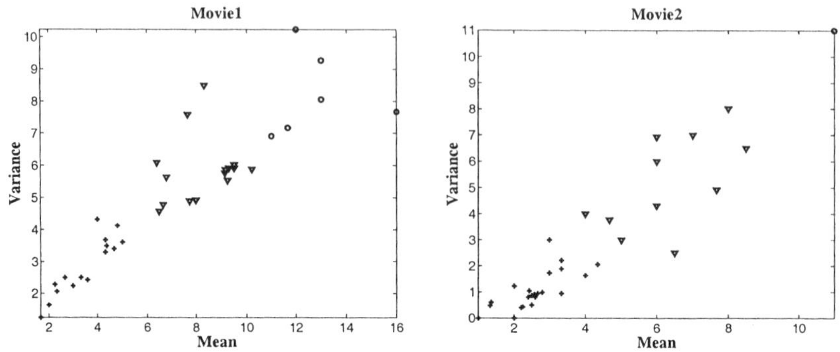

(a) (b)

Figure 5.2. Clustering shot sinks into three classes with (a) Movie1, (b) Movie2, where crosses, triangles and circles stand for sinks in the periodic, partly periodic and non-periodic classes, respectively.

1. If an event contains at least two periodic sinks, at most one partly-periodic, and no non-periodic shot sinks, it is declared as a two-speaker dialog. This rule is quite intuitive as the camera will basically track the speakers back and forth during a typical movie conversation, thus producing a series of alternating close-up shots.
2. If the event contains several partly-periodic sinks, or if the periodic and non-periodic shot sinks coexist, we label it as a multiple-speaker dialog. The reason to tolerate non-periodic sinks is that, when there are multiple speakers present, we have no control of who will be the next speaker since everyone has an equal opportunity to talk.
3. All remaining events are labeled as the hybrid.

5.3.4 Integrating Speech and Face Information

Due to the limitation of pure color information, we have observed the following two types of false alarms in the coarse-level event results: 1) Type I: *Misdetected conversation-like montage presentation as a spoken dialog.* In these scenarios, the camera usually shuttles back and forth between two silent objects (or humans), thus resulting in a series of repeated shots. Nevertheless, since no dialog actually goes on in these events, we shall not detect them as two-speaker dialogs; 2) Type II: *Misclassified multiple-speaker dialog as a two-speaker dialog.* This type of false alarm usually occurs when the camera frequently switches between two couples instead of among individual speakers. Errors will also occur in the scenario where one person dominates the dialog while the rest of the speakers talk less.

To reduce the Type I false alarm, we integrate the embedded audio information into the detection scheme. Specifically, we first classify every shot in the candidate dialog into one of the four audio classes described in Section 5.2. Then, we calculate the ratio of its contained speech shots. If the ratio is above a certain threshold, we confirm the event to be a dialog; otherwise, we label it as a hybrid event.

To reduce the Type II false alarm, we include the facial cue into the detection scheme. Specifically, for each shot in a two-speaker dialog, we first perform a face detection on its underlying frames and output the average of its detected face counts. Then, we go check if more than half of these values are larger than one. If yes, we re-label this event as a multiple-speaker dialog.

5.4 Adaptive Speaker Identification

In this module, we aim at identifying target movie cast members for content indexing purposes. Figure 5.3 shows the proposed system framework that consists of the following six major blocks: (1) shot detection and audio classification, (2) face detection, recognition and mouth tracking, (3) speech segmentation and clustering, (4) initial speaker modeling, (5) audiovisual (AV)-based speaker identification, and (6) unsupervised speaker model adaptation. As shown, given an input video, we first split it into audio and visual streams, then a shot detection is performed on the visual source. Following this, a shot-based audio classification is carried out. Next, with non-speech shots being discarded, all speech shots are further processed in the speech segmentation and clustering module where the same person's speeches are grouped into a homogeneous cluster. Meanwhile, a face detection/recognition and mouth tracking process is performed on speech shots to recognize talking faces. The speech and face cues are then effectively integrated to finalize the speaker identification in the AV-based identification module. Finally, the identified speaker's model is updated in the unsupervised model adaptation module, which comes into effect in the next round of the identification process.

5.4.1 Face Detection, Recognition and Mouth Tracking

The goal of this module is to detect and recognize talking faces in speech shots. Some previous work on talking face detection can be found in [Tsekeridou *et al.*, 2001] which used a mouth template to track the

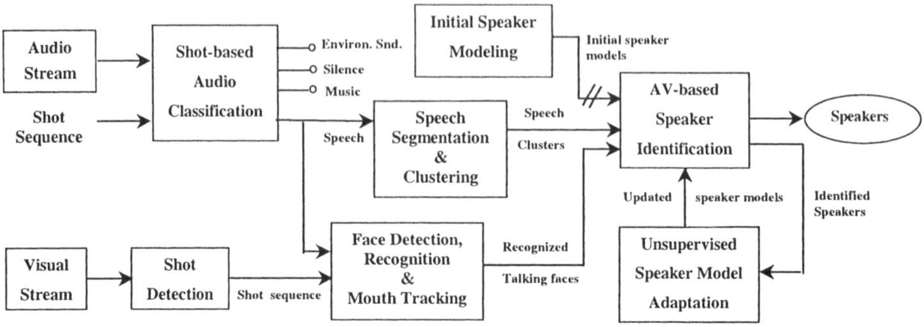

Figure 5.3. Block diagram of the proposed adaptive speaker identification system.

face, as well as in [Li *et al.*, 2003] where a mathematical framework which incorporated the face-and-speech correlation with the latent semantic indexing approach was applied.

Face Detection and Recognition. To speed up the face detection process, here we will only carry out the detection on speech shots as shown in Figure 5.3. Figure 5.4(a) shows one detection example where the detected face is boxed by a rectangle and eyes are indicated by crosses. Also, to facilitate the subsequent recognition process, we organize the detection results into a set of face sequences, where all frames within each sequence contain the same nonzero number of human faces.

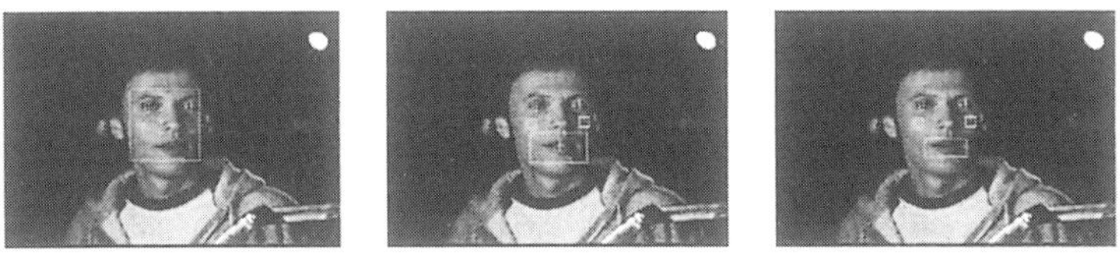

(a) (b) (c)

Figure 5.4. (a) A detected human face, (b) the coarse mouth center, the mouth search area, and two small squares for skin-color determination, and (c) the detected mouth region.

To construct the face database for recognition, we first ask users to select N target cast members or speakers during the system initialization by choosing frames that contain their faces. These faces are then detected, associated with the cast members' names, and stored into the face database.

During the face recognition process, each detected face in the first frame of each face sequence is recognized. The result is returned as a face vector $\vec{f} = [f_1, \ldots, f_N]$, where f_i is a value in [0, 1] which indicates the confidence of being target cast i.

Mouth detection and tracking. In this step, we first apply a weighted block matching approach to detect the mouth for the first frame of a face sequence, then we track the mouth for the rest of the frames. Note that if more than two faces are present in the sequence, we will virtually split it into a number of sub-sequences, with each focusing on one face.

1. Mouth detection

According to the facial biometric analogies, we know that there is a certain ratio between the inter-ocular distance and the distance *dist* between eyes and mouth. Thus, once we obtain the eyes' positions from the face detector, which are denoted by (x_1, y_1) and (x_2, y_2), we can subsequently locate the coarse mouth center (x, y) for an upright face. However, when a face is rotated, we need to recalculate its mouth center, as in [Li *et al.*, 2003]

$$x = \frac{x_1 + x_2}{2} \pm dist \times \sin(\theta), \quad y = \frac{y_1 + y_2}{2} + dist \times \cos(\theta), \tag{5.2}$$

where θ is the head rotation angle.

We then expand the coarse mouth center (x, y) into a rectangular mouth search area as shown in Figure 5.4(b), and perform a weighted block-matching process to locate the target mouth. The criterion we used to detect the mouth is that the mouth area should present the largest color difference from the skin color, which is determined from the average pixel color in the two small under-eye squares as shown in Figure 5.4(b). An example of a correctly detected mouth is shown in Figure 5.4(c). For the rest of the discussion, we denote the detected mouth center by (cx, cy).

2. Mouth tracking

To track the mouth for the rest of frames, we assume that for each subsequent frame, the centroid of its mouth mask can be derived from that of the previous frame as well as from its eye positions. Moreover, we assume that the distance between the coarse mouth center (x, y) and the detected mouth center (cx, cy) remains the same for all frames. Figure 5.5 shows the mouth detection and tracking results on a face sequence containing ten consecutive frames.

Finally, a color histogram-based approach is applied to determine if the tracked mouth is talking. Particularly, if the normalized accumulated

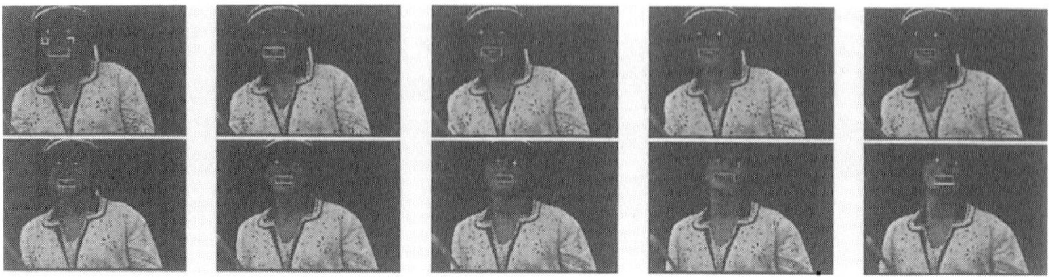

Figure 5.5. Mouth detection and tracking results on ten consecutive video frames.

histogram difference in the mouth area of the entire or part of the face sequence f exceeds a certain threshold, we label it as a talking mouth; and correspondingly, we mark sequence f as a talking face sequence.

5.4.2 Speech segmentation and clustering

For each speech shot, the two major speech processing tasks are speech segmentation and speech clustering. In the segmentation step, all individual speech segments are separated from the background noise (also called silence). In the clustering step, we group the same speaker's segments into homogeneous clusters so as to facilitate the successive identification process.

Speech segmentation. A two-step process is applied to separate speech from the background: 1) given the audio signal of a speech shot, we first sort all audio frames into an array based on their energies, then we quantize all frames into N bins. Next, the threshold T which separates speech and silence is determined from the average energies in the first and last three bins [Li *et al.*, 2003]; 2) a four-state transition diagram [Li and Zheng, 2001] is then employed to extract the speech segments as shown in Figure 5.6. Particularly, the transition conditions between two states are labelled on each edge, and the corresponding actions are described in parentheses. As we can see, this state machine basically groups blocks of continuous silence/speech frames as silence/speech segments while removing impulsive noises at the same time.

Speech clustering. Speech clustering has been studied for a long time, and many approaches have been proposed. In this work, we use Bayesian Information Criterion (BIC) to measure the similarity between two speech segments [Chen *et al.*, 1998].

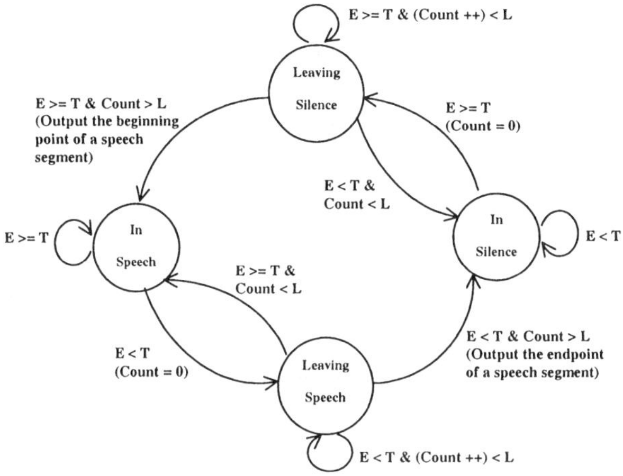

Figure 5.6. A state transition diagram for speech-silence segmentation where T stands for the derived adaptive threshold, E denotes the frame energy, *count* is a frame counter and L indicates the minimum speech/silence segment length.

When comparing two segments using the BIC, the distance measure can be stated as a model selection criterion where one model is represented by two separate segments X_1 and X_2, and the other model represents the joined segment $X = \{X_1, X_2\}$. The difference between these two modeling approaches equals

$$\begin{aligned}\Delta BIC(X_1, X_2) &= \frac{1}{2}(M_{12}\log|\Sigma| - M_1 \log|\Sigma_1| \\ &- M_2 \log|\Sigma_2|) - \frac{1}{2}\lambda(d + \frac{1}{2}d(d+1))\log M_{12},\end{aligned}$$

where Σ_1, Σ_2, Σ are X_1, X_2 and X's covariance matrices, and M_1, M_2, M_{12} are their respective feature vector numbers. λ is a penalty weight and equals 1 in this case. d gives the dimension of the feature space. According to the BIC theory, if $\Delta BIC(X_1, X_2)$ is negative, the two speech segments, X_1 and X_2, can be considered from the same speaker.

Now, assume cluster C contains n homogeneous speech segments, then given a new speech segment X, we compute $Dist(X, C)$ as $Dist(X, C) = \sum_{i=1}^{n} w_i \times \Delta BIC(X, X_i)$, where $w_i = M_i / \sum_{j=1}^{n} M_j$. Finally, if $Dist(X, C)$ is less than 0, we merge X to cluster C; otherwise, if none of the existing clusters is matched, a new cluster will be initialized.

5.4.3 Initial speaker modeling

To bootstrap the identification process, we need initial speaker models as shown in Figure 5.3. This is achieved by exploiting the inter-relations between the face and speech cues. Specifically, for each target cast member A, we first find a speech shot where A is talking based on the face detection and recognition result. Then, we collect all of its speech segments as described in Section 5.4.2.0, and build A's initial model. The Gaussian Mixture Model (GMM) has been employed here for modeling. Note that at this stage, the initial model will only contain one Gaussian component with its mean and covariance computed as global due to the limited amount of training data.

5.4.4 Likelihood-based speaker identification

At this stage, we will identify speakers based on pure speech information. Specifically, given a speech signal, we first decompose it into a set of overlapped audio frames; then 14 Mel-frequency cepstral coefficients [Reynolds *et al.*, 1995] are extracted from each frame to form an observation sequence X. Finally, we calculate the likelihood $L(X; M_i)$ between X and all speaker models M_i, and obtain a speaker vector $\vec{v}$.

Likelihood calculation. Because the Gaussian mixture density can provide a smooth approximation to the underlying long-term sample distribution of a speaker's utterances [Reynolds *et al.*, 1995], we choose to use GMM to model speakers in this work. Particularly, a GMM model M can be represented by the notation $M = \{p_j, \vec{\mu}_j, \Sigma_j\}, j = 1, \ldots, m$. where m is the number of components in M, and p_j, $\vec{\mu}_j$, Σ_j are the weight, mean vector and covariance matrix of the ith component, respectively.

Now, let M_i be the GMM model corresponding to the ith enrolled speaker with $M_i = \{p_{ij}, \vec{\mu}_{ij}, \Sigma_{ij}\}$, and let X be the observation sequence consisting of T cepstral vectors $\vec{x}_t$, $t = 1, \ldots, T$, under the assumption that all observation vectors are independent, the log likelihood $\ell(X; \vec{\mu}_{ij}, \Sigma_{ij})$ between X and M_i is computed as [Mardia *et al.*, 1979]

$$\ell(X; \vec{\mu}_{ij}, \Sigma_{ij}) = -\frac{T}{2}\log|2\pi\Sigma_{ij}| - \frac{T}{2}tr(\Sigma_{ij}^{-1}S) - \frac{T}{2}(\bar{X} - \vec{\mu}_{ij})'\Sigma_{ij}^{-1}(\bar{X} - \vec{\mu}_{ij}), \tag{5.3}$$

where S and $\bar{X}$ are X's covariance and mean, respectively.

Based on this identification scheme, a speaker vector $\vec{v} = [v_1, \ldots, v_N]$ can be obtained for each cluster C, where v_i is a value in [0, 1] which equals the normalized log likelihood value $\ell(X; \vec{\mu}_{ij}, \Sigma_{ij})$, and indicates the confidence of being target speaker i.

5.4.5 Audiovisual integration for speaker identification

This step aims at finalizing the speaker identification task for cluster C (in shot S) by integrating the audio and visual cues obtained in Sections 5.4.1, 5.4.2 and 5.4.4. Specifically, given cluster C and all recognized talking face sequences F in S, we examine if there is a temporal overlap between C and any sequence F_i. If yes, we assign F_i's face vector $\vec{f}$ to C if the overlap ratio exceeds a threshold. Otherwise, we set C's face vector to null. However, if C is overlapped with multiple F_i due to speech clustering or talking face detection errors, we choose the one with the highest overlap ratio.

Now, we determine the speaker's identity in cluster C as

$$speaker(C) = \arg \max_{1 \leq j \leq N} (w_1 \cdot f[j] + w_2 \cdot v[j]), \tag{5.4}$$

where $\vec{f}$ and $\vec{v}$ are C's face and speaker vectors, respectively. N is the total number of target speakers. w_1 and w_2 are two weights that sum up to 1.0. Currently we set them to be equal.

5.4.6 Unsupervised Speaker Model Adaptation

Now, after we identify speaker P for cluster C, we will update his model using C's data in this step. Meanwhile, a background model will be either initialized or updated to account for all non-target speakers. Specifically, when there is no a priori background model, we use C's data to initialize it if the minimum of $L(C; M_i), i = 1, \ldots, N$ is less than a preset threshold. Otherwise, if the background model produces the largest likelihood, we denote the identified speaker as "unknown" and use C's data to update the background model.

The following three approaches are investigated to update the speaker model: *Average-based* model adaptation, *MAP-based* model adaptation, and *Viterbi-based* model adaptation.

Average-based Model Adaptation. In this approach, P's model is updated in the following three steps.

Step 1: Compute BIC distances between cluster C and all of P's mixture component b_i. Denote the component that gives the minimum distance d_{min} by b_0.

Step 2: If d_{min} is less than an empirically determined threshold, we consider C to be acoustically close to b_0, and use C's data to update this component. Specifically, let $N(\mu_1, \Sigma_1)$ and $N(\mu_2, \Sigma_2)$ be C and b_0's

Gaussian models, respectively, we update b_0's mean and covariance as

$$\mu_2' = \frac{N_1}{N_1+N_2}\mu_1 + \frac{N_2}{N_1+N_2}\mu_2, \tag{5.5}$$

$$\Sigma_2' = \frac{N_1}{N_1+N_2}\Sigma_1 + \frac{N_2}{N_1+N_2}\Sigma_2 + \frac{N_1 N_2}{(N_1+N_2)^2}(\mu_1-\mu_2)(\mu_1-\mu_2)^T, \tag{5.6}$$

where N_1 and N_2 are the numbers of feature vectors in C and b_0, respectively [Mokbel, 2001].

Otherwise, if d_{min} is larger than the threshold, we will initialize a new mixture component for P with its mean and covariance equaling to μ_1 and Σ_1. However, once the total number of P's components reaches a certain value (which is set to 32 in our experiments), only component adaptation is allowed. This is adopted to avoid having too many Gaussian components in each model.

Step 3: Update the weight for each of P's mixture component.

MAP-based Model Adaptation. MAP adaptation has been widely and successfully used in speech recognition, yet it has not been well explored in speaker identification. In this work, due to the limited speech data, only Gaussian means will be updated. Specifically, given P's model M_p, we update component b_i's mean μ_i via

$$\mu_i' = \frac{L_i}{L_i+\tau}\bar{\mu} + \frac{\tau}{L_i+\tau}\mu_i, \tag{5.7}$$

where τ defines the "adaptation speed" and is currently set to 10.0. L_i gives the occupation likelihood of the adaptation data to component b_i, and is defined as

$$L_i = \sum_{t=1}^{T} p(i|\vec{x}_t, M_p), \tag{5.8}$$

where $p(i|\vec{x}_t, M_p)$ is the *a posteriori* probability of $\vec{x}_t$ to b_i. Finally, $\bar{\mu}$ gives the mean of the observed adaptation data, and is defined as

$$\bar{\mu} = \frac{\sum_{t=1}^{T} p(i|\vec{x}_t, M_p)\vec{x}_t}{\sum_{t=1}^{T} p(i|\vec{x}_t, M_p)}. \tag{5.9}$$

Unlike the previous method, this MAP adaptation is applied to every component of P based on the principle that every feature vector has a certain possibility of occupying every component. Thus, MAP adaptation provides a soft decision on which feature vector belongs to which component.

Viterbi-based Model Adaptation. Similar to the MAP-based approach, this approach also allows different feature vectors belonging to different components. Nevertheless, while the MAP approach provides a soft decision, this approach implies a hard decision, *i.e.* for any one particular feature vector $\vec{x}_t$, it can either occupy component b_i or not. Therefore, the probability function $p(i|\vec{x}_t, M_p)$ in Equation 5.8 is now replaced by an indicator function which is either 0 or 1. Now, given any feature vector $\vec{x}_t$, the mixture component it occupies will be determined by

$$m_0 = \arg \max_{1 \leq i \leq m} p(i|\vec{x}_t, M_p). \tag{5.10}$$

Finally, Equations 5.5 and 5.6 are used to update P's components after we assign every feature vector to its belonged component. As one can see, this approach is actually a compromise between the previous two methods.

To summarize, based on the proposed model adaptation approaches, a speaker model will grow from one Gaussian mixture component up to 32 components as we go through the entire movie sequence.

5.5 Event-based Movie Skimming

An ideal video skim should retain most of the original informative parts which are critical for content understanding. From the discussion in Section 5.3, we know that the movie content can be compactly represented by a set of events, where each event contains an individual thematic topic and has continuous audiovisual contents. These events thus form the candidate skimming components.

The proposed movie skimming system consists of the following two steps: 1) *event feature extraction*, where six types of mid- to high-level features are extracted from each event. These features are then used to evaluate the event importance when integrated with user's preference; 2) *movie skim generation*, where selected important events are assembled to generate the final skim. Each of these two steps are detailed below.

5.5.1 Event Feature Extraction

For simplicity, we denote the current event by EV, and the number of its component shots by L.

1. Music ratio. The music ratio MR of an event is computed as

$$MR = \frac{\sum_{l=1}^{L} M_l}{L}, \tag{5.11}$$

where M_l is a binary number. M_l equals 1 when shot l in EV contains music; otherwise, it equals 0. Based on this ratio, we say EV has a low music ratio if $MR < 0.3$. When $0.3 \leq MR \leq 0.6$, EV has a medium music ratio; and otherwise, it has a high music ratio.

2. Speech ratio. The speech ratio SR is computed as

$$SR = \frac{\sum_{l=1}^{L} S_l}{L}, \tag{5.12}$$

where S_l is a binary number. S_l equals 1 when shot l in EV contains speech signals; otherwise, it is set to 0. Similarly, EV is also classified into three levels based on this ratio.

3. Sound loudness. The sound loudness SL is used to indicate if the current event has a loud and noisy background. Currently, SL is computed as the average sound energy for all environmental shots. Also, for the sake of event importance ranking, SL is normalized to the range of [0,1] by dividing the largest SL value of the entire movie. The sound loudness is classified into low, medium and high three levels as well.

4. Action level. The action level AL is used to indicate the amount of motions involved in the current event, and is computed as the average motions contained in its component frames [Li and Kuo, 2003]. For the same purpose, the value of AL is also normalized to the range of [0,1] over the entire movie. The action level is also classified into low, medium and high three levels.

5. Present cast. The present cast PC not only includes the identified speakers but also contains the cast whose face is recognized in the current event. This parameter is used to indicate if certain movie characters of interest to the user are present in the event. Currently, we only provide the top five user-selected cast members.

6. Theme topic. The theme topic TT in the current event corresponds to one of the three event types: the two-speaker dialog, the multiple-speaker dialog, and other general event.

Based on these features, we define an attribute matrix A for an incoming movie as

$$A = \begin{bmatrix} a_{1,1} & a_{1,2} & \cdots & a_{1,M} \\ a_{2,1} & \cdots & \cdots & a_{2,M} \\ \cdots & \cdots & a_{i,j} & \cdots \\ a_{N,1} & a_{N,2} & \cdots & a_{N,M} \end{bmatrix},$$

where M is the total number of features extracted from each event (which equals 6 in this case), N is the total number of events in the current movie, and $a_{i,j}$ is the value of the jth feature in the ith event. For example, if $a_{1,1} = 0.4$, $a_{2,5} = 0 \cup 1 \cup 3$, and $a_{4,6} = 0$, it means that the first event has a music ratio of 0.4, the second one has three present cast members, and the fourth one is a two-speaker dialog.

5.5.2 Movie Skim Generation

At this step, we generate the movie skim by choosing important events from the candidate set based on user's preference. Specifically, a preference vector $\vec{P} = [p_1, p_2, \ldots, p_M]^T$ is used to collect user's feature preference. For instance, the use can specify his preference on the music ratio, speech ratio, sound loudness and action level as "low", "medium", "high", or "no preference". For feature PC, p_5 can be a combination of all desired cast members such as "$0 \cup 3$", or simply "no preference". As for feature p_6 on the theme topic, it can be a union of the numbers between 0 to 3, with 0 being "the two-speaker dialog", and 3 being "no preference". Finally, the user can also specify his desired skim length.

Given the preference vector $\vec{P}$, we then compute the event importance vector $\vec{E}$ as

$$\begin{aligned} \vec{E} &= A \odot \vec{P} \\ &= \begin{bmatrix} a_{1,1} & a_{1,2} & \cdots & a_{1,M} \\ a_{2,1} & \cdots & \cdots & a_{2,M} \\ \cdots & \cdots & a_{i,j} & \cdots \\ a_{N,1} & a_{N,2} & \cdots & a_{N,M} \end{bmatrix} \bigodot \begin{bmatrix} p_1 \\ \cdots \\ \cdots \\ p_M \end{bmatrix} = \begin{bmatrix} e_1 \\ \cdots \\ \cdots \\ e_N \end{bmatrix} \end{aligned}$$

where

$$e_i = a_{i,1} \odot p_1 + \ldots + a_{i,M} \odot p_M, \tag{5.13}$$

and "$\odot$" is a mathematical operator that functions as a logical "AND". For instance, if $a_{1,1} = 0.2$, and $p_1 = low$, we have $a_{1,1} \odot p_1 = 1$ since $a_{1,1}$ denotes a low music ratio which is consistent with p_1. Otherwise, we set $a_{1,1} \odot p_1 = 0$. Similarly, we can define the operations for $a_{i,2} \odot p_2$, $a_{i,3} \odot p_3$, and $a_{i,4} \odot p_4$. For the operation between $a_{i,5}$ and p_5, we set $a_{i,5} \odot p_5 = 1$ if $a_{i,5}$ contains at least one of the cast members listed in p_5; otherwise, it equals 0. The same definition applies to $a_{i,6} \odot p_6$ as well. Finally, in case that p_j equals "no preference", we set $a_{i,j} \odot p_j = -1$.

Now, given that each event has a score e_i that ranges from -6 to 6, we can generate the final skim by selecting the events that maximize the cumulative score in the resulting skim while preserving the desired skim length at the same time. This could be viewed as an example of the 0-1 knapsack problem defined as [Martello *et al.*, 1990]

$$\text{To maximize} \quad \sum_{i=1}^{N} w_i e_i, \qquad \text{subject to} \quad \sum_{i=1}^{N} w_i T_i \leq T, \tag{5.14}$$

where w_i is a binary variable, which equals 1 when the ith event is selected for the skim and 0, otherwise. T_i is its temporal duration and T is the target skim length.

To solve this knapsack problem, we first sort all events based on their scores in a descending order. Then, a greedy selection algorithm is applied which starts by selecting the topmost event. We then keep on selecting the next event as long as its duration is less than the remaining skim time. Finally, all selected events are concatenated to form the final skim. Note that, to include as many events as possible, we could remove silence shots from the selected dialogs since they usually do not contain important messages.

5.5.3 Discussion

Up to now, a theoretically complete movie skimming system has been proposed. There are, however, practical considerations to be included for more satisfactory results.

When More Judging Rules Are Needed. Since the event score e_i has a relatively narrow value range, different events may share the same score. Consequently, new rules are needed to distinguish their importance. Moreover, when the user has no preference on any features, we need a set of default rules to generate the skim. Possible solutions are given below to each of these two scenarios: 1) ask for more user preferences such as the cast members' ranking from the most interesting to the least, or apply more strict rules so that no event that contains uninteresting actors is selected; 2) derive judging rules from the movie genres. We know that different types of movies have different, yet stereotyped ways to attract their audience. For instance, romances usually contain many dialog and music scenes while action movies tend to use many thrilling scenes with deafening sounds and crazy actions.

Sub-sampling the Video Skim. To include as much content as possible within a limited skimming time, we have added a video subsampling scheme to this work. Specifically, given the desired skim length L, we first generate an αL-long skim with α greater than 1. Then, we play one out of every α frames during the skim playback, thus to maintain its target duration. Currently, we choose α between 1 and 2.5.

To maintain a comprehensible audio quality in the skim, and also to make the accompanying audio track be consistent with the compressed image sequence, we also compress the audio track with the same compression ratio. The MWSOLA (Modified Waveform Similarity Overlap-And-Add) technique proposed in [Liu *et al.*, 2001] is employed for this purpose.

Discovering the Story Structure. Because feature films have elaborately designed and meticulously edited story structures, we are able to generate more informative movie skims if we could discover them. According to [Block, 2001], a movie's story structure is usually instantiated by its visual structure, which could be detected and extracted by analyzing seven basic visual components including space, line, shape, tone, color, movement, and rhythm.

A typical movie story contains three basic parts: the beginning (*exposition*), the middle (*conflict*), and the end (*resolution*). The exposition part gives the facts that are needed to begin the story such as the main characters. The middle part contains the rising actions or the conflict, and as the story develops, the conflict increases in intensity. The most intense part of the movie is the climax where the conflict gets resolved. The resolution part wraps up incomplete story elements and ends the story.

It is apparent that, if we can extract the story structure as discussed, we are able to include more informative contents in the skim. For instance, when the plot reaches the climax, we shall give it more weight since it usually attracts audience attention. In contrast, for the exposition and resolution parts, a brief explanation would be enough.

5.6 Experimental Results

5.6.1 Event Detection Results

For all the experiments reported in this section, video streams are compressed in MPEG-1 format with a frame rate of 29.97 frames/sec. To validate the effectiveness of the proposed approach, representatives of various movie genres were tested. Specifically, the test set includes Movie1 ("*The Legend of the Fall*", a tragic romance), Movie2 ("*When Harry Met Sally*", a comedic drama) and Movie3 ("*Braveheart*", an action movie). Each movie clip is approximately onehour long.

Due to the inherent subjectivity of the event definition, we do not attempt to discuss the appropriateness of extracted events since people's opinions may differ. Instead, we will only examine the correctness of the event classification results, for which it is easier to reach a consensus. Experimental results are shown in Table 5.1 for all 3 movies which contain 80 events in total. Also, because the hybrid class contains all events excluding the dialogs, it is omitted from the table. Precision and recall rates are computed to evaluate the system performance, where

$$\text{Precision} = \frac{\text{hits}}{\text{hits} + \text{false alarms}}, \qquad \text{Recall} = \frac{\text{hits}}{\text{hits} + \text{misses}}.$$

Table 5.1. Event detection results for Movie1, Movie2 and Movie3

Movie1 – Tragic Romance										
	Combining Speech & Face Cues					Without Speech/Face Cues				
Event	Hit	Miss	False	Prec.	Recall	Hit	Miss	False	Prec.	Recall
M-spk	4	0	0	100%	100%	4	0	1	80%	100%
2-spk	6	1	0	100%	86%	6	1	3	67%	86%
Movie2 – Comedic Drama										
	Combining Speech & Face Cues					Without Speech/Face Cues				
Event	Hit	Miss	False	Prec.	Recall	Hit	Miss	False	Prec.	Recall
M-spk	7	0	0	100%	100%	5	2	0	100%	72%
2-spk	14	0	0	100%	100%	14	0	2	88%	100%
Movie3 – Action										
	Combining Speech & Face Cues					Without Speech/Face Cues				
Event	Hit	Miss	False	Prec.	Recall	Hit	Miss	False	Prec.	Recall
M-spk	5	1	0	100%	83%	5	1	0	100%	83%
2-spk	13	0	0	100%	100%	13	0	3	81%	100%

As shown in this table, encouraging event extraction results have been achieved. When the speech and facial cues are integrated, both precision and recall rates reach 83% in all three movies, which is very encouraging. Regarding the misses observed in the table, the missed two-speaker dialog in Movie1 was misclassified as a hybrid event, where one of the speakers was walking all time, which resulted in a frequent background change and therefore an irregular periodicity. In Movie3, a multiple-speaker dialog was misdetected due to the reason that people were talking in a too random fashion in that event, which resulted in an irregular shot repeat pattern.

5.6.2 Speaker Identification Results

To evaluate the performance of the proposed adaptive speaker identification system, studies have been carried out on above three movies. However, due to the space limit, only the results on Movie2 will be reported here.

Three characters were selected for Movie2, and a total of 952 speech clusters were generated. An average of 90% clustering purity was achieved which is defined as the ratio between the number of segments from the dominant speaker and the total number of segments in the cluster. Regarding the talking face detection, we have achieved 83% precision and 88% recall rates on 425 detected face sequences. However, the ratio of talking-face frames over the total number of video frames is as low as 11.5%. This is because movie cast members are in constant motion, thus making it difficult to detect their faces.

The identification results for all obtained speech clusters are reported in the form of a confusion matrix as shown in Table 5.2. The three cast members are indexed by A, B, C, and their corresponding movie characters are denoted by A', B' and C'. "Unknown" is used for all non-target speakers. The number in each grid, say grid (A', B), indicates the number of speech segments where character A' is talking yet actor B is identified. Three parameters, namely, *false acceptance* (FA), *false rejection* (FR) and *identification accuracy* (IA) are calculated to evaluate the system performance. Particularly, for each cast or character, we have

$$FR = \frac{\text{sum of off-diagonal numbers in the row}}{\text{sum of all numbers in the row}},$$

$$FA = \frac{\text{sum of off-diagonal numbers in the column}}{\text{sum of all numbers in the column}},$$

$$IA = 1 - FR.$$

Table 5.2(a) gives the identification result when the average-based model adaptation is applied. An average of 75.3% IA and 22.3% FA are observed. Result obtained from the MAP-based approach is given in Table 5.2(b) where we have an average 78.6% IA and 21% FA. This result is slightly better than that in (a), yet at the cost of a higher computation complexity. Table 5.2(c) shows the result for the Viterbi-based approach. As we can see, this table presents the best performance with an average 82% IA and 20% FA. The fact that this approach outperforms the MAP approach may imply that, for speaker identification, a hard decision would be good enough.

By carefully studying the results, we found two major factors that degrade the system performance: (a) imperfect speech segmentation and clustering, and (b) inaccurate facial analysis results. Due to the various sounds/noises existing in movies, it is extremely difficult to achieve perfect speech segmentation and clustering. Besides, incorrect facial data can result in mouth detection and tracking errors, which further affect the identification accuracy.

To determine the upper limit of the number of mixture components in each speaker model, we examined the average identification accuracy in terms of 32 and 64 components for all three adaptation methods and plotted them in Figure 5.7(a). As shown, except for the average-based method where a similar performance is observed, the use of 32 Gaussian mixture components has produced a better performance.

Table 5.2. Adaptive speaker identification results for Movie2 using: (a) the average-based, (b) the MAP-based, and (c) the Viterbi-based model adaptation approaches.

	A	B	C	Ukwn	FR	IA
A'	228	42	6	32	26%	74%
B'	37	281	35	24	25%	75%
C'	10	9	115	16	23%	77%
Ukwn	10	15	4	88		
FA	20%	19%	28%			

(a)

	A	B	C	Ukwn	FR	IA
A'	239	25	22	22	22%	78%
B'	59	302	5	11	20%	80%
C'	10	8	117	15	22%	78%
Ukwn	19	16	7	75		
FA	27%	14%	22%			

(b)

	A	B	C	Ukwn	FR	IA
A'	246	29	13	20	20%	80%
B'	41	317	13	6	16%	84%
C'	10	10	123	7	18%	82%
Ukwn	18	22	14	63		
FA	22%	14%	24%			

(c)

Finally, the average identification accuracy obtained by using or without using face cues is compared in Figure 5.7(b). Clearly, without the assistance of face cue, the system performance has been significantly degraded, especially for the average-based adaptation approach. This indicates that the face cue plays an important role in model adaptation.

Figure 5.8 gives a detailed description of a speaker identification example. Specifically, the upper part shows the waveform of an audio signal recorded from a speech shot where two speakers take turns to talk. The superimposed pulse curve illustrates the speech-silence separation result where all detected speech segments are bounded by the passband of the curve. Two segments which are indicated by the circles, are discarded due to their short lengths. The true speaker identity for each speech segment is given right below this sub-figure, where speakers A and B are represented by dark- and light-colored blocks, respectively. The likelihood-based speaker identification result is given in the next sub-figure. As shown, there are two false alarms in this result where

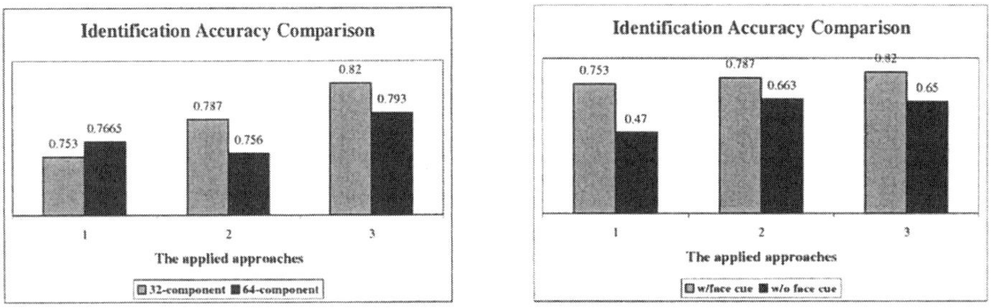

(a) (b)

Figure 5.7. Identification accuracy comparison for the average-based, the MAP-based, and the Viterbi-based approaches with: (a) 32-component vs. 64-component for speaker models, and (b) using vs. without using face cues.

B is falsely recognized as A by twice. The talking face detection result is shown in the next sub-figure. Finally, the last sub-figure shows the ultimate identification result obtained by integrating both speech and face cues. As shown, although the first error still exists as the face cue cannot offer help, the second error has been corrected.

5.6.3 Movie Skimming Results

Since it is difficult to qualitatively evaluate a video skimming system, we carried out a preliminary user study to quantitatively measure the system performance. Specifically, we have designed the following six statements, and asked each participant to assess them on a 5-point scale (1-5), where 1 stands for "strongly disagree", and 5 for "strongly agree": 1) *visual comprehension*: "the visual quality of the skim is good, no jerky motions"; 2) *audio comprehension*: "the audio quality is good, no staccato speeches"; 3) *semantic continuity*: "the embedded semantic flow is continuous and understandable"; 4) *good abstraction*: "the generated skim can summarize the movie well"; 5) *quick browsing*: "this skimming system can help me browse the movie content quickly"; and 6) *video skipping*: "I can skip watching the original movie by only viewing the skim".

We performed the study on both Moviel and Movie2. Two graduate students were invited to participate in the experiment; they were familiar with one movie. The survey results are reported in Table 5.3.

From this table, we see that encouraging results have been obtained in the study. Both participants were very satisfied with the skim's audio and visual quality. Moreover, since thc event forms the building block of the skim, the underlying semantic flow is continuous and understandable.

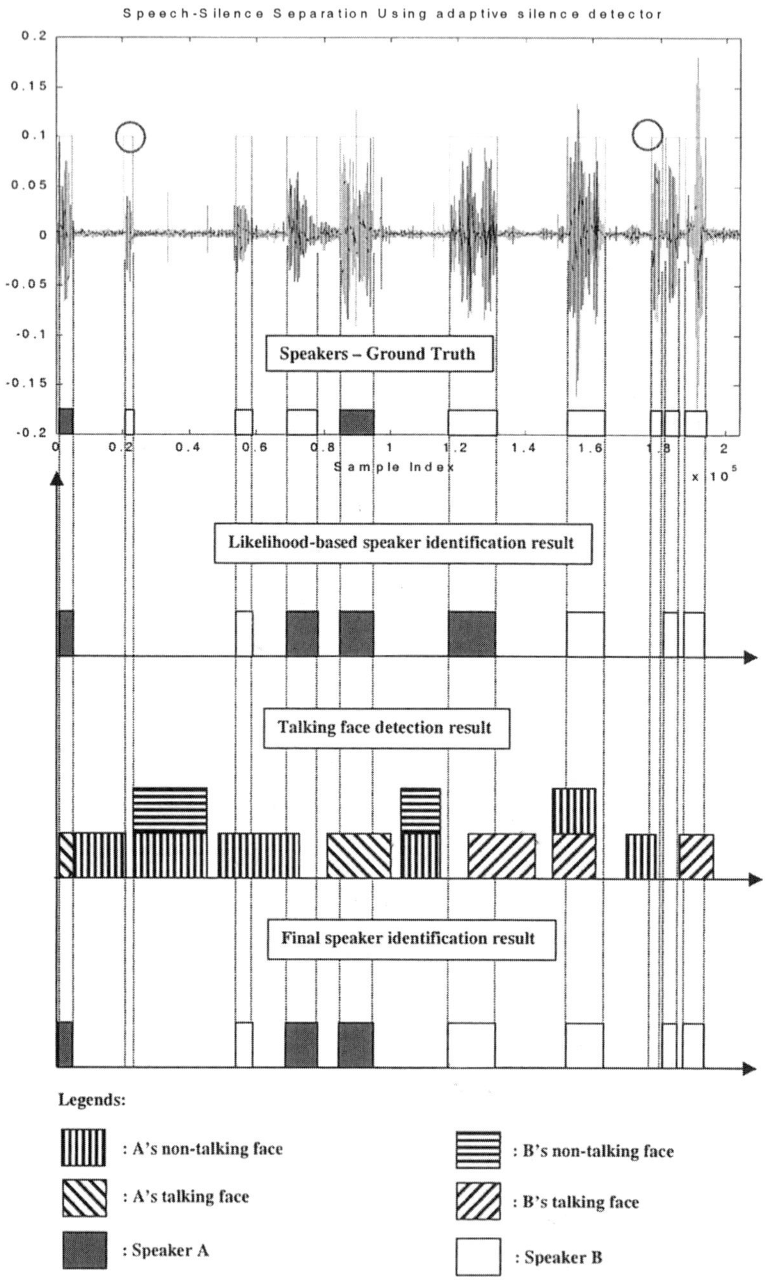

Figure 5.8. A detailed description of a speaker identification example.

Table 5.3. User study result.

Questions	Score	
	Mean	Std. deviation
Visual comprehension	4.8	0.21
Audio comprehension	4.62	0.17
Semantic continuity	4.4	0.36
Well abstraction	4.5	0.22
Quick browsing	4.7	0.15
Video skipping	4.15	0.78

Overall, they agreed that the generated skims have well summarized the movie content, and can help them quickly browse the sequence. However, when we asked them if they would skip watching the original movie by only viewing the skim, they were not affirmative since the original movie definitely contains much richer information. Finally, they suggested that a scalable video skim would be interesting and helpful.

5.7 Conclusion

A content-based movie analysis, indexing and skimming system is presented in this chapter. Various media cues such as face, audio and visual information, are employed to extract high-level video semantics (event and speaker identity), as well as to represent movie contents in a compact, yet meaningful manner. Although feature films have been our major focus, the methodology presented here could be easily extended to other types of generic video. More robust results could be achieved by integrating other image/video processing techniques such as human/object tracking. Finally, representing the system output in MPEG-7 standardized description format would be another interesting topic.

Acknowledgments

This research has been partially funded by the Integrated Media Systems Center, a National Science Foundation Engineering Research Center in USC, under Cooperative Agreement Number EEC-9529152 and partially funded by the Hewlett-Packard Company. The authors would also like to acknowledge the HP Labs for providing the face detection and recognition library.

References

Block, B., *The Visual Story: Seeing the Structure of Film, TV, and New Media.* Massachusetts, Focal Press, 2001.

Chen, S., and Gopalakrishnan, P., "Speaker, Environment and Channel Change Detection and Clustering Via the Bayesian Information Criterion," Proc. of DARPA Broadcast News Transcription and Understanding Workshop, 1998.

Hauptmann, A.G., and Smith, M.A., "Text, Speech, and Vision For Video Segmentation: The Informedia Project," Proc. of the AAAI Fall Symposium on Computer Models for Integrating Language and Vision, 1995.

HP Labs, Computational Video Group, "The HP Face Detection and Recognition Library," User's Guide and Reference Manual, Version 2.2, December 1998.

Huang, J., Liu, Z., and Wang, Y., "Integration of Audio and Visual Information For Content-based Video Segmentation," ICIP'98, October 1998.

Johnson, S.E., "Who Spoke When? - Automatic Segmentation and Clustering For Determining Speaker Turns," Eurospeech'99, 1999.

Li, D., Wei, G., Sethi, I.K., and Dimitrova, N., "Person Identification in TV Programs," Journal of Electronic Imaging, 10(4):930-938, 2001.

Li, M., Li, D., Dimitrova, N., and Sethi, I., "Audio-visual Talking Face Detection," ICME'03, July 2003.

Li, Q., Zheng, J., Zhou, Q., and Lee, C., "A Robust, Real-time Endpoint Detector With Energy Normalization For ASR in Adverse Environments,", ICASSP'01, May 2001.

Li, Y., and Kuo, C.-C., *Content-based Video Analysis, Indexing and Representation Using Multimodal Information*, Ph.D Thesis, University of Southern California, 2003.

Li, Y., Narayanan, S., and Kuo, C.-C., "Identification of Speakers in Movie Dialogs Using Audiovisual Cues," ICASSP'02, Orlando, May 2002.

Li, Y., Narayanan, S., and Kuo, C.-C., "Adaptive Speaker Identification with AudioVisual Cues For Movie Content Analysis," Invited Paper in Pattern Recognition Letters with special issue on Recent Trends in Video Computing, 2003.

Liu, F., Kim, J., and Kuo, C.-C., "Adaptive Delay Concealment For Internet Voice Applications with Packet-based Time-scale Modification," ICASSP'01, 2001.

Mardia, K., Kent, J., and Bibby, J., *Multivariate Analysis.* Academic Press, San Diego, 1979.

Martello, S., and Toth, P., *Knapsack Problems: Algorithms and Computer Implementations.* Chichester, NY, Wiley and Sons, 1990.

Mokbel, C., "Online Adaptation of HMMs to Real-life Conditions: A Unified Framework," IEEE Transactions on Speech and Audio Processing, 9(4):342-357, May 2001.

Monaco, J., *How To Read A Film: The Art, Technology, Language, History and Theory of Film and Media*, New York, Oxford University Press, 1982.

MPEG Requirements Group, "MPEG-7 Context, Objectives and Technical Roadmap," Doc. ISO/MPEG N2861, MPEG Vancouver Meeting, July 1999.

Pfeiffer, S., Lienhart, R., Fischer, S., and Effelsberg, W., "Abstracting Digital Movies Automatically," Journal of Visual Communication and Image Representation, 7(4):345-353, December 1996.

Reisz, K., and Millar, G., *The Technique of Film Editing.* New York: Hastings House, Publishers, 1968.

Reynolds, D., and Rose, R., "Robust Text-independent Speaker Identification Using Gaussian Mixture Speaker Models," IEEE Transactions on Speech and Audio Processing, 3(1):72-83, 1995.

Rui, Y., Huang, T.S., and Mehrotra, S., "Constructing Table-of-content For Video," ACM Journal of Multimedia Systems, 7(5):359-368, 1998.

Smith, M., and Kanade, T., "Video Skimming and Characterization Through the Combination of Image and Language Understanding Techniques," Proc. of the IEEE Computer Vision and Pattern Recognition, pages 775-781, 1997.

Sundaram, H., and Chang, S.F., "Determining Computable Scenes in Films and Their Structures Using Audio-visual Memory Models," ACM Multimedia'00, Marina Del Rey, November 2000.

Tarkovsky, A., *Sculpting in Time – Reflections on the Cinema*, Austin, University of Texas Press, 1986.

Taskiran, C.M., Amir, A., Ponceleon, D., and Delp, E.J., "Automated Video Summarization Using Speech Transcripts," Proc. of SPIE, 4676:371-382, January 2002.

Toklu, C., Liou, S.P., and Das, M., "Videoabstract: A Hybrid Approach To Generate Semantically Meaningful Video Summaries," ICME'00, New York, 2000.

Tsekeridou, S., and Pitas, I., "Content-based Video Parsing and Indexing Based on Audio-visual Interaction," IEEE Transactions on Circuits and Systems for Video Technology, 11(4):522-535, 2001.

Tseng, B.L., Lin, C.Y., and Smith, J.R., "Video Summarization and Personalization For Pervasive Mobile Devices," Proc. of SPIE, 4676:359-370, January 2002.

Wan, V., and Campbell, w., "Support Vector Machines for Speaker Verification and Identification," Proc. of the IEEE Signal Processing Society Workshop on Neural Networks, 2:775-784, 2000.

Yeung, M., Yeo, B.L., and Liu, B., "Extracting Story Units From Long Programs For Video Browsing and Navigation," IEEE Proceedings of Multimedia, pages 296-305, 1996.

Yeung, M., and Yeo, B.L., "Video Content Characterization and Compaction For Digital Library Applications," Proc. of SPIE, 3022:45-58, February 1997.

Zhang, T., and Kuo, C.-C., "Audio Content Analysis For On-line Audiovisual Data Segmentation," IEEE Transactions on Speech and Audio Processing, 9(4):441-457, 2001.

Chapter 6

VIDEO OCR: A SURVEY AND PRACTITIONER'S GUIDE

Rainer Lienhart
Intel Labs, Intel Corporation,
Santa Clara, California, USA
Rainer.Lienhart@intel.com

Abstract This survey strives to present the core concepts underlying the different *texture-based* approaches to automatic detection, segmentation and recognition of visual text occurrences in complex images and videos. It emphasizes the different approaches to attack the many issues in this space. For each kind of approach only a few representative references are given. This survey does not try to give an exhaustive listing of all relevant work, but to help practitioners and engineers new in the field to get a thorough overview of the state-of-the-art principles, methods, and systems in Video OCR. To this end, the approaches of the various researchers are broken up into constituents and presented as a design choice in a hypothetical image and video OCR system.

Keywords: Video OCR, text detection, text segmentation, text recognition, text tracking, texture, survey, guide, scene text, overlay text, font attributes, pixel classification, non-Roman languages, edge detection, wavelets, scale integration.

Introduction

Video text detection algorithms are sometimes classified by whether the originally proposed detection algorithm was designed to operate on uncompressed or compressed video streams. In this survey, we do not make this distinction, since almost all texture-based detection algorithms can be applied to the compressed as well as to the uncompressed domain. The issue of compressed versus uncompressed processing is orthogonal to the task of Video OCR. It only determines the space in which to perform the text texture detection. Text segmentation is usually performed in the uncompressed domain.

From a bird's eye view, the task of detecting, segmenting and recognizing text visually appearing in complex images and/or video seems to be well defined. However, many design decisions have to be taken based on the overall goal.

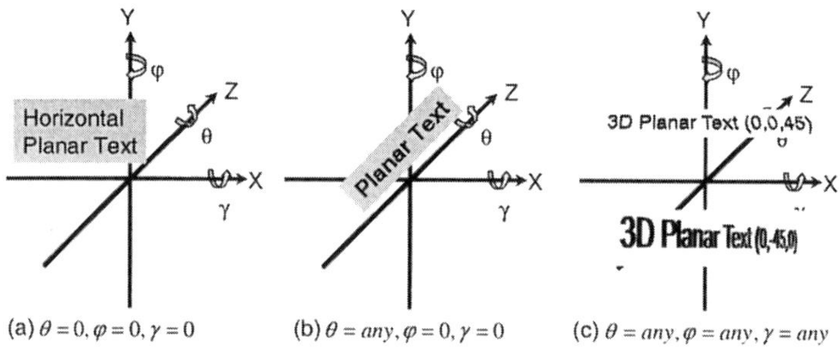

Figure 6.1. Different degrees of freedom in the placement of planar text in the 3D space.

Typical design choices are:

- *What kind of text occurrences should be considered?*

  Based on its origin there exist two different kinds of text in videos and images [Lienhart *et al.*, 1996; Lienhart, 1996]. ***Scene text*** is text that was recorded as part of scene such as street names, shop names, and text on T-shirts. It mostly appears accidentally and is seldom intended. Due to its incidental and the thus resulting unlimited variety of its appearance, it is hard to detect, extract and recognize. It can appear with any slant, tilt, in any lighting and upon straight or wavy surfaces. It may also be partially occluded.

  In contrast, the appearance of ***overlay text*** is carefully directed. It is often an important carrier of information and herewith suitable for indexing and retrieval. For instance, embedded captions in TV programs represent a highly condensed form of key information on the content of the video [Yeo *et al.*, 1996]; in commercials, the product and company name are often part of the text shown. Here, the product name is often scene text but used like artificial text. Most research work concentrates on artificial text occurrences, where it is implicitly implied that text lies in a plane roughly perpendicular to the optical axis of the camera. Only little work can be found on scene text [Myers *et al.*, 2001; Ohya *et al.*, 1994; Clark *et al.*, 2000; Clark *et al.*, 2001].

A different classification scheme of text occurrences is based on the constraints in the placement of planar text in the 3D space. Typical classes with increasing degree of freedom are

1 horizontal overlay text in the plane parallel to the camera plane ($\theta = 0, \varphi = 0, \gamma = 0$) (see Fig. 6.1(a)),

2 planar overlay text in the plane parallel to the camera plane ($\theta = any, \varphi = 0, \gamma = 0$) (see Fig. 6.1(b)),

3 Unconstrained planar 3D text ($\theta = any, \varphi = any, \gamma = any$) (see Fig. 6.1(c)), and

4 Unconstrained 3D text ($\theta = any, \varphi = any, \gamma = any$, text on any surface).

- *With what font attributes?*

  Text occurrences can differ significantly in font size, type, style, and color. Some research work has been tailored to very specific domains with limited variations in these attributes. For instance, the Video OCR module in the Informedia project is tailored to CNN Headline News video encoded in MPEG-1 [Sato *et al.*, 1998; Sato *et al.*, 1999]. It explicitly exploits the domain knowledge about the tight restrictions in font attributes such as font types and font sizes. Other Video OCR systems avoid any attribute restrictions [Lienhart *et al.*, 2002a; Wu *et al.*, 1999]. Text can be of any size, type, style and color.

- *In what kind of media data?*

  Should the underlying text detection, segmentation and recognition approach be image-based (i.e., treating a video as a set of independent images) or should it exploit the fact that the same text line occurs in videos for some time and that, therefore, the multiple instances of the same text line can be utilized to achieve better detection, segmentation and recognition performance.

- *How will the output of the Video OCR system be used?*

  Different usages have different levels of tolerance against errors. For instance, if the Video OCR output is only used for image/video indexing based on the transcribed text, pixel errors in the localization and segmentation steps as well as recognition errors can be tolerated and compensated. If, however, the output is used for object-based video encoding, the system must minimize the errors in pixel classification. The system in [Lienhart *et al.*, 2002a], for example, was explicitly designed to label each pixel in a video as whether it belongs to text or not. This information was used to encode text occurrences then as a high-fidelity foreground MPEG-4 video objects (VOPs) at low frame per second (fps) rate, while re-coloring the pixels behind the text pixels with colors efficient for compression. A gain of about 1.5dB in PSNR at low bit rates were reported (see Fig. 6.2).

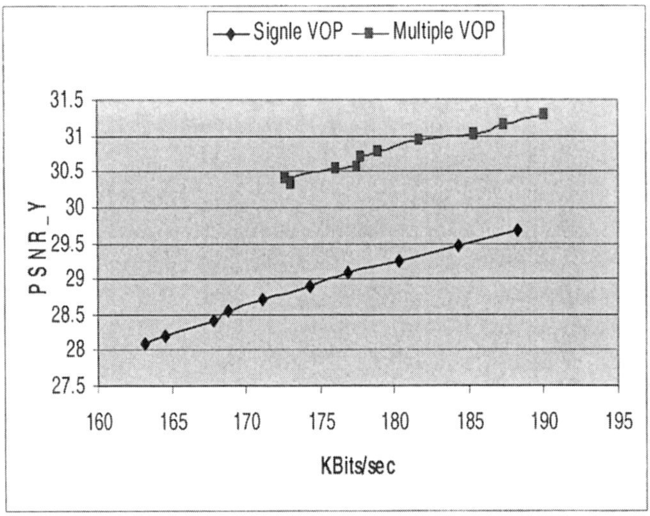

Figure 6.2. PSNR comparison of same video encoded as a single VOP MPEG-4 video and a multiple VOP MPEG-4 video with one additional VOP for each detected text line.

Other usage scenarios are the visual removal of text from videos and the automatic translation of detected text from one language into another language.

This paper focuses on texture-based Video OCR algorithms. It does not address the many connected-component-based approaches such as [Lienhart *et al.*, 1996; Lienhart, 1996; Lienhart *et al.*, 2000; Shim *et al.*, 1998]. This choice was purely made to present the vast research in this field in a more structured way. In no sense it should be understood as a judgment of the connected-component-based approach. In fact, connected-component based approaches are working surprisingly well and successful in practice and can compete in performance with the best texture-based approaches.

The remainder of this paper is organized at follows. In Section 6.1 we address texture-based text detection—the task of finding the locations of text occurrences in images and videos. Starting with general observations about text, a set of suitable texture features are listed in Section 6.1.1. Then, Section 6.1.2 details how to use the texture features to achieve image-based text detection. For video specialized extensions exist to further improve text detection performance. They are explained in Section 6.1.3. An overview of common performance measures and the performance of existing systems is given in Section 6.1.4. Section 6.2 addresses text segmentation—the task of preparing bitmaps of localized text occurrences for optical character recognition (OCR). It is subdivided into three parts. Section 6.2.1 and 6.2.2 discuss pre-processing steps helping to improve text segmentation performance. The former Section focuses on approaches in the image domain, while the latter investigates the unique possibilities with videos. Finally Section 6.2.3 introduces the segmentation algorithms. An overview of common performance measures and the performance of existing text segmentation systems is given in Section 6.2.4. Section 6.3 concludes the paper with a summary and outlook.

6.1 Detection

Text detection is the task of finding the locations of text occurrences in images and videos. Dependent on the subsequent task the circumscribing shapes of the text locations either comprise whole text columns or individual text lines. For planar text occurrences in the plane parallel to the camera plane the circumscribing shape is a rectangle, while for scene text it usually is a rotated parallelogram ignoring the foreshortening under fully perspective projection.

Text detection has many applications. It is the prerequisite for text segmentation. It, however, can also be used to rectify documents captured with still image or wearable cameras for improved readability. In most cases without rectification the text would exhibit significant perspective distortions.

6.1.1 Text Features

General Observations. Humans can quickly identify text regions without having to search for individual characters. Even text too far to be legible can easily be identified as such. This is due to the stationary pattern text lines and text columns exhibit at different scales.

In Roman languages text regions consist of text lines of the same orientation with roughly the same spacing in between. Each text line is composed of characters of approximately the same size, placed next to each other. A text line contrasting with the background shows a large intensity variation vertically to the writing direction as well as horizontally at its upper and lower boundaries.

The mainstream of overlay text in Roman languages is characterized by the following features [Lienhart *et al.*, 1996; Lienhart, 1996]. Only a few exceptions may be observed in practice:

- Characters are in the foreground. They are never partially occluded.
- Characters are monochrome.
- Characters are rigid. They do not change their shape, size or orientation from frame to frame.
- Characters have size restrictions. A letter is not as large as the whole frame. Nor are letters smaller than a certain number of pixels as they would otherwise be illegible to viewers.
- Characters are mostly upright.
- Characters are either stationary or linearly moving. Moving characters also have a dominant translation direction: horizontally from right to left or vertically from bottom to top.
- Characters contrast with their background since artificial text is designed to be read easily.
- The same characters appear in multiple consecutive frames.
- Characters appear in clusters at a limited distance aligned to a virtual line. Most of the time the orientation of this virtual lines is horizontal since that is the natural writing direction.

Most of these features also hold for ***non-Roman languages***, but some need to be adapted to the characteristics of the particular language system. For instance, the minimal readable font size of Roman languages is about 7 to 8 pt. In contrast, Chinese characters due to their complex structure require at least twice the size. In Roman languages meaningful words are built from multiple characters. Therefore, a semantically meaningful text line should be composed of at least three or more characters. In Chinese, however, each character has a meaning voiding this constraint. Roman languages are most readable with justified characters. Justified text lines in turn result in homogenous stroke densities that can easily be detected. Chinese characters, in contrast, have a fixed block size letting its spatial stroke density vary significantly. Every character occupies the same space. At the same time the number of strokes can vary from 1 to 20 [Cai *et al.*, 2002]. Some texture-based features might therefore not be applicable to Chinese character detection.

In this survey we will concentrate on texture-based approaches for Roman languages. Other language system as well as its dual approach, text detection and text segmentation based on connected component analysis, will not be addressed here.

All approaches should keep the following general challenges in mind:

- The contrast of text in complex backgrounds may vary in different areas of the image. Complex background usually requires strong contrast to make text still readable, while for simple background even a small contrast is sufficient [Cai *et al.*, 2002].

- The color of text is not uniform due to color bleeding, noise, compression artifacts, and applied anti-aliasing. Colour homogeneity should therefore not be strictly assumed [Lienhart, 1996; Loprestie *et al.*, 2000].

Texture-based Features. Text exhibits unique features at many scales. Researchers have developed many statistical features based on the local neighborhood to capture certain texture aspects of text. Some features operate at different text scales and are designed to identify individual text lines, while others measure certain attributes of text paragraphs. In this subsection, the most important features are listed. None of them will uniquely identify text regions. Each individual feature will still confuse text with non-text areas, but models one or several important aspects of text versus non-text regions. A society of features will complement each other and allow identifying text unambiguously.

Gray Levels of Raw Pixels. Shin et al. suggest the use of grey levels of raw pixels as features. The input feature vector size is reduced by taking only a structured subset of all pixels in a neighborhood. For instance, they suggest the use of a star pattern mask as shown in Fig.6.3 [Shin *et al.*, 2000].

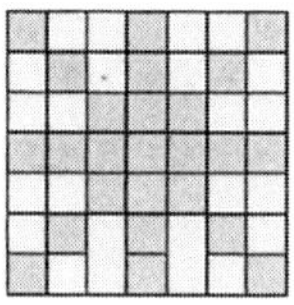

Figure 6.3. Star-like pixel pattern

Local Variance. The observed local variance in text regions depends on the scale. For small and medium text medium values are expected, since text in such areas undergoes aliasing at the boundaries. Very high variance region indicate single sharp edges and not text. In [Clark *et al.*, 2000] a circular disk filter S of radius 3 is applied to measure local variance V :

$$V = S * (I - S * I)^2 .$$

S is the area mask of the local neighborhood and I the input image.

Local Edge Strength. Characters consist of strokes. Text regions thus have a high density of edges. The local edge strength E is defined as the average edge magnitude in a neighborhood:

$$E = S * |D * I| .$$

I is the input image, D an edge filter (e.g., gradient or Sobel filter), and S some averaging filter (e.g., box, binomial, or Gaussian filter). In [Cai *et al.*, 2002] and [Clark *et al.*, 2000] a Sobel filter is applied, followed by a circular disk filter of radius 6. The local edge strength responds to text of any orientation.

If only horizontal in plane text should be detected, it is favorable to consider primarily only the horizontal edge strength:

$$E_h = S * |D_x * I| ,$$

where D_x is some horizontal edge detector. In [Zhong *et al.*, 2000] the horizontal edge strength is directly derived from DCT-encoded JPEG-images and MPEG-based I-frames by means of the sum of the absolute amplitude of the horizontal harmonics in each DCT block(i, j):

$$E_h(i,j) = \sum_{v=v_1}^{v_2} |c_{ov}(i,j)|$$

$c_{ov}(i,j)$ are the horizontal harmonics of 8×8 DCT block (i, j). The boundaries v_1 and v_2 have to be chosen according to the character size. [Zhong *et al.*, 2000] uses 2 and 6 for v_1 and v_2, respectively. The DCT coefficients capture the spatial periodicity and directionality in a local block and are therefore a shortcut to edge detection. Such a compressed domain edge detector, however, covers only a small part of the many resolutions of a frame posing a problem to scale-independent text extraction. This is especially true for high resolution videos such as HDTV video sequences.

Cai et al. suggest using an adaptive edge strength threshold [Cai *et al.*, 2002]. They observed that for text embedded in simple background low contrast suffices to render text readable, and that this can also be observed in practice. However, for text embedded in complex-background a high-contrast is always required and used. In a first step a low threshold is applied to the edge strength map. The threshold is selected to accommodate for low-contrast text in simple background. Based on a sliding window, the number of edge-free rows is counted. A high count suggests simple background and no threshold adjustments, while higher counts suggest choosing a higher adaptive threshold in that area to remove more edge pixels. One might argue that a more efficient continuous classifier can be built by using machine learning algorithms.

Edge Density. Text density is usually evaluated by opening/closing operations applied to binarized edge maps. In [Cai *et al.*, 2002] specific filters are designed, however, it is not clear why they should perform better than standard opening/closing operations. In general, the optimization criterion would be to learn a filter or morphological operation that keeps text regions of certain edge density, while removing non-text regions based on their diverging edge density.

Symmetric Edge Distribution. In areas of clearly readable text one expects—besides high local edge strength—to find edges at all angles and that in most cases an edge of a certain angle is accompanied by an edge in the opposite direction [Lienhart *et al.*, 2000] Clearly visible and readable text should have an edge on both sides of a stroke. Thus

$$1 - \sum_{\theta=0}^{\pi} \left(A\left(\theta\right) - A\left(\theta + \pi\right)\right)^2$$

is a measure of symmetry using local edge angle histograms [Clark *et al.*, 2001]. $A(\theta)$ is the total magnitude of edges in direction θ. This feature is scale invariant. Fig. 6.4 shows an example taken from [Clark *et al.*, 2000].

Edge Angle Distribution. For text regions we expect edge angles to be well distributed, i.e., almost all edge angles will occur. An appropriate measure is:

$$EAD = \sum_{\theta=0}^{2\pi} \left(A(\theta) - \overline{A}\right)$$

$\overline{A}$ represents the average magnitude over all directions. The EAD measure has its lowest value for homogenous edge distributions and will increase for skewed ones. Unlike most other features, this feature allows to distinguish straight ramps, canals, or ridges from text [Clark *et al.*, 2000]. In other words, at the appropriate scale text areas are isotropic. Alternatively, this attribute could be measured by Jaehne's Inertia tensor [Jaehne *et al.*, 1995].

Wavelets. Wavelet decomposition naturally captures directional frequency content at different scales. Li et al. suggest using the mean, second order (variance) and third-order central moments of the LH, HL, and HH component of the first three levels of each 16×16 window [Li *et al.*, 2000b].

Derivatives. In [Lienhart *et al.*, 2002a] the gradient image of the RGB input image $I = (I_r, I_g, I_b)$ is used to calculate the complex-values

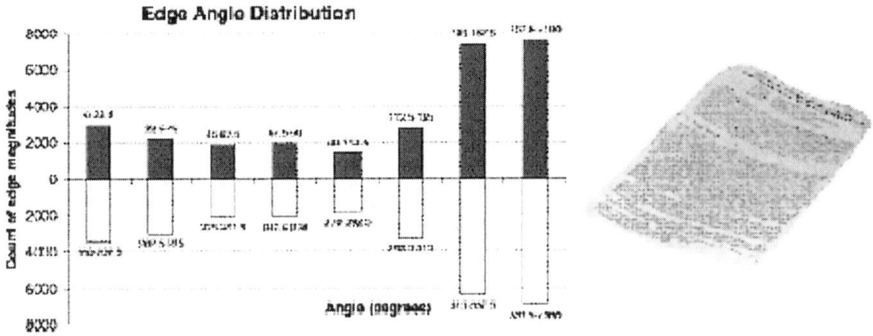

Figure 6.4. Histogram of edge angle values between 0 ° and 360 ° for the text shown on the right (Figure taken from [Clark *et al.*, 2000]).

edge orientation image E:

$$E(x,y) = \left(\sum_{c \in \{r,g,b\}} \left| \frac{I_c(x,y)}{dx} \right| + \left| \frac{I_c(x,y)}{dy} \right| \cdot i \right).$$

E maps all edge orientations between 0 ° and 90 °, and thus distinguishes only between horizontal, diagonal and vertical orientations.

6.1.2 Detection

The most common and generic form of feature-based text detection is based on a ***fixed scale and fixed position text classifier*** on some feature image F. A feature image F is a multi-band image where each band can be one of the features described in Section 6.1.1 computed at a given scale from the input image I. Given a $W \times H$ window region in a multi-band feature image F, a fixed size fixed position text detector classifies the window as containing text if and only if text of a given size is completely contained in the window. Often the window height is chosen to be one or two pixels larger than the largest targeted font height, and the width is chosen based on the width of the shortest possible, but still semantically meaningful word. For instance, in [Lienhart *et al.*, 2002a] a window of 20×10 was used.

Many different supervised machine learning techniques have been used to train a fixed scale fixed position text classifier such as Decision Trees, Neural Networks, complex Neural Networks, Boosting, Support Vector Machines, GMs, and handcrafted methods. An important design consideration at this stage is the amount of scale and location independence that should be trained into the fixed size fixed position classifier. Common choices for scale independence range from $\pm$ 10% to $\pm$50% of some reference font size, while for position independence ± 1 to $\pm W$*10% pixels are common.

Location independence is achieved by sliding the $W \times H$ window pixel by pixel over the whole feature image and recording the probability of having text at that location in a scale-dependent saliency map (see Fig. 6.4, single row). ***Scale independence*** is achieved by applying the fixed scale detection scheme to rescaled input images of different resolution [Li *et al.*, 2000b; Lienhart *et al.*, 2002a; Wu *et al.*, 1999]. Alternatively the features instead of the image can be rescaled to achieve a multi-scale search [Lienhart *et al.*, 2002b; Viola *et al.*, 2001].

As one can observe from the forth column in Fig. 6.4, where confidence in text locations is encoded by brightness, text locations stick out as correct hits at multiple scales, while false alarms appear less consistent over multiple scales. Similar results have been observed by Rowley et al. for their neural network-based face detector [Rowley *et al.*, 1998] and by Laurent Itti in his work on models of saliency-based visual attention [Itti *et al.*, 1998].

In order to recover initial text bounding boxes, the response images at the various scales must be integrated into a consistent text detection result. Different approaches are used for ***scale integration***. Examples are:

- Extract and refine initial text boxes at each scale from its associated saliency map in parallel before integrating them into the final detection result. Each scale might also take into account the response of nearby scales (3).
- Extract and refine initial text boxes sequentially—from the saliency maps at lower scales to the saliency maps at higher scales. Remove

all regions in the higher scale response maps which have already been detected at lower scales.

- Project the confidence of being text back to the original scale of the input image and extract and refine initial text boxes from the scale-integrated saliency map. Fig. 6.4 column 5 gives an example [Lienhart *et al.*, 2002a].

There are two principal ways of extracting initial text boxes: bottom-up and top-down approaches. ***Bottom-up approaches*** are region growing algorithms. Starting with seed pixels of highest text probability, text regions are grown iteratively. While this works well for Roman languages due to their low-variance stroke density property, it might cause problems for Chinese characters due to their large variance in stroke density [Cai *et al.*, 2002]. ***Top-down approaches*** split images regions alternately in horizontal and vertical directions based on texture features [Cai *et al.*, 2002]. Sometimes both approaches are used simultaneously. For instance in [Lienhart *et al.*, 2002a] a bottom-up approach is used to find text columns, while a top-down approach is used to partition these text columns into individual text lines.

The overall multi-scale search procedure is summarized in Fig. 6.5. Note that the raw scale and scale independent saliency maps are often smoothed by some morphological operations such as opening and closing.

6.1.3 Exploiting Temporal Redundancy

Videos differ from images by temporal redundancy. Each text line appears over several contiguous frames. This temporal redundancy can be exploited to

- increase the chance of localizing text since the same text may appear under varying conditions from frame to frame,
- remove false text alarms in individual frames since they are usually not stable throughout time,
- interpolate the locations of 'accidentally' missed text lines in individual frames, and
- enhance text segmentation by bitmap/stroke integration over time.

Early approaches used tracking primarily to remove false alarms. Therefore, potential text lines or text stroke segments were only tracked over a few frames (e.g., five frames) [Lienhart *et al.*, 1996; Shim *et al.*, 1998]. Dependent on whether the tracking was successful or not, a text

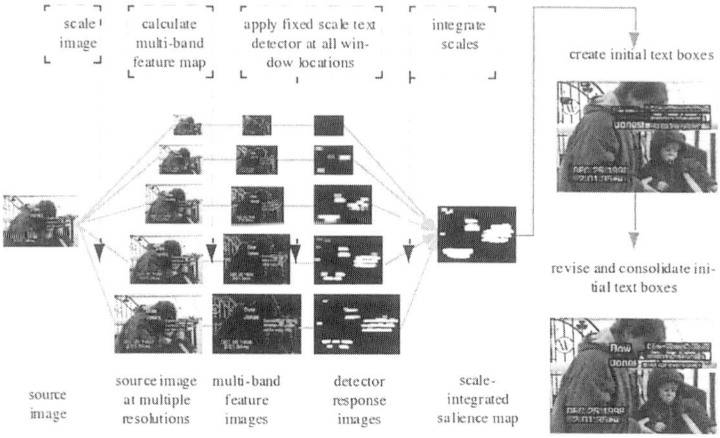

Figure 6.5. Scale and position independent text localization

candidate box or text stroke region was either preserved or discarded. Short term tracking also put fewer requirements on the quality of the tracking module.

More recent approaches summarize text boxes and character strokes of the same content in contiguous frames into a single text object. A text object describes a text line over time by its text bitmaps or connected-components, their sizes and their positions in the various frames as well as their temporal range of occurrence.

Text objects are extracted in a two-stage process in order to reduce computational complexity: In stage 1, a **video** is **monitored** at a coarse temporal resolution (see Fig. 6.6 and [Li *et al.*, 2000b; Lienhart *et al.*, 2002a]). For instance, the image-based text localizer of Section 6.1.2 is only applied to every second (i.e., every 30th and 25th frame in NTSC and PAL, respectively). The maximum possible step size is given by the assumed minimum temporal duration of text line occurrences. It is known from vision research that humans need between 2 and 3 seconds to process a complex scene. Thus, it is safe to assume that text appears clearly for at least one second.

If text is detected, the second stage of **text tracking** will be entered. In this stage text lines found in the monitoring stage are tracked backwards and forwards in time up to their first and last frame of occurrence.

We will restrict our description to forward tracking only since backward tracking is identical to forward tracking except in the direction you go through the video. Also the tracking description will be biased towards the feature based approach, although most can be directly applied to the stroke-based text detection approaches, too.

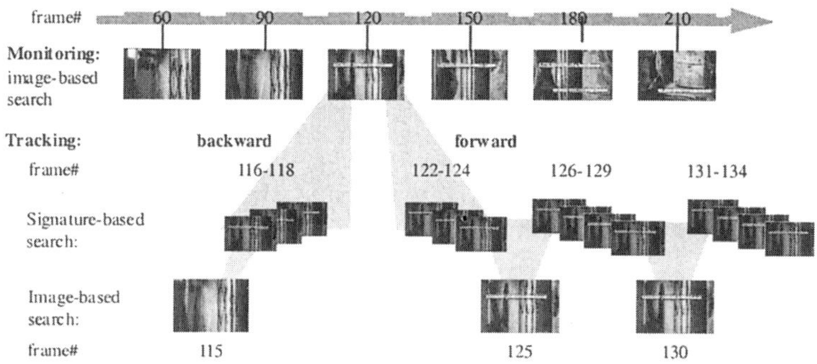

Figure 6.6. Relationship between video monitoring and text tracking stage

A fast text tracker takes the text line in the current video frame, calculates a characteristic signature, which allows discrimination of this text line from text lines with other contents, and searches in the next video frame for a region of the same dimension, which best matches the reference signature. If the best match exceeds a minimal required similarity, the text line is declared to be found and added to the text object. If the best match does not exceed a minimal required similarity, a signature-based drop-out is declared. The size of the search radius depends on the maximal assumed velocity of text. Heuristically text needs at least 2 seconds to move from left to right in the video. Given the frame size and the playback rate of the video this translates directly to the search radius in pixels. In principle, the search space can be narrowed down by predicting the location of text in the next frame based on the information contained in the text object so far.

The signature-based text line search cannot detect a text line fading out slowly since the search is based on the signature of the text line in the previous frame and not on a fixed master/prototype signature. The frame to frame changes are likely to be too small to be detectable. Further, the signature-based text line search can track zooming in or zooming out text only over a very short period of time. To overcome these limitations, the signature-based search is replaced every x-th frame by the image-based text localizer in order to re-calibrate locations and sizes of the text lines.

Often continuous detection and tracking of text objects is not possible due to imperfections in the video signal such as high noise, limited bandwidth, text occlusion, and compression artifacts. Therefore tracking should be terminated only if for a certain number of contiguous frames no corresponding text line could be found. For this, two thresholds $\max_{DropOut}^{signature-based}$ and $\max_{DropOut}^{image-based}$ are used. Whenever a text object cannot be extended to the next frame, the drop-out counter of the respective localization technique is incremented. The respective counter is reset to zero whenever the search succeeds. The tracking process is finished as soon as one of both counters exceeds its threshold.

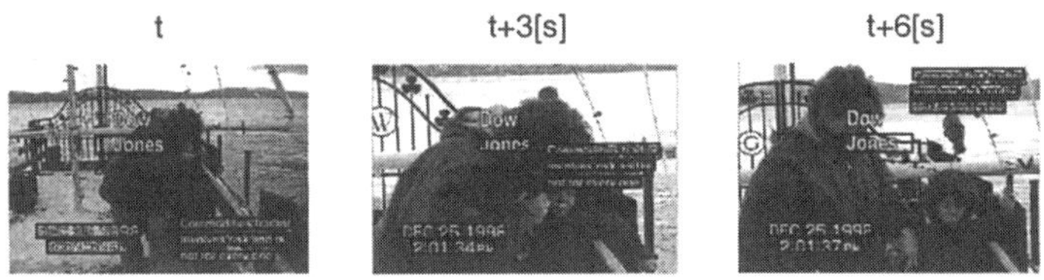

Figure 6.7. Example of text tracking of located text lines. All text lines except 'Dow' could be successfully tracked. The line 'Dow' is missed during text localization due to its difficult background (iron gate and face border).

Post-Processing. In order to prepare a text object for text segmentation, it must be trimmed down to the part which has been detected with high confidence: the first and last frame in which the image-based text localizer detected the text line. Text objects with a high drop-out rate and/or short duration (e.g., less than a second) should be discarded. The first condition rests on our observation that text lines are usually visible for at least one second. The second condition removes text objects resulting from unstable tracking which cannot be handled by subsequent processing. Unstable tracking is usually caused by strong compression artifacts or non-text objects.

Finally, a few attributes should be determined for each text object:

- **Text color**: Assuming that the text color of the same text line does not change over the course of time, a text object's color is determined as the median of the text colors per frame.

- **Text position**: The position of a text line might be static in one or both coordinates. If static, all text bounding boxes are replaced by the median text bounding box. The median text bounding box is the box whose left/right/top/bottom border is the median over all left/ right/top/bottom borders. If the position is only fixed in one direction such as the x or y axes, the left and right or the top and bottom are replaced by the median value, respectively. Temporally changing coordinate components may be smoothed by linear regression over time.

Fig. 6.7 shows the result of text tracking of located text lines for a sample sequence. All text lines except 'Dow' could be successfully tracked. The line 'Dow' is missed due to its partially difficult background such as the iron gate and face border. The iron gate's edge pattern is very similar to text in general. It also contains individual characters, thus confusing the image-based text localization system, which in turn renders tracking impossible.

6.1.4 Experimental Results

Two different kinds of performance measure have been used by the researchers in the field:

- Pixel-based performance measures and

- Text box-based performance measures.

Both performance measures require ground truth knowledge, i.e., precise knowledge about the text positions in each image/frame. Such ground truth knowledge usually has to be created by hand.

Pixel-based performance numbers calculate the hit rate, false hit rate and miss rate based on the percentage of pixels the ground truth and the detected text bounding boxes have in common:

$$hitrate_{pixel-based} = \frac{100}{|G|} \sum_{g \in G} \frac{1}{|g|} \max_{a \in A} \{|a \cap g|\}$$

$$missrate_{pixel-based} = 100 - hitrate_{pixel-based}$$

$$falsehits_{pixel-based} = \frac{100}{|G|} \sum_{g \in G} \frac{1}{|g|} \left(\left| \arg\max_{a \in A} \{|a \cap g|\} \right| - \max_{a \in A} \{|a \cap g|\} \right)$$

where $A = \{a_1, \ldots, a_N\}$ and $G = \{g_1, \ldots, g_M\}$ are the sets of pixel sets representing the automatically created text boxes and the ground truth text boxes of size $N = |A|$ and $M = |G|$, respectively. $|a|$ and $|g|$ denote the number of pixels in each text box, and $a \cap g$ the set of joint pixels in a and g.

In contrast, the ***text box-based performance numbers*** refer to the number of detected boxes that match with the ground truth. An automatically created text bounding box A is regarded as matching a ground truth text bounding box G if and only if the two boxes overlapped by at least $x\%$. Typical values for x are 80% or 90%:

$$hitrate_{box-based} = \frac{100}{M} \sum_{g \in G} \max_{a \in A} \{\delta(a, g)\}$$

$$missrate_{box-based} = 100 - hitrate_{box-based}$$

$$falsehits_{box-based} = \frac{100}{M} \left(N - \sum_{a \in A} \max_{g \in G} \{\delta(a, g)\} \right),$$

where

$$\delta(a, b) = \begin{cases} \min\left(\frac{|a \cap g|}{|a|}, \frac{|a \cap g|}{|g|}\right) & if \ \min\left(\frac{|a \cap g|}{|a|}, \frac{|a \cap g|}{|g|}\right) \geq 0.8 \\ 0 & else \end{cases}.$$

Alternatively, often recall and precision values are reported:

$$recall = \frac{hits}{hits + missed} \quad , \quad precision = \frac{hits}{hits + falsealarms}$$

The most important text detection approaches and their reported performance numbers are listed and compared in Table 6.1.

Commonly reported sources of text misses are due to weak text contrast with the background, large spacing between the characters, or too large fonts. Non-text regions with multiple vertical structures often result in false alarms.

Table 6.1. Comparison of text detection approaches; H=hit rate; F_1 = false alarm rate with respect to text regions; F_2 = false alarm rate with respect to patches tested; F = false alarm rate with unknown basis; P/R=precision/recall values.

Work	Scope				Domain	Performance	Comments
	Image	Exploit video	Captions	Scene text	Compressed/ Uncompr.		
[Cai *et al.*, 2002]	x		x		U	H: 98.2% F: 6.5%	Detection of horizontal English & Chinese text
[Jeong *et al.*, 1999]	x		x		U	H: 92.2% F: 5.1%	NN-based text detection for news video; English and Chinese
[Li *et al.*, 2000b]	x	x	x	(x)	U	R: 92.8% P: 91.0%	Tracking system sensitive to complex background; multi-scale search
[Lienhart *et al.*, 2002a]	x	x	x	(x)	U	H: 94.7% F_1: 18%	Complete NN-based system; multi-scale search
[Mariano *et al.*, 2000]	x		x		U	H: 94% F: 39%	Designed for horizontal, uniformed colored text
[Ohya *et al.*, 1994]	x		x	x	U	H: 95.0%	Detection, segmentation and recognition tightly integrated; focus on upright scene text
[Sato *et al.*, 1999]	x	x	x		U	H: 98.6%	Complete innovative system for CNN Headline News; designed for very small font sizes
[Shim *et al.*, 1998]	x	x	x	x	U	H: 98.8%	Designed for horizontal text only; similar to [Lienhart, 1996]
[Shin *et al.*, 2000]	x		x		U	H: 94.5% F: 4.2%	Uses SVM on raw pixel inputs; multi-scale search
[Wu *et al.*, 1999]	x		x	x	U	H: 93.5%	Complete system for video, newspapers, ads, photos, etc.; multi-scale search
[Zhong *et al.*, 1999]	x	x	x		C	H: 96% F_1: 6.07%	Very fast pre-filter for text detection
[Zhong *et al.*, 2000]	x		x		C	H: 99.1% F_1: 36% F_2:: 1.58%	Very fast pre-filter for text detection

6.2 Segmentation

Text segmentation is the task of preparing the bitmaps of localized text occurrences for optical character recognition (OCR). Often standard commercial OCR software packages, which are optimized for scanned documents, are used for recognition due to their high level of maturity.

Text segmentation is commonly performed in two steps: In a first step, the image quality is enhanced in the still image and/or video domain, before in a second step a binary image is derived from the visually enhanced image by means of standard binarization algorithms [Ohya *et al.*, 1994; Otsu, 1979].

6.2.1 Enhancements in the Image Domain

Resolution Enhancement. The low resolution of video (typically 72 ppi) is a major source of problems in text segmentation and text recognition. Individual characters in MPEG-1 encoded videos often have a height of less than 11 pixels. Although such text occurrences are still recognizable for humans, they challenge today's standard OCR systems due to anti-aliasing, spatial sampling and compression artifacts [Loprestie *et al.*, 2000; Lienhart *et al.*, 2000; Sato *et al.*, 1998]. Today's OCR systems have been designed to recognize text in documents, which were scanned at a resolution of at least 200dpi to 300dpi resulting in a minimal text height of at least 40 pixels. In order to obtain good results with standard OCR systems it is necessary to enhance the resolution of segmented text lines.

A common pre-processing step is to obtain higher resolution text bitmaps by sub-pixel accurate rescaling of the original text bitmaps to a fixed target height, while preserving the aspect ratio. Typical values for the target height range from 40 to 100 pixels, and cubic interpolation or better up-sampling filters are used for rescaling. Fixing a target height is computationally efficient, because text with a larger height neither improves segmentation nor OCR performance [Li *et al.*, 2000b; Lienhart *et al.*, 2002a; Sato *et al.*, 1998]. In addition, the fixed target height effectively normalizes the stroke widths to a narrow range for Roman characters, which in turn can be used later for additional refinement operations.

Character Stroke Enhancement. Sato et al. propose to use 4 directional stroke filters of 0°, +45°, -45°, and 90° trained by fixed English fonts. These filters calculate the probability of each pixel being on a text stroke of that direction. By integrating the four filter results an enhance text stroke bitmap is formed (see Fig. 6.8 taken from [Sato *et al.*, 1998])

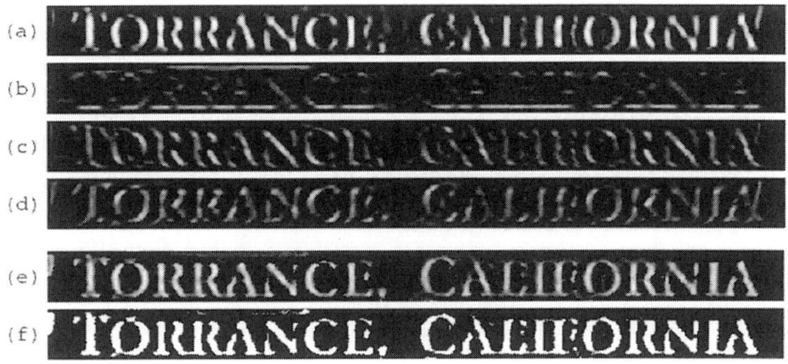

Figure 6.8. "Result of character extraction filters: (a) 0 ° (b) 90 ° (c)-45 ° (d) 45 ° (e) Integration of four filters (f) Binary image" (Figure taken from [Sato *et al.*, 1998]).

6.2.2 Enhancements in the Video Domain

Temporal Integration. Text objects in videos consist of many bitmaps of the same text line in contiguous frames. This redundancy can be exploited in the following way to remove the complex background surrounding characters: Suppose the bitmaps of a text object are piled up over time such that the characters are aligned perfectly with each other. Looking through a specific pixel in time, one may notice that pixels belonging to text vary only slightly, while background pixels often change tremendously through time. Since a text line's location is static due to its alignment its pixels are not supposed to change. In contrast, background pixels are very likely to change due to motion in the background or motion of the text line (see Table 6.2(a)).

A temporal maximum/minimum operator applied to all or a subset of perfectly aligned greyscale bitmaps of a text object for normal/inverse text is generally capable to separate text pixels from background pixels. This temporal maximum/minimum operation was first proposed by Sato et al. for static text [Sato *et al.*, 1999], but can also applied to moving text if the text segmentation system supports sub-pixel accurate text line alignment [Lienhart *et al.*, 2002a]. An alternative approach to the min/max operation is to calculate a pixel's temporal mean and variance and reject pixels with large standard deviations or a few outliers.

Sub-pixel Accurate Text Alignment. Two similar proposals have been developed by Li [Li *et al.*, 2000b] and Lienhart [Lienhart *et al.*, 2002a]. The latter approach, though, is more robust since it exploits the estimated text color during tracking and, therefore, does not have problems with complex background as reported by [Li *et al.*, 2000b].

The sub-pixel accurate text alignment is achieved as follows: In a first step, the bounding boxes of detected text locations are slightly increased to ensure that text is always 100% contained in the enlarged bounding boxes (see Figure 10). Let $B_1(x,y), ..., B_N(x,y)$ denote the N bitmaps of the enlarged bounding boxes of a text object and $B^r(x,y)$ the representative bitmap, which is to be derived and initialized to $B^r(x,y) = B_1(x,y)$. Then, for each bitmap $B_i(x,y)$, $i \in \{2,\ldots,N\}$, the algorithm searches for the best displacement vector $\left(dx_i^{opt}, dy_i^{opt}\right)$, which minimizes the difference between $B^r(x,y)$ and$B_i(x,y)$ with respect to pixels having text color, i.e.,

$$(dx_i^{opt}, dy_i^{opt}) = \arg\min \sum_{(x,y)\in B^r \cap B^r(x,y) \subseteq textColor} |B^r(x,y) - B_i(x+dx, y+dy)|$$

A pixel is defined to have text color if and only if it does not differ more than a certain amount from the greyscale text color estimated for the text object.

At each iteration,$B^r(x,y)$is updated to

$$B^r(x,y) = op(B^r(x,y), B_i(x + dx_i^{opt}, y + dy_i^{opt})),$$

where $op{=}max$ for normal text and $op{=}min$ for inverse text. Table 6.2(b) shows an example of the min/max operation.

6.2.3 Segmentation

Different segmentation techniques have been used for text segmentation. Sometimes several of them are combined to achieve better and more reliable segmentation results.

Seed-Filling from Border Pixels. Text occurrences are supposed to have enough contrast with their background in order to be easily readable. This feature can be exploited to remove large parts of the complex background. The basic idea is to increase the text bounding boxes such

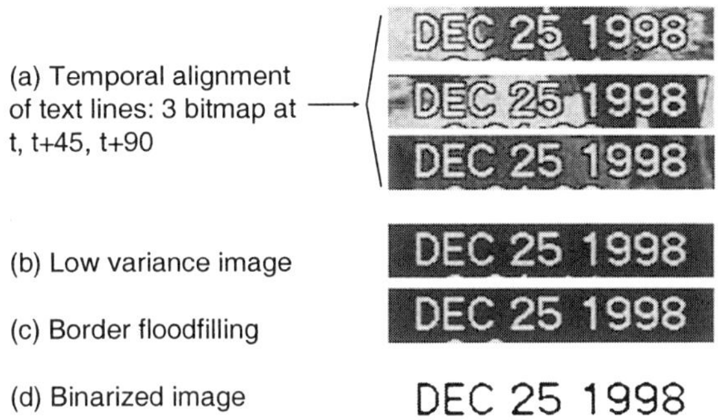

Figure 6.9. Example of the various text segmentation steps

that no text pixels fall onto the border and then to take each pixel on the boundary of the text bounding box as a seed to a virtual seed-fill procedure, which is tolerant to small color changes. Pixels which differ not more than $threshold_{seedfill}$ from the seed will be regarded as pixels of the same color as the seed. In theory the virtual seed-fill procedure should never remove character pixels since the pixels on the boundary do not belong to text and text contrasts with its background. We attributed the seed-fill procedure with "virtual" since the fill operation is only committed after the seed-fill procedure has been applied to all pixels on the border line in order to avoid side effects between different seeds [Lienhart *et al.*, 2002a].

In practice, however, text segmentation sometimes has to deal with low contrast, which may cause the seed-fill algorithm to leak into a character. A stop criterion may be defined based on the expected stroke thickness. Regions which over a large extent comply with the stroke thickness range of characters in one dimension should not be deleted.

Not all background pixels are eliminated by this procedure, since the sizes of the regions filled by the seed-fill algorithm are limited by the maximum allowed color difference between a pixel and its border pixel seed. In addition, some regions are not connected to the border such as the interior of closed stroke characters 'o' and 'p'. Therefore, a hypothetical 8-neighborhood seed-fill procedure with $threshold_{seedfill}$ is applied to each non-background pixels in order to determine the dimension of the region that can hypothetically be filled. Background regions should be smaller then text character regions. Therefore, all hypothetical regions violating the typical range of width and height values for characters are deleted.

Figure 6.10. Relationship between bounding boxes and their enlarged counterparts

Thresholding. The simplest form of thresholding rests on a single, global threshold. Many different variants of global thresholding have been designed—ranging from bi-level schemes to tri-level schemes. More sophisticated variants also exploit the estimated text color [Lienhart *et al.*, 2002a; Sato *et al.*, 1998; Wu *et al.*, 1999].

For text on complex background a global threshold may not be appropriate since background pixels can have similar greyscale values as the text, or it be brighter and darker than the text at different locations.

In these cases an adaptive threshold should be applied. Commonly used adaptive binarization algorithms are derivatives of Otsu's [Otsu, 1979] and Ohya's work [Ohya *et al.*, 1994].

6.2.4 Experimental Results

For text segmentation no generally accepted performance measure has emerged in the literature. The three most common performance measures are:

- ***Manual visual inspection***: Correctness is determined by manual visual inspection of all created binary bitmaps.

- ***OCR Accuracy***: Segmentation performance is evaluated indirectly by means of the resulting OCR error rate with a given OCR engine making the results dependent on the OCR engine and its peculiarities.

- **Probability of Error**: The probability of error measure requires pixel maps of the ground truth data, which in most cases is very hard to provide. The probability of error (PE) is defined as follows [Lee *et al.*, 1990]:

$$PE = P(O)P(B|O) + P(B)P(O|B),$$

where P(B|O) and P(O|B) are the probability of error in classifying a text/background pixel as background/text pixel, P(O) and P(B) are the a priori probabilities of text/background pixels in the test images.

Table 6.2 reports the important text segmentation approaches and their performance numbers. Only OCR accuracy is reported for comparability.

6.3 Conclusion

Text localization and text segmentation in complex images and video have reached a high level of maturity. In this survey we focused on texture-based approaches for artificial text occurrences. The different core concepts underlying the different detection and segmentation schemes were presented together with guidelines for practitioners in video processing. Future research in Video OCR will focus more on scene text as well as on further improvements of the algorithms for localization and segmentation of artificial text occurrences.

Table 6.2. Comparison of text segmentation approaches; RC = character recognition rate, RW = word recognition rate

Work	Analysis scope		Performance	Comments
	Image/ video	Captions/ Scene text		
Li'99	I,V	C, S	RC: 88%	Addresses overlay text and scene text with two separate approaches; resolution and video enhancement
[Lienhart *et al.*, 2002a]	I,V	C, (S)	RC: 69.9%	Uses standard OCR software; resolution and video enhancement; difficult and large test set
[Lienhart, 1996]	I,V	C, (S)	RC: 80.3%	Connected-component based approach; simple self-trained OCR engine
[Lienhart *et al.*, 2000]	I,V	C, (S)	RC: 60.7%	Connected-component based approach; standard OCR engine
[Loprestie *et al.*, 2000]	I	C	RC: 87.9	Developed for text in web images
[Ohya *et al.*, 1994]	I	C,S	RC: 34.3%	Detection, Segmentation and Recognition are tightly integrated into each other
[Sato *et al.*, 1999]	I,V	C	RC: 83.5%	Integrated domain specific (CNN Headline News) text segmentation and recognition approach; achieved further recognition improvements by using domain-specific dictionaries; resolution and video enhancement
[Wu *et al.*, 1999]	I	C, (S)	RC: 83.8% RW: 72.4%	Uses standard OCR software; segmentation is based on global threshold per text region

References

Lalitha Agnihotri and Nevenka Dimitrova. Text Dection for Video Analysis. *IEEE Workshop on Content-Based Access of Image and Video Libraries*, 22 June 1999, Fort Collins, Colorado, 1999.

Min Cai, Jiqiang Song, and Michael R. Lyu. A New Approach for Video Text Detection. *IEEE International Conference on Image Processing*, pp. 117-120, 2002.

P. Clark and M. Mirmehdi. Finding Text Regions Using Localised Measures. Proceedings of the *11th British Machine Vision Conference*, pp. 675-684, BMVA Press, September 2000.

P. Clark and M. Mirmehdi. Estimating the orientation and recovery of text planes in a single image. Proceedings of the *12th British Machine Vision Conference*, pp. 421-430, BMVA Press, September 2001.

L. Itti, C. Koch, and E. Niebur. A Model of Saliency-Based Visual Attention for Rapid Scene Analysis. IEEE Transactions on Pattern Analysis and Machine Intelligence, 20(11):1254-1259, 1998.

S. U. Lee, S. Y. Chung, and R. H. Park. A Comparative Performance Study of Several Global Thresholding Techniques for Segmentation. *Computer Vision, Graphics, and Image Processing*, Vol. 51, pp. 171-190, 1990.

Huiping Li, O. Kia and David Doermann. Text Enhancement in Digital Videos. *Proc. SPIE Vol. 3651: Document Recognition and Retrieval VI*, p. 2-9, 1999.

Huiping Li and David Doermann. Superresolution-Based Enhancement of Text in Digital Video. 15th *Pattern Recognition Conference*, Vol.1, pp. 847-850, 2000.

H. Li, D. Doermann and O. Kia. Automatic Text Detection and Tracking in Digital Video. *IEEE Transactions on Image Processing.* Vol. 9, No. 1, pp. 147-156, Jan. 2000.

Bernd Jaehne. Digital Image Processing. Springer-Verlag Berlin Heidelberg, 1995.

Anil K. Jain and Bin Yu. Automatic Text Localication in Images and Video Frames. *Pattern Recognition*, 31(12), pp. 2055-2076, Dec. 1998.

Ki-Young Jeong, Keechul Jung, Eun Yi Kim, and Hang Joon Kim. Neural Network-based Text Location for News Video Indexing. *IEEE International Conference on Image Processing*, Vol. 3, pp. 319-323, 1999.

Rainer Lienhart and Frank Stuber. Automatic Text Recognition in Digital Videos. *Proc. SPIE 2666: Image and Video Processing IV*, pp. 180-188, 1996.

Rainer Lienhart. Automatic Text Recognition for Video Indexing. *Proc. ACM Multimedia 96*, Boston, MA, pp. 11-20, Nov. 1996.

Rainer Lienhart and Wolfgang Effelsberg. Automatic Text Segmentation and Text Recognition for Video Indexing. *ACM/Springer Multimedia Systems*, Vol. 8, pp. 69-81, Jan. 2000.

Rainer Lienhart and Axel Wernicke. Localizing and Segmenting Text in Images, Videos and Web Pages. *IEEE Transactions on Circuits and Systems for Video Technology*, Vol.12, No. 4, pp. 256 -268, April 2002.

Rainer Lienhart and Jochen Maydt. An Extended Set of Haar-like Features for Rapid Object Detection. *IEEE International Conference on Image Processing*, Vol. 1, pp. 900-903, Sep. 2002.

Daniel Loprestie and JiangYing Zhou. Locating and Recognizing Text in WWW Images. *Information Retrieval*, Kluwer Academic Publishers, pp. 177-206, 2000.

Vladimir Y. Mariano and Rangachar Kasturi. Locating Uniform-Colored Text in Video Frames. *15th Int. Conf. on Pattern Recognition*, Vol.4, pp. 539-542, 2000.

G. Myers, R. Bolles, Q.-T. Luong, and J. Herson. Recognition of Text in 3-D Scenes. 4th *Symposium on Document Image Understanding Technology*, Columbia, Maryland, pp. 23-25, April 2001.

Jun Ohya, Akio Shio, and Shigeru Akamatsu. Recognizing Characters in Scene Images. *IEEE Transaction on Pattern Analysis and Machine Intelligence*, Vol. 16, No. 2, Febr. 1994.

N. Otsu. A Threshold Selection Method From Gray-Level Histograms. *IEEE Transactions on Systems, Man, and Cybernetics*, Vol. 9, No. 1, pp. 62-66, 1979.

Henry A. Rowley, Shumeet Baluja, and Takeo Kanade. Neural Network-Based Face Detection. *IEEE Transactions on Pattern Analysis and Machine Intelligence*, Vol. 20, No. 1, pp. 23-38, January 1998.

T. Sato, T. Kanade, E. Hughes, M. Smith. Video OCR for Digital News Archives. *IEEE Workshop on Content-Based Access of Image and Video Databases*, Bombay, India, January, pp. 52-60, 1998.

T. Sato, T. Kanade, E. K. Huges, M. A. Smith, and S. Satoh. Video OCR: Indexing Digital News Libraries by Recognition of Superimposed Caption. *ACM Multimedia Systems*, Vol. 7, No. 5, pp. 385-395, 1999.

Jae-Chang Shim, Chitra Dorai, and Ruud Bolle. Automatic Text Extraction from Video for Content-based Annotation and Retrieval. *IBM Technical Report*, RC21087, IBM Thomas J. Watson Research Center, Yorktown Heights, New York, January 1998.

C.S. Shin, K.I. Kim, M.H. Park, H.J. Kim. Support Vector Machine-based Text Detection in Digital Video. *Proceedings of the IEEE Signal Processing Society Workshop on Neural Networks for Signal Processing X*, Vol. 2, pp. 634-641, 2000.

Paul Viola and Michael J. Jones. Rapid Object Detection using a Boosted Cascade of Simple Features. *IEEE Computer Vision and Pattern Recognition*, Vol. 1, pp. 511-518, 2001.

V. Wu, R. Manmatha, E.M. Riseman. Textfinder: An Automatic System to Detect and Recognize Text in Images. *IEEE Transactions on Pattern Analysis and Machine Intelligence,* Vol. 21, Issue 11, pp. 1224-1229, Nov. 1999.

Boon-Lock Yeo and Bede Liu. Visual Content Highlighting via Automatic Extraction of Embedded Captions on MPEG Compressed Video. in Digital Video Compression: Algorithms and Technologies, Proc. SPIE 2668-07 (1996).

Yu Zhong, Hongjiang Zhang, and A.K. Jain. Automatic Caption Localization in Compressed Videos. *IEEE International Conference on Image Processing*, Vol. 2, pp. 96-100, 1999.

Yu Zhong, Hongjiang Zhang, and A.K. Jain. Automatic Caption Localization in Compressed Videos. *IEEE Transactions on Pattern Analysis and Machine Intelligence*, Vol. 22, Issue 4, pp. 385-392, April 2000.

Chapter 7

VIDEO CATEGORIZATION USING SEMANTICS AND SEMIOTICS

Zeeshan Rasheed and Mubarak Shah
Computer Vision Lab
School of Electrical Engineering and Computer Science
University of Central Florida
Orlando, FL 32816
zrasheed@cs.ucf.edu

Abstract This chapter discusses a framework for segmenting and categorizing videos. Instead of using a direct method of content matching, we exploit the semantic structure of the videos and employ domain knowledge. There are general rules that television and movie directors often follow when presenting their programs. In this framework, these rules are utilized to develop a systematic method for categorization that corresponds to human perception. Extensive experimentation was performed on a variety of video genres and the results clearly demonstrate the effectiveness of the proposed approach.

Keywords: Video categorization, segmentation, shot detection, key-frame detection, shot length, motion content, audio features, audio energy, visual disturbance, shot connectivity graph, semantics, film grammar, film structure, preview, video-on-demand, game shows, host detection, guest detection, categorization, movie genre, genre classification, human perception, terabytes, film aesthetics, music and situation, lighting.

Introduction

The amount of audio-visual data currently accessible is staggering; everyday, documents, presentations, homemade videos, motion pictures and television programs augment this ever-expanding pool of information. Recently, the Berkeley "How Much Information?" project [Lyman and Varian, 2000] found that 4,500 motion pictures are produced annually amounting to almost 9,000 hours or half a terabyte of data *ev-*

ery year. They further found that 33,000 television stations broadcast for twenty-four hours a day and produce eight million hours per year, amounting to 24,000 terabytes of data! With digital technology becoming inexpensive and popular, there has been a tremendous increase in the availability of this audio-visual information through cable and the Internet. In particular, services such as *video on demand* allow the end users to interactively search for content of their interest. However, to be useful, such a service requires an intuitive organization of data available. Although some of the data is labelled at the time of production, an enormous portion remains un-indexed. Furthermore, the provided labeling may not contain sufficient context for locating data of interest in a large database. Detailed annotation is required so that users can quickly locate clips of interest without having to go through entire databases. With appropriate indexing, the user could extracting relevant content and navigate effectively in large amounts of available data. Thus, there is great incentive for developing automated techniques for indexing and organizing audio-visual data, and for developing efficient tools for browsing and retrieving contents of interest.

Digital video is a rich medium compared to text material. It is usually accompanied by other information sources such as speech, music and closed captions. Therefore, it is important to fuse this heterogenous information intelligently to fulfill the users' search queries. Conventionally, the data is often indexed and retrieved by directly matching homogeneous types of data. Multimedia data, however, also contains important information related to the *interaction* between heterogenous types of data, such as video and sound, a fact confirmed through human experience. We often observe that a scene may not evoke the same response of horror or sympathy, if the accompanying sound is muted. Conventional methods fail to utilize these relationships since heterogenous data types cannot be compared directly. The challenge is to develop sophisticated techniques that fully utilize the rich source of information contained in multimedia data.

7.1 Semantic Interpretation of Videos

We believe that the categorization of videos can be achieved by exploring the concepts and meanings of the videos. This task requires bridging the gap between low-level contents and high-level concepts. Once a relationship is developed between the computable features of the video and its semantics, the user would be allowed to navigate through videos by ideas instead of the rigid approach of content matching. However, this relationship must follow the norms of human perception and abide by the rules that are most often adhered to by the creators (directors) of these videos. These rules are generally known as *Film Grammar* in video production literature. Like any natural language, this grammar also has several dialects, but is fortunately, more or less universal. For example, most television game shows share a common pattern of transitions among the shots of host and guests, governed by the grammar of the show. Similarly, a different set of rules may be used to film a dialogue between two actors as compared to an action scene in a feature movie. In his landmark book "Grammar of the Film Language", Daniel Arijon writes:

> *"All the rules of film grammar have been on the screen for a long time. They are used by film-makers as far apart geographically and in style as Kurosawa in Japan, Bergman in Sweden, Fellini in Italy and Ray in India. For them and countless others this common set of rules is used to solve specific problems presented by the visual narration of a story"* [Arijon, 1976], p. 4.

The interpretation of concepts using this grammar first requires the extraction of appropriate features. Secondly, these features or *symbols* need to be semiotically (symbolic as opposed to semantic) explored as in natural languages. However, the interpretation of these symbols must comply with the governing rules for video-making of a particular genre. An important aspect of this approach is to find a suitable mapping between low-level video features and their bottom-line semantics. These steps can be summarized as:

- Learn the video making techniques used by the directors. These techniques are also called *Film Grammar.*
- Learn the theories and practices of film aesthetics, such as the effect of color on the mood, the effect of music on the scene situation and the effect of postprocessing of the audio and video on human perception.
- Develop a model to integrate this information to explore concepts.
- Provide users with a facility to navigate through the audio-visual data in terms of concepts and ideas.

This framework is represented in Fig. 7.1. In the next section, we will define a set of computable features and methods to evaluate them. Later, we will demonstrate that by combining these features with the semantic structure of talk and game shows, interview segments can be separated from commercials. Moreover, the video can be indexed as *Host-shots* and *Guest-shots.* We will also show that by employing cinematic principles, Hollywood movies can be classified into different genres such as *comedy*, *drama* and *horror* based on their previews. We will present the experimental results obtained, which demonstrate the appropriateness of our methodology.

We now discuss the structure of a film, which is an example of audio-visual information, and define associated *computable features* in the next section.

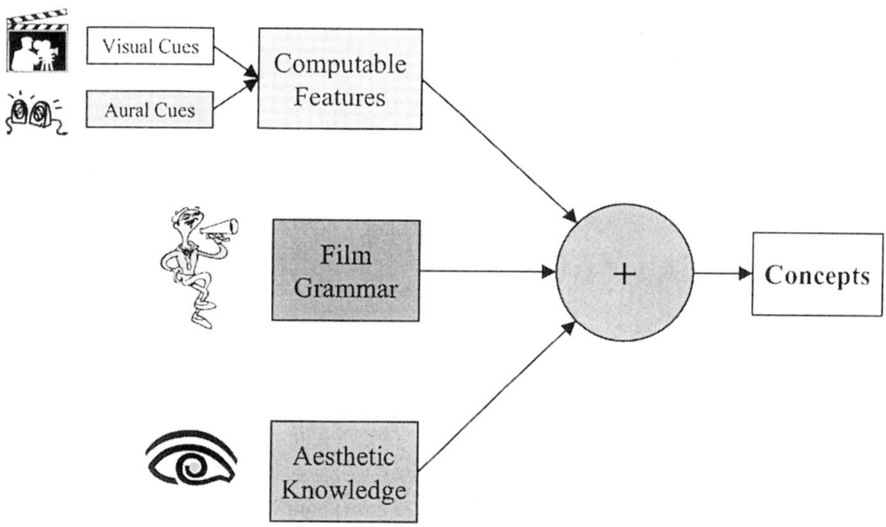

Figure 7.1. Our approach.

7.1.1 Film Structure

There is a strong analogy between a film and a novel. A *shot*, which is a collection of coherent (and usually adjacent) image frames, is similar to a *word.* A number of words make up a *sentence* as shots make visual thoughts, called *beats.* Beats are the representation of a subject and are collectively referred to as a *scene* in the same way that sentences collectively constitute a *paragraph.* Scenes create *sequences* like paragraphs make *chapters.* Finally, sequences produce a *film* when combined together as the chapters make a *novel* (see Fig.7.2). This final audio-visual product, i.e. the film, is our input and the task is to extract the concepts within its small segments in a bottom-up fashion. Here, the ultimate goal is to decipher the meaning as it is perceived by the audience.

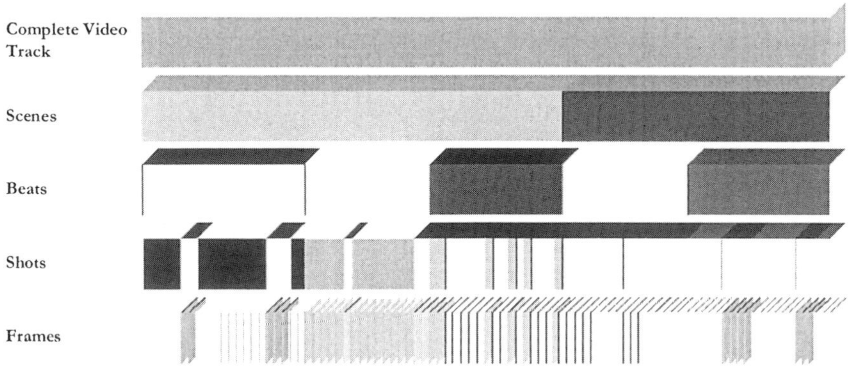

Figure 7.2. A film structure; frames are the smallest unit of the video. Many frames constitute a shot. Similar shots make scenes. The complete film is the collection of several scenes presenting an idea or concept.

7.2 Computable Features of an Audio-Visual Data

We define computable features of an audio-visual data as a set of attributes that can be extracted using image/signal processing and computer vision techniques. This set includes, but is not limited to, shot boundaries, shot length, shot activity, camera motion, color characteristics of image frames (for example histogram, color-key using brightness and contrast) as video features. The audio features may include amplitude and energy of the signal as well as the detection of speech and music in the audio stream. Following, we discuss these features and present methods to compute them.

7.2.1 Shot Detection

A shot is defined as a sequence of frames taken by a single camera with no major changes in the visual content. We have used a modified version of the color histogram intersection method proposed by [Haering, 1999]. For each frame, a 16-bin HSV normalized color histogram is estimated with 8 bins for hue and 4 bins each for saturation and value. Let $S(i)$ represent the histogram intersection of two consecutive frames i and $j = i - 1$ that is:

$$S(i) = \sum_{k \in bins} min(H_i(k) - H_j(k)), \tag{7.1}$$

where H_i and H_j are the histograms and $S(i)$ represents the maximum color similarity of frames. Generally, a fixed threshold is chosen empirically to detect the shot change. This approach works quite well [Haering, 1999] if the shot change is abrupt without any shot transition effect. However, a variety of shot transitions occur in videos for example *wipes* and *dissolves*. Applying a fixed threshold to $S(i)$ when the shot transition occurs with a dissolve generates several outliers because consecutive frames differ from each other until the shot transition is completed. To improve the accuracy, an iterative smoothing of the one dimensional function S is performed first. We have adapted the algorithm proposed by [Perona and Malik, 1990], based on anisotropic diffusion. This is done in the context of scale-space. S is smoothed iteratively using a Gaussian kernel such that the variance of the Gaussian function varies with the signal gradient:

$$S^{t+1}(i) = S^t(i) + \lambda \left[c_E \cdot \nabla_E S^t(i) + c_W \cdot \nabla_W S^t(i) \right], \tag{7.2}$$

where t is the iteration number and $0 < \lambda < 1/4$ with:

$$\begin{aligned} \nabla_E S(i) &\equiv S(i+1) - S(i), \\ \nabla_W S(i) &\equiv S(i-1) - S(i). \end{aligned} \tag{7.3}$$

The conduction coefficients are a function of the gradients and are updated for every iteration as:

$$\begin{aligned} c_E^t &= g\left(\mid \nabla_E S^t(i) \mid \right), \\ c_W^t &= g\left(\mid \nabla_W S^t(i) \mid \right), \end{aligned} \tag{7.4}$$

where $g(\nabla S) = e^{-(\frac{|\nabla_E|}{k})^2}$. In our experiments, the constants were set to $\lambda = 0.1$ and $k = 0.1$. Finally, the shot boundaries are detected by finding the local minima in the smoothed similarity function S. Thus, a shot boundary will be detected where two consecutive frames have minimum color similarity. This approach reduces the false alarms produced by fixed threshold methods. Figure 7.3 presents a comparison of the two methods. The similarity function S is plotted against the frame numbers. Only 400 frames are shown for convenient visualization. There are several outliers in (a) because gradually changing visual contents from frame to frame (dissolve effect) are detected as shot changes. For example, there are multiple shots detected around frame numbers 50, 150 and 200. However, in (b), a shot is detected when the similarity between consecutive frames is minimum. Compare the detection of shots with (a). Figure 7.4 also shows improved shot detection for the preview of

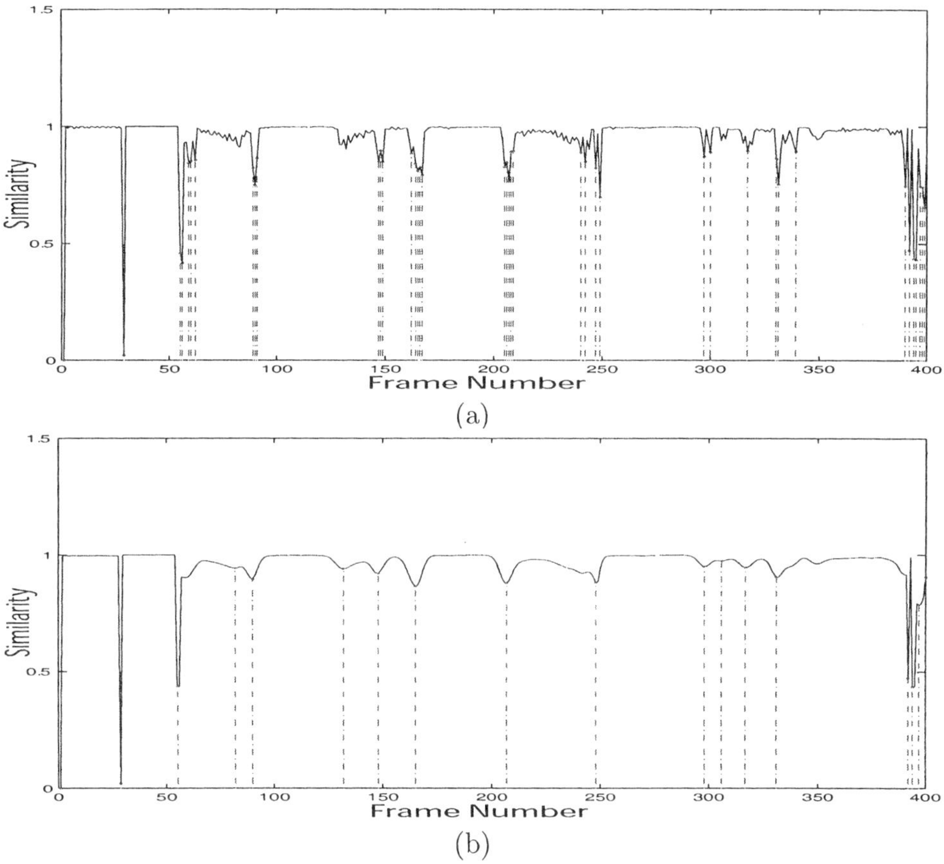

Figure 7.3. Shot detection results for the movie preview of "Red Dragon". There are 17 shots identified by a human observer. (a) Fixed threshold method. Vertical lines indicate the detection of shots. Number of shots detected: 40, Correct: 15, False-positive: 25, False-negative: 2 (b) Proposed method. Number of shots detected: 18, Correct: 16, False-positive: 2, False-negative: 1.

the movie "Road Trip". In our experiments, we achieved about 90% accuracy for shot detection in most cases.

Once the shot boundaries are known, each shot S_i is represented by a set of frames, that is:

$$S_i = \left\{ f^a, f^{a+1}, ..., f^b \right\}, \tag{7.5}$$

where a and b are the indices of the first and the last frames of the ith shot respectively. In the next section, we describe a method for compact representation of shots by selecting an appropriate number of key frames.

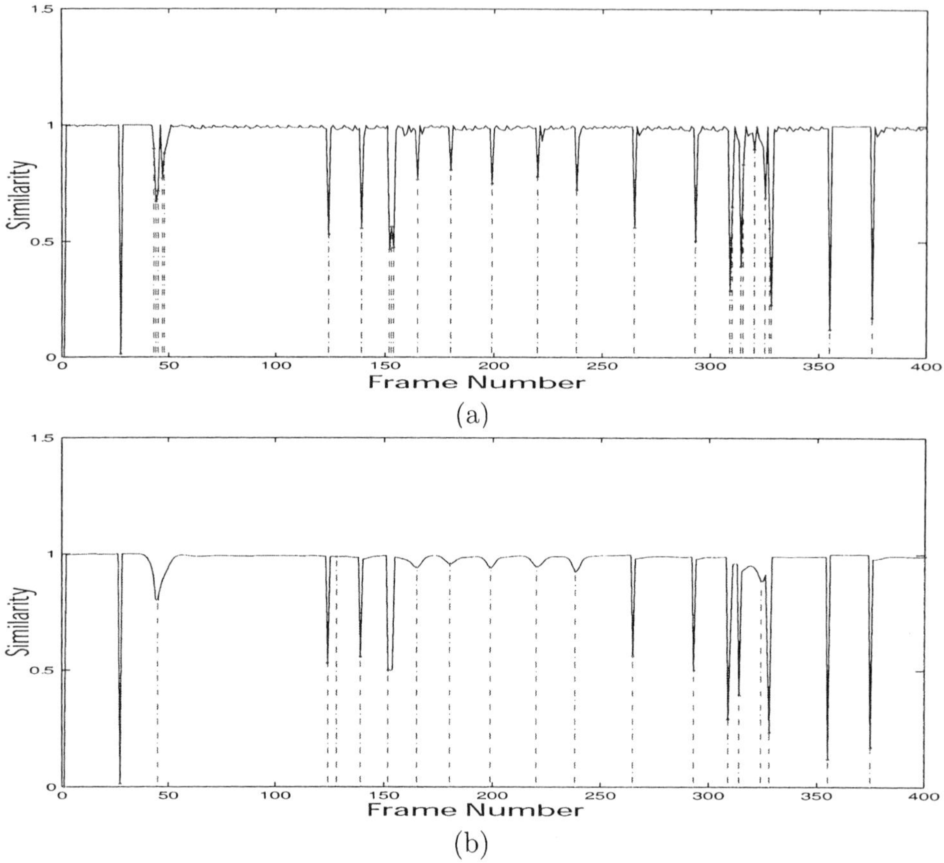

Figure 7.4. Shot detection results for the movie preview of "Road Trip". There are 19 shots identified by a human observer. (a) Fixed threshold method. Vertical lines indicate the detection of shots. Number of shots detected: 28, Correct: 19, False-positive: 9, False negative: 0. (b) Proposed method. Number of shots detected: 19, Correct: 19, False-positive: 0, False negative: 0.

7.2.2 Key frame Detection

Key frames are used to represent the contents of a shot. Choosing an appropriate number of key frames is difficult since we consider a variety of videos including feature movies, sitcoms and interview shows, which contain both action and non-action scenes. Selecting one key frame (for example the first or middle frame) may represent a static shot (a shot with little actor/camera motion) quite well, however, a dynamic shot (a shot with higher actors/camera motion) may not be represented adequately. Therefore, we have developed a method to select variable number of key frames depending upon the shot activity. Each shot, S_i, is represented by a set of key frames, K_i, such that all key frames are distinct. Initially, the middle frame of the shot is selected and added to the set K_i (which is initially empty) as the first key frame. The reason for taking the middle frame instead of the first frame is to make sure that the frame is free from shot transition effects, for instance, a *diffusion* effect. Next, each frame within a shot is compared to every frame in the set K_i. If the frame differs from all previously chosen key frames by a fixed threshold, it is added in the key frame set, otherwise it is ignored. This algorithm of key frame detection can be summarized as:

STEP 1: Select middle frame as the first key frame
$K_i \leftarrow \{f^{\lfloor (a+b)/2 \rfloor}\}$
STEP 2: for $j = a$ to b
if $max\left(S\left(f^j, f^k\right)\right) < Th \; \forall f^k \in K_i$
Then $K_i \leftarrow K_i \cup \{f^j\}$

where Th is the minimum frame similarity threshold that declares two frames to be similar. Using this approach, multiple frames are selected for the shots which have higher dynamics and temporally changing visual contents. For less dynamic shots, fewer key frames are selected. This method assures that every key frame is distinct and, therefore, prevents redundancy.

7.2.3 Shot Length and Shot Motion Content

Shot length (the number of frames present in a shot) and *shot motion content* are two interrelated features. These features provide cues to the nature of the scene. Typically, dialogue shots are longer and span a large number of frames. On the other hand, shots of fight and chase scenes change rapidly and last for fewer frames [Arijon, 1976]. In a similar fashion, the motion content of shots also depends on the nature of the shot. The dialogue shots are relatively calm (neither actors nor the camera exhibit large motion). Although camera pans, tilts and zooms are common in dialogue shots, they are generally smooth. In fight and chase shots, the camera motion is jerky and haphazard with higher movements of actors. For a given scene, these two attributes are generally consistent over time to maintain the pace of the movie.

Computation of Shot Motion Content. Motion in shots can be divided into two classes; *global motion* and *local motion.* Global motion in a shot occurs due to the movements of the camera. These may include *pan* shots, *tilt* shots, *dolly/truck* shots and *zoom in/out* shots [Reynertson, 1970]. On the other hand, local motion is the relative

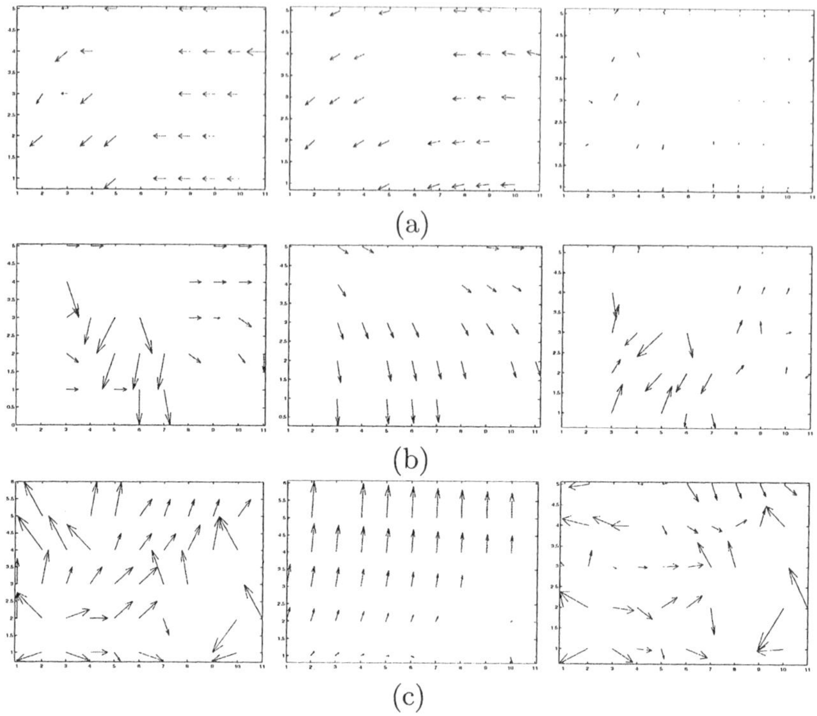

Figure 7.5. Estimation of shot motion content using motion vectors in three different shots of the movie "Golden Eye". The first column shows encoded motion vectors from the MPEG file. The second column shows the reprojected flow vectors after a least squares fit using an affine model. The third column shows the difference between the actual and the reprojected flow vectors. The shot motion content values computed by our algorithm for three shots are (a) a dialogue between two people, ϵ=9.8, (b) a person using a fire arm, ϵ=46.64 and (c) a moving tank, ϵ=107.03. These values are proportional to the shot activity.

movement of objects with respect to the camera, for example, an actor walking or running. We define shot motion content as the amount of local motion in a shot and exploit the information encoded in MPEG-1 compressed video to compute it. The horizontal and vertical velocities of each block are encoded in the MPEG stream. These velocity vectors reflect the global or local motion. We estimate the global affine motion using a least squares method. The goodness of the fit is measured by examining the difference between the actual and reprojected velocities of the blocks. The magnitude of this error is used as a measure of shot motion content. An affine model with six parameters is represented as follows:

$$u = a_1 \cdot x + a_2 \cdot y + b_1$$
$$v = a_3 \cdot x + a_4 \cdot y + b_2, \tag{7.6}$$

where u and v are horizontal and vertical velocities obtained from the MPEG file, a_1 through a_4 capture the camera rotation, shear and scaling, b_1 and b_2 represent the global translation in the horizontal and vertical directions respectively, and $\{x, y\}$ are the coordinates of the block's centroid. Let u_k and v_k be the encoded velocities and u'_k and v'_k be the reprojected velocities of the kth block in the jth frame using the affine motion model, then the error ϵ_j in the fit is measured as:

$$\epsilon_j = \sum_{k \in motionblocks} \sqrt{(u'_k - u_k)^2 + (v'_k - v_k)^2}. \tag{7.7}$$

The shot motion content of shot i is the aggregation of ϵ of all P frames in the shot:

$$SMC_i = \sum_{j \in S_i} \epsilon_j, \tag{7.8}$$

where SMC is the *shot motion content.* Figure 7.5 shows the shot motion content for three different cases. The SCM in the shot is normalized by the total number of frames in the shot.

7.2.4 Audio Features

Music and nonliteral sounds are often used to provide additional energy to a scene. They can be used to describe a situation, such as whether the situation is stable or unstable. In movies, the audio is often correlated with the scene. For example, shots of fighting and explosions are mostly accompanied by a sudden change in the audio level. Therefore, we detect events when the peaks in the audio energy are relatively high. The energy of an audio signal is computed as

$$E = \sum_{i \in interval} (A_i)^2, \tag{7.9}$$

where A_i is the audio sample indexed by time i and *interval* is a small window which is set to 50ms for our experiments. See Figure 7.6 for plots of audio signal and its energy for the movie preview of "The World Is Not Enough".

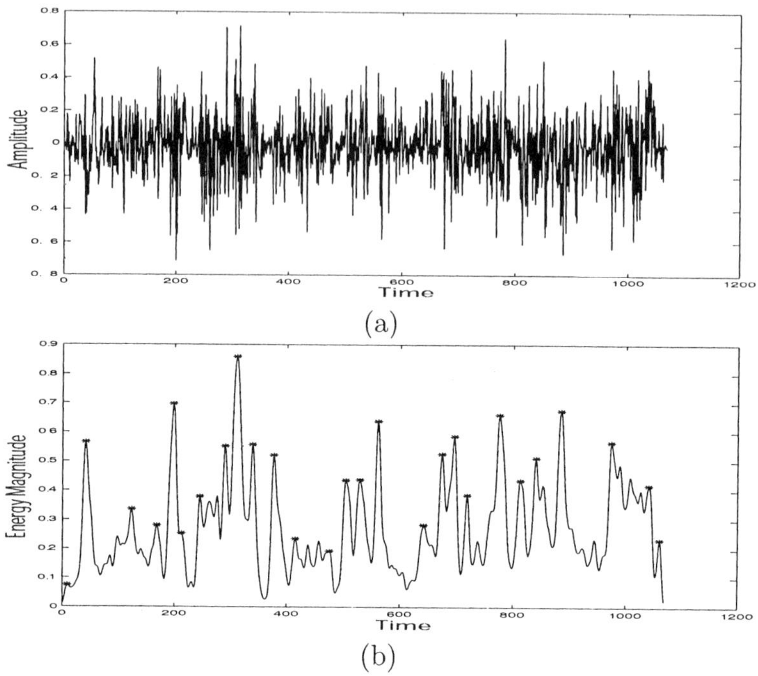

Figure 7.6. Audio processing: (a) the audio waveform of the movie "The World Is Not Enough", (b) Energy plot of the audio: Good peaks detected by our test are indicated by asterisks.

7.3 Segmentation of News and Game Shows using Visual Cues

Talk show videos are significant components of televised broadcast. Several popular prime-time programs are based on the host and guests concept, for example "Crossfire", "The Larry King Live", "Who Wants To Be A Millionaire", "Jeopardy" and "Hollywood Squares". In this section, we address the problem of organizing such video shows. We assume that the user might be interested in looking only at interview segments without commercials. Perhaps the user wants to view only clips that contain the questions asked during the show or only the clips which contain the answers of the interviewee. For example, the user might be motivated to watch only the questions in order to get a summary of the topics discussed in a particular interview. Therefore, we exploit the *Film Grammar* of such shows and extract interview segments by separating commercials. We further classify interview segments between shots of the host and the guests. The programs belonging to this genre in which a host interacts with guests share a common grammar. This grammar can be summarized as:

- The camera switches back and forth between the host and the guests.
- Frequent repetitions of shots
- *Guests'* shots are lengthier than *Hosts* shots.

On the other hand, commercials are characterized by the following grammar:

- More colorful shots than talk and game shows
- Fewer repetitions of shots
- Rapid shot transitions and small shot durations.

In the next section we describe a data structure for videos which is used for the extraction of program sections and for the detection of program host and guests.

7.3.1 Shot Connectivity Graph

We first find the shot boundaries and organize the video into a data-structure, called a *Shot Connectivity Graph, G*. This graph links similar shots over time. The vertices V represent the shots and edges represent the relationship between the nodes. Each vertex is assigned a label indicating the serial number of the shot in time and a weight w equal to the shot's length. In order to connect a node with another node we test the key frames of respective shots for three conditions:

- *Shot similarity constraint*: Key frames of two shots should have similar distribution of HSV color values.
- *Shot proximity constraint*: A shot may be linked with a recent shot (within the last T_{mem} shots)
- *Blank shot constraint*: Shots may not be linked across a blank in the shot connectivity graph. Significant story boundaries (for example, between the show and the commercials) are often separated by a short blank sequence.

Eq.7.10 is used to link two nodes in the Shot Connectivity Graph.

$$\sum_{j \in bins} min(H_q(j), H_{q-k}(j)) \geq T_{color} \quad \text{forsome } k \leq T_{mem}, \qquad (7.10)$$

where T_{color} is a threshold on the intersection of histograms. Thus two vertices v_p and v_q, such that $v_p, v_q \in V$ and $p < q$, are adjacent, that is they have an edge between them if and only if

- v_p and v_q represent consecutive shots or
- v_p and v_q satisfy the shot similarity, shot proximity and blank-shot constraints.

The shot connectivity graph exploits the structure of the video selected by the directors in the editing room. Interview videos are produced using multiple cameras running simultaneously, recording the host and the guest. The directors switch back and forth between them to fit these parallel events on a sequential tape. Examples of shot connectivity graphs automatically computed by our method are shown in Fig. 7.7 and 7.8.

7.3.2 Story Segmentation and Removal of Commercials

Shots in talk shows have strong visual correlation, both backwards and forwards in time, and this repeating structure can be used as a key cue in segmenting them from commercials, which are non-repetitive and rapidly changing. There may still be repetitive shots in a commercial sequence, which appear as cycles in the shot connectivity graph. However, these shots are not nearly as frequent, or as long in duration, as those in the interview. Moreover, since our threshold of linking shots back in time is based on the *number of shots*, and not on the total time elapsed, commercial segments will have less *time memory* than talk shows.

To extract a coherent set of shots, or *stories*, from the shot connectivity graph G, we find all strongly connected components in G. A strongly connected component $G'(V', E')$ of G has the following properties:

- $G' \subseteq G$
- There is a path from any vertex $v_p \in G'$ to any other vertex $v_q \in G'$.
- There is no $V_z \in (G - G')$ such that adding V_z to G' will form a strongly connected component.

Each strongly connected component $G' \in G$ represents a story. We compute the likelihood of all such stories being part of a program segment. Each story is assigned a weight based on two factors; the *number of frames in a story* and the *ratio of number of repetitive shots* to the *total number of shots in a story.* The first factor follows from the observation that long stories are more likely to be program segments than commercials. Stories are determined from strongly connected components in the shot connectivity graph. Therefore, a long story means that we have observed multiple overlapping cycles within the story since the length of each cycle is limited by T_{mem}. The second factor stems

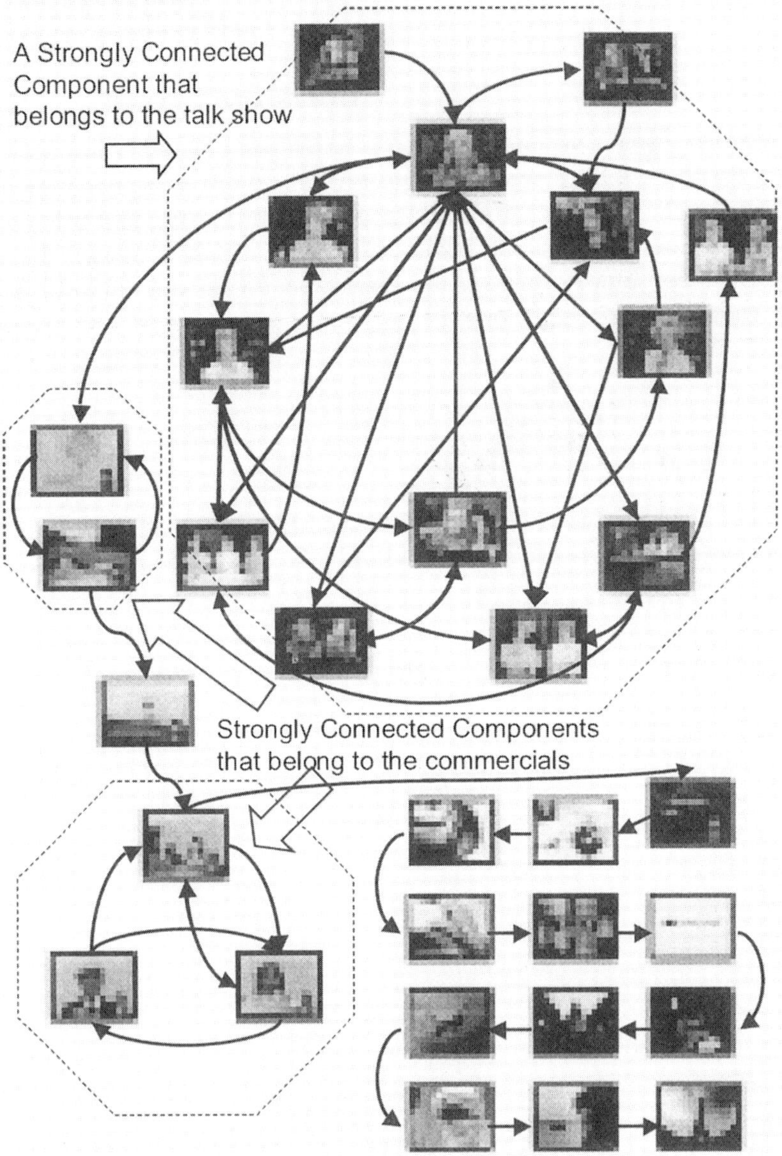

Figure 7.7. A Shot Connectivity Graph of a talk show with commercials. Strongly connected components of the graph are enclosed by dotted lines. The interview section of the video has more connected components and many repetitions. Commercials have smaller cycles and fewer repetitions.

from the observation that programs have a large number of repetitive shots in proportion to the total number of shots. Commercials, on the other hand, have a high shot transition rate. Even though commercials

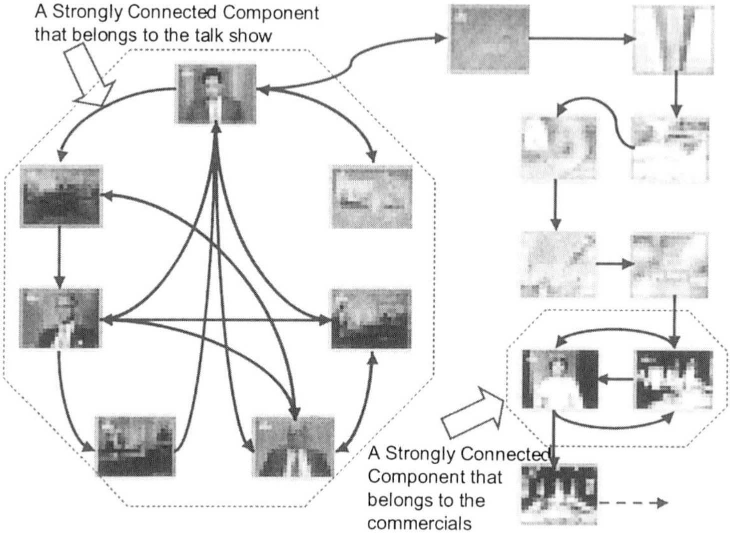

Figure 7.8. A Shot Connectivity Graph of a Pakistani talk show followed by commercials. Strongly connected components of the graph are enclosed by dotted lines. The interview section has more connected components and many repetitions. Commercials have smaller cycles and fewer repetitions.

may have repetitive shots, this repetition is small compared to the total number of shots. Thus, program segments will have more repetition than commercials, relative to total number of shots. Both of these factors are combined in the following likelihood of a story being a program segment:

$$L(G') = \sum_{\forall j \in G'} w_j \cdot \frac{\sum_{\forall E'_{ji} \in G' | j > i} 1}{\sum_{\forall j \in G'} 1} \cdot \Delta t, \qquad (7.11)$$

where G' is the strongly connected component representing the story; w_j is weight of the jth vertex i.e. the number of frames in the shot; E' are the edges in G'; Δt is the time interval between consecutive frames. Note that the denominator represents the total number of shots in the story. This likelihood forms a weight for each story, which is used to determine the label for the story. Stories with $L(story)$ higher than a certain threshold are labelled as program stories, whereas those that fall below the threshold are labelled as commercials. This scheme is robust and yields accurate results, as shown in Section 7.3.4.

7.3.3 Host Detection: Analysis of Shots Within an Interview Story

We perform further analysis of program stories to differentiate host shots from those of guests. Note that in most talk shows a single person is host for the duration of program but the guests keep changing. Also the host asks questions which are typically shorter than the guests' answers. These observations can be utilized for successful segmentation. Note that no specific training is used to detect the host. Instead, the host is detected from the pattern of shot transitions, exploiting the semantics of scene structure.

For a given show, we first find the N shortest shots in the show containing only one person. To determine whether a shot has one or more persons, we use the skin detection algorithm presented by [Kjedlsen and Kender, 1996], using RGB color space. The key frames of the N shortest shots containing only one person are correlated in time to find the most repetitive shot. Since questions are typically much shorter than answers, host shots are typically shorter than guest shots. Thus it is highly likely that most of the N shots selected will be host shots. An $N \times N$ correlation matrix C is computed such that each term of C is given by:

$$C_{ij} = \frac{\sum_{r \in rows} \sum_{c \in cols} (I_i(r,c) - \mu_i)'(I_j(r,c) - \mu_j)}{\sqrt{\left(\sum_{r \in rows} \sum_{c \in cols} (I_i(r,c))^2\right) \left(\sum_{r \in rows} \sum_{c \in cols} (I_j(r,c))^2\right)}} \tag{7.12}$$

where I_k is the gray-level intensity image of frame k and μ_k is its mean. Notice that all the diagonal terms in this matrix are 1 (and therefore do not need to be actually computed). Also, C is symmetric, and therefore only half of the non-diagonal elements need to be computed. The frame which returns the highest sum for a row is selected as the key frame representing the host. That is,

$$HostID = argmax_r \sum_{c \in allcols} C_{rc} \quad \forall r. \tag{7.13}$$

Table 7.1 demonstrates the detection of the host for the game show, "Who Wants To Be A Millionaire". Six candidates are picked for the host. Note that of the six candidates, four are shots of the host. The bottom row shows the summation of correlation values for each candidate. The sixth candidate has the highest correlation sum and is automatically

Candidates	Cand. 1	Cand. 2	Cand. 3	Cand. 4	Cand. 5	Cand. 6
Cand. 1	1	0.3252	0.2963	0.3112	0.1851	0.3541
Cand. 2	0.3252	1	0.5384	0.6611	0.3885	0.7739
Cand. 3	0.2963	0.5384	1	0.5068	0.3487	0.6016
Cand. 4	0.3112	0.6611	0.5068	1	0.3569	0.6781
Cand. 5	0.1851	0.3885	0.3487	0.3569	1	0.4036
Cand. 6	0.3541	0.7739	0.6016	0.6781	0.4036	1
Sum	2.4719	3.6871	3.2918	3.5141	2.6828	**3.8113**

Table 7.1. Detection of host shots in the game show "Who Wants To Be A Millionaire". Six shots were selected as host candidate. Candidates 2, 3, 4 and 6 belong to the actual host shot whereas candidates 1 and 5 are guests. The table shows the correlation values. Note that candidate 6 has the highest correlation sum and therefore is correctly identified as a show-host shot.

selected as the host. Guest-shots are the shots which are non-host. The key host frame is then correlated against key frames of all shots to find all shots of the host.

Show	Frames	Shots	Story Segments		Recall	Precision
			Ground Truth	Found		
Larry King 1	34,611	733	8	8	0.96	0.99
Larry King 2	12,144	446	6	6	0.99	0.99
Larry King 3	17,157	1,101	8	9	0.86	0.99
Larry King 4	13,778	754	6	6	0.97	0.99
Millionaire 1	19,700	1,496	7	7	0.92	0.99
Millionaire 2	17,442	1,672	7	7	0.99	0.99
Meet The Press	32,142	561	2	2	0.99	1.00
News Night (Pakistani)	9,729	501	1	1	1.00	1.00
News Express (Taiwanese)	16,472	726	4	4	1.00	0.92

Table 7.2. Results of story detection in a variety of videos. Precision and recall values are also mentioned. Video 1 was digitized at 10fps. All other videos were digitized at 5fps.

7.3.4 Experimental Results

The test suite was four full-length "Larry King Live" shows, two complete "Who Wants To Be A Millionaire" episodes, one episode of "Meet The Press", one Pakistani talk show, "News Night" and one Taiwanese

Show	Correct Host ID ?	Host Detection Accuracy
Larry King 1	Yes	99.32%
Larry King 2	Yes	94.87%
Larry King 3	Yes	96.20%
Larry King 4	Yes	96.85%
Millionaire 1	Yes	89.25%
Millionaire 2	Yes	95.18%
Meet the Press	Yes	87.7%
News Night	Yes	62.5 %

Table 7.3. Host detection results. All hosts are detected correctly.

show, "News Express". The results were compared with the ground truth obtained by a human observer i.e. classifying frames as either belonging to a commercial or a talk show. Table 7.2 shows that the correct classification rate is over 95% for most of the videos. The classification results for "Larry King 3" are not as good as the others. This particular show contained a large number of outdoor video clips that did not conform to the assumptions of the talk show model. The overall accuracy of talk show classification results is about the same for all programs, even though these shows have quite different layout and production styles. Table 7.3 contains host detection results with the ground truth established by a human observer. The second column shows whether the host identity was correctly established. The last column shows the overall rate of misclassification of host shots. Note that for all videos, very high accuracy and precision are achieved by the algorithm.

7.4 Movie Genre Categorization By Exploiting Audio-Visual Features Of Previews

Movies constitute a large portion of the entertainment industry. Currently several web-sites host videos and provide users with the facility to browse and watch online. Therefore, automatic genre classification of movies is an important task, and with the trends in technology, likely to become far more relevant in the near future. Due to the commercial nature of movie productions, movies are always preceded by previews and promotional videos. From an information point of view, previews contain adequate context for genre classification. As mentioned before, movie directors often follow general rules pertaining to the film genre. Since previews are made from the actual movies, these rules are reflected in them as well. In this section we establish a framework which exploits these cues for movie genre classification.

7.4.1 Approach

Movie previews are initially divided into *action* and *non-action* classes using the shot length and *visual disturbance* features. In the next step, the audio information and color features of key frames are analyzed. These features are combined with cinematic principles to subclassify non-action movie into *comedy, horror* and *drama/other.* Finally, action movies are classified into the *explosion/fire* and *other-action* categories. Figure 7.4.1 shows the proposed hierarchy.

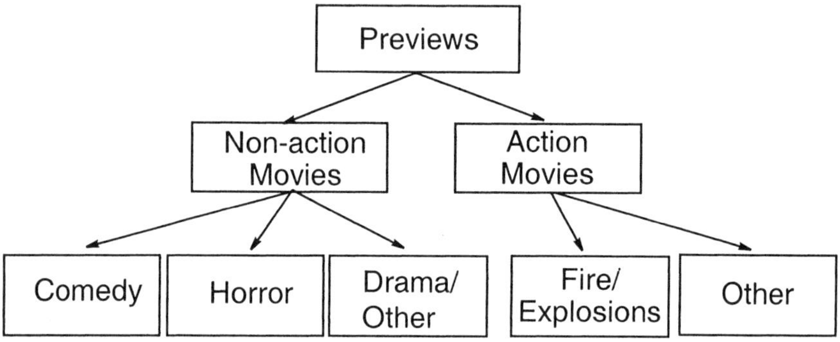

Figure 7.9. Proposed hierarchy of movie genres.

7.4.2 Visual Disturbance in the Scenes

We use an approach based on the structural tensor computation introduced in [Jahne, 1991], to find the visual disturbance. The frames contained in a video clip can be thought of as a volume obtained by combining all the frames in time. Thus $I(x, y, t)$ represents the gray scale value of a pixel located at the coordinate (x, y) in an image at time t. This volume can be decomposed into a set of two 2D temporal slices such that each is defined by planes (x, t) and (y, t) for horizontal and vertical slices respectively. We analyze only the horizontal slices and use only four rows of images in the video sequences to reduce computation. The structure tensor of the slices is expressed as:

$$\mathbf{\Gamma} = \begin{bmatrix} J_{xx} & J_{xt} \\ J_{xt} & J_{tt} \end{bmatrix} = \begin{bmatrix} \sum_w H_x^2 & \sum_w H_x H_t \\ \sum_w H_x H_t & \sum_w H_t^2 \end{bmatrix}, \qquad (7.14)$$

where H_x and H_t are the partial derivatives of $I(x, t)$ along the spatial and temporal dimensions respectively, and w is the window of support (3x3 in our experiments). The direction of gray level change in w, which is expressed by angle θ of Γ, is expressed as:

$$\mathbf{\Gamma} = \begin{bmatrix} \lambda_x & 0 \\ 0 & \lambda_t \end{bmatrix} = R \begin{bmatrix} J_{xx} & J_{xt} \\ J_{xt} & J_{tt} \end{bmatrix} R^T, \tag{7.15}$$

where λ_x and λ_y are the eigenvalues and R is the rotation matrix defined as

$$R = \begin{bmatrix} cos\theta & sin\theta \\ -sin\theta & \cos\theta \end{bmatrix}.$$

With the help of the above equations we can solve for the value of θ as

$$\theta = \frac{1}{2} tan^{-1} \frac{2J_{xt}}{J_{xx} - J_{tt}}. \tag{7.16}$$

Now the local orientation, ϕ, of the window, w, in a slice can be computed as

$$\phi = \begin{cases} \theta - \frac{\pi}{2} & \theta > 0 \\ \theta + \frac{\pi}{2} & otherwise \end{cases} \tag{7.17}$$

such that $-\frac{\pi}{2} < \phi \leq \frac{\pi}{2}$.

When there is no motion in a shot, ϕ is constant for all pixels. In the case of global motion (for example, camera translation) the gray levels of all pixels in a row change in the same direction. This results in equal or similar values of ϕ. However, in case of local motion, pixels that move independently will have different values of ϕ. Thus, this angle can be used to identify each pixel in a column of a slice as a moving or non-moving pixel.

We analyze the distribution of ϕ for every column of the horizontal slice by generating a nonlinear histogram. Based on experiments, we divide the histogram into 7 nonlinear bins which are [-90, -55, -35, -15, 15, 35, 55, 90]. The first and the last bins accumulate the higher values of ϕ, whereas the middle ones capture the smaller values. In case of a static scene or a scene with global motion, all pixels have similar values of ϕ and therefore they fall into one bin. On the other hand, pixels with motion other than global motion have different values of ϕ and they fall into different bins. We locate the peak in the histogram and mark the pixels in that bin as the *static* pixels, whereas the remaining ones are marked as *moving*. Next, we generate a binary mask for the whole video clip separating static pixels from moving pixels. The overall *visual disturbance* is the ratio of moving pixels to the total number of the pixels in a slice.

We use the average of the visual disturbance of four equally separated slices for each movie trailer as a disturbance measure. Shots with large

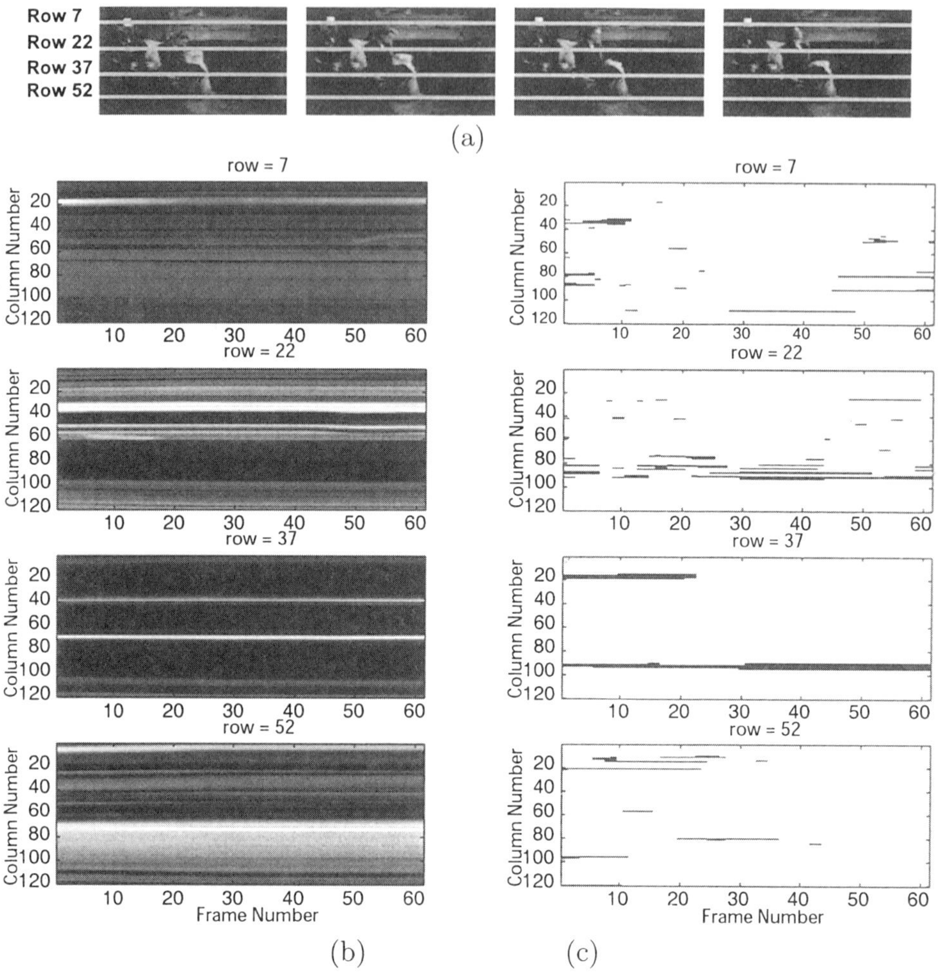

Figure 7.10. Plot of visual disturbance for a static shot. (a) Four frames of a dialogue shot. (b) Horizontal slices for four fixed rows of the shot from the preview. Each column in the horizontal slice is a row of the image. (c) Active pixels (black) in corresponding slices.

local motion cause more pixels to be labelled as moving. This measure is, therefore, proportional to the amount of action occurring in a shot. Fig 7.10 and 7.11 show this measure for shots of two different movies. It is clear that the density of visual disturbance is much smaller for a non-action scene than for an action scene. The computation of visual disturbance is very efficient and computationally inexpensive. Our method processes only four rows per image compared to [Vasconcelos

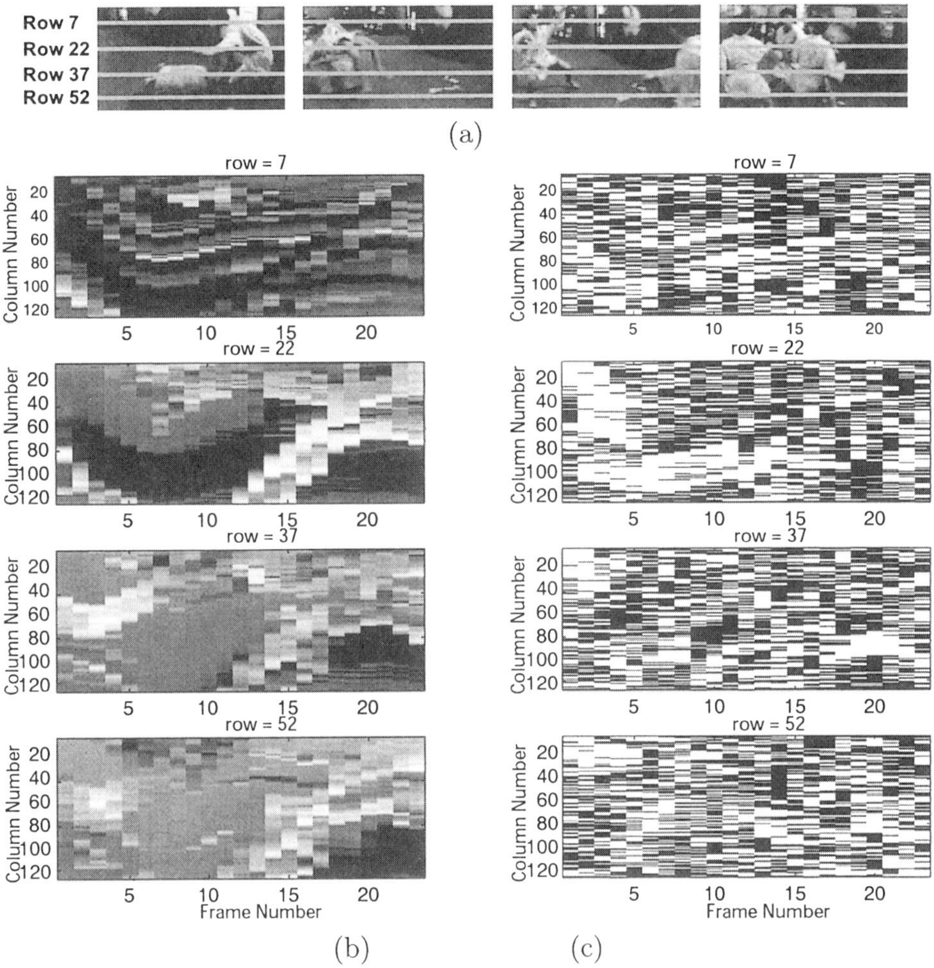

Figure 7.11. Plot of visual disturbance for a dynamic shot. (a) Four frames of a fight shot. (b) Horizontal slices for four fixed rows of a shots from the preview. Each column in the horizontal slice is a row of image. (c) Active pixels (black) in corresponding slices.

and Lippman, 1997] who estimate affine motion parameters for every frame.

7.4.3 Initial Classification

We have observed that action movies have more local motion than drama or horror movies. The former class exhibits a denser plot of visual disturbance and the latter has fewer active pixels. We have also noticed that in action movies, shots change more rapidly than in other genres like drama and comedy. Therefore, by plotting visual disturbance against average shot length, we can separate action from non-action movies.

7.4.4 Sub-classification of Non-action Movies

Key-Lighting. Light intensity in the scene is controlled and changed in accordance with the scene situation. In practice, the movie directors use multiple light sources to balance the amount and direction of light while filming a shot. The purpose of using several light sources is to provide a specific perception of the scene as it influences how the objects appear on the screen. Similarly the nature and size of objects' shadows are also used by maintaining a suitable proportion of intensity and direction of light sources. Reynertson comments on this issue: *"The amount and distribution of light in relation to shadow and darkness and the relative tonal value of the scene is a primary visual means of setting mood."* [Reynertson, 1970], p.107. In other words, lighting is used in the scene not only to provide good exposure but also to create a dramatic effect of light and shade consistent with the scene. Debating on this, Wolf Rilla says *"All lighting, to be effective, must match both mood and purpose. Clearly, heavy contrasts, powerful light and shade, are inappropriate to a light-hearted scene, and conversely a flat, front-lit subject lacks the mystery which back-lighting can give it."* [Rilla, 1970], p. 96.

Using the gray scale histogram, we classify images into two classes:

- **High-key lighting:** A *high-key* lighting means that the scene has an abundance of bright light. It usually has lesser contrast and the difference between the brightest light and the dimmest light is small. Practically, this configuration is achieved by maintaining a low *key-to-fill* ratio i.e. a low contrast between the dark and light. *High-key* scenes are usually happy or less dramatic. Many situation comedies also have high-key lighting ([Zettl, 1990], p. 32.)
- **Low-key lighting:** In this lighting, the background and part of the scene are generally predominantly dark. In *low-key* scenes, the contrast ratio is high. *Low-key* lighting is more dramatic and often used in *film noir* and horror films.

We have observed that most of the shots in *horror* movies are *low-key* shots, especially in the case of previews, as previews contain the most important and interesting scenes from the movie. On the other hand, comedy movies tend to have more *high-key* shots. To exploit this information we consider all key frames of the preview in the gray

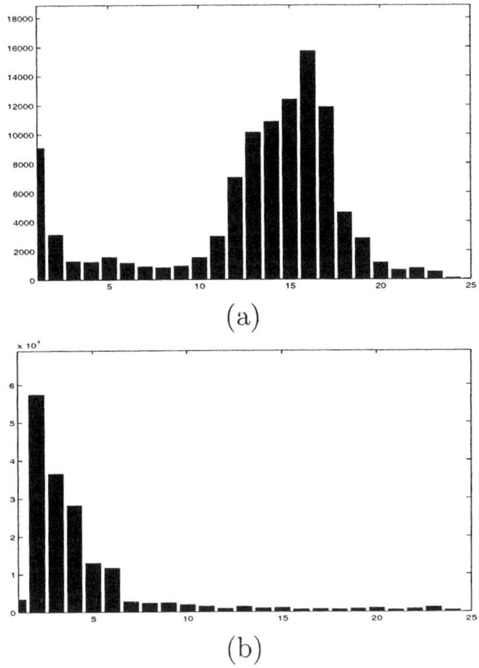

Figure 7.12. Distribution of gray scale pixel values. (a) The histogram of a *high-key* shot of the movie "The One". (b) The histogram of a *low-key* shot of the movie "The Others".

scale space and compute the distribution of the gray level of the pixels (Figures 7.4.4.0 and 7.13).

Our experiments show the following trends:

- *Comedy* : Movies belonging to this category have a gray-scale mean near the center of the gray-scale axis, with a large standard deviation, indicating a rich mix of intensities in the movie.
- *Horror* : Movies of this type have a mean gray-scale value towards the dark end of the axis, and have low standard deviation. This is because of the frequent use of dark tones and colors by the director.
- *Drama/other* : Generally, these types of movies do not have any of the above distinguishing features.

Based on these observations, we define a scheme to classify an unknown movie as one of these three types. We compute the mean, μ, and standard deviation, σ, of the gray-scale values of the pixels in all key frames. For each movie, i, we define a quantity $\zeta_i(\mu, \sigma)$ which is the product of μ_i and σ_i, that is:

$$\zeta_i = \mu_i \cdot \sigma_i, \tag{7.18}$$

where μ_i and σ_i are normalized to the maximum values in the data set. Since horror movies have more *low-key* frames, both mean and standard deviation values are low resulting in a small value of ζ. Comedy movies, on the other hand will return a high ζ because of high mean and high standard deviation.

We therefore define two thresholds, τ_c and τ_h, and assign a category to each movie i based on the following criterion.

$$L(i) = \begin{cases} Comedy & \zeta_i \geq \tau_c \\ Horror & \zeta_i \leq \tau_h \\ Drama/Other & \tau_h < \zeta_i < \tau_c \end{cases} \tag{7.19}$$

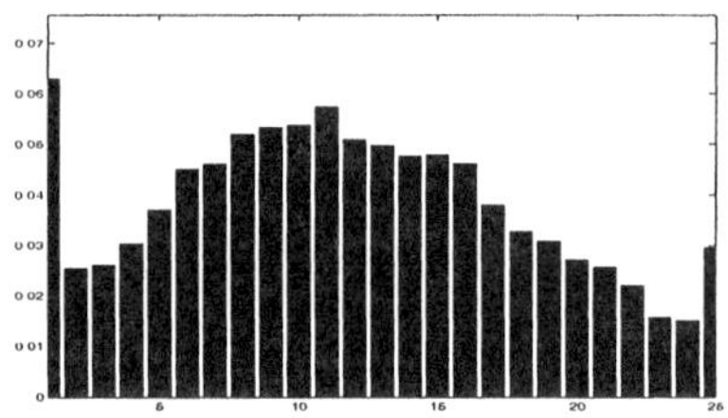

(a)

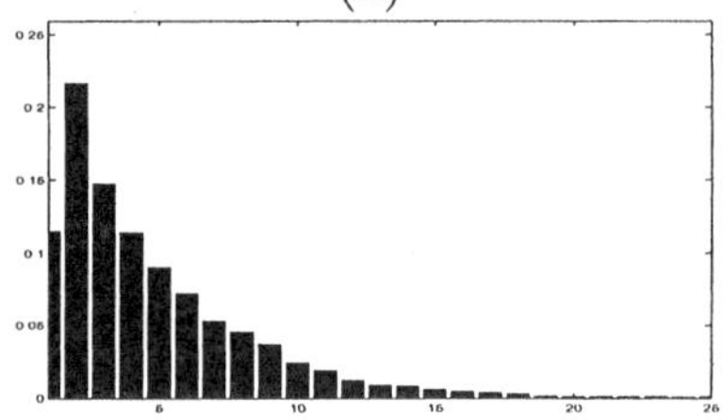

(b)

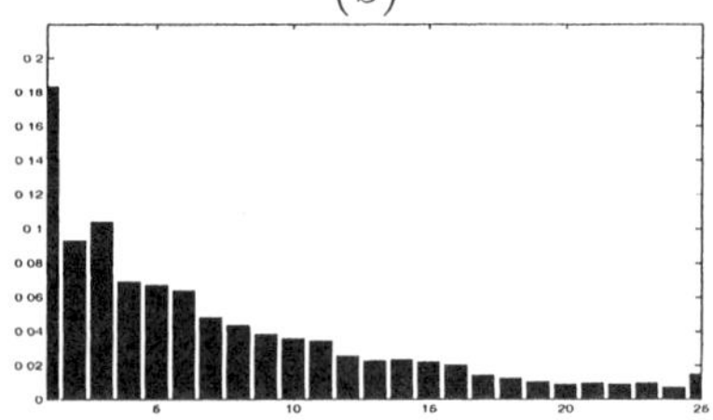

(c)

Figure 7.13. Average intensity histogram of key frames (a) "Legally Blonde", a *comedy* movie (b) "Sleepy Hollow", a *horror* movie and (c) "Ali", an example of *drama/other.*

7.4.5 Sub-classification Within Action Movies Using Audio and Color

Action movies can be classified as *martial arts, war* or *violent* such as those containing gunfire and explosions. We further rate a movie on the basis of the amount of fire/explosions by using both audio and color information.

7.4.6 Audio Analysis

In action movies, the audio is always correlated with the scene content. For example, fighting and explosions are usually accompanied by a sudden change in the audio level. To identify events with an unusual change in the audio, the energy of the audio can be used. We, therefore, first compute the energy in the audio track and then detect the presence of fire/explosion.

7.4.7 Fire/Explosion Detection

After detecting the occurrence of important events in the movie by locating the peaks in the audio energy plot, we analyze the corresponding frames to detect fire and/or explosions. In such cases there is a gradual change in the intensity of the images in the video. We locate the beginning and the end of the scene from the shot boundary information and process all corresponding frames for this test. Histograms with 26 bins are computed for each frame and the index of the bin with the maximum number of votes is plotted against time. During an explosion, the scene shows a gradual increase in the intensity. Therefore, the gray levels of the pixels move from lower intensity values to higher intensity values and the peak of the histogram moves from a lower index to a higher index.

Using this heuristic, a camera flash might be confused with an explosion. Therefore we further test the stability of the peak as a function of time. We exclude shots that show stability for less than a threshold since a camera flash does not last for more than a few frames. Figure 7.14 shows plots of the index of the histogram peak of the color histogram against time. Each shot has an abrupt change in audio; therefore our algorithm successfully differentiates between explosion and non-explosion shots.

7.4.8 Experimental Results

We have experimented with previews of 19 Hollywood movies downloaded from Apple's website (http://www.apple.com/trailers/). Video was analyzed at the frame rate of 24 Hz and at a resolution of 120 ×

(a) (b) (c)

Figure 7.14. Detection of *fire/explosion* in a shot of the movie "The Fast And The Furious". (a) and (b) are two frames of the shot. (c) the plot of the index of the histogram peak against time. The shot was successfully identified as *fire/explosion* (images courtesy of "Original Film".)

68 whereas the audio was processed at 22 KHz and with 16-bit precision. Figure 7.15 shows the distribution of movies on the feature plane obtained by plotting the visual disturbance against the average shot length. We use a linear classifier to separate these two classes. Movies with more action contents exhibit shorter average shot length. On the other hand *comedy/drama* movies have low action content and longer shot length.

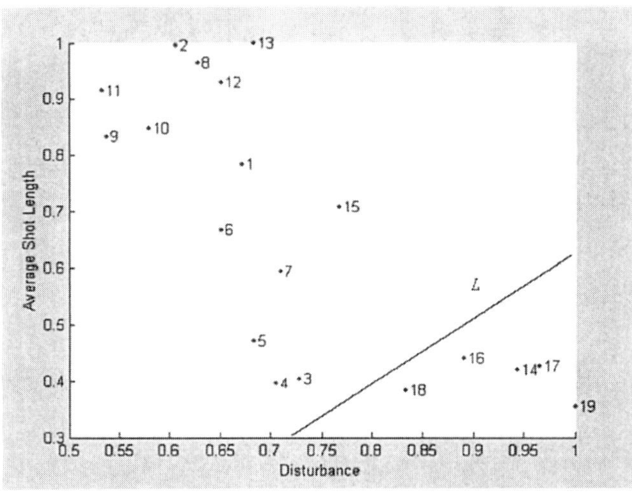

1. Ali
2. Jackpot
3. Mandolin
4. What Lies Beneath
5. Dracula
6. Hannibal
7. Sleepy Hollow
8. The Others
9. Legally Blonde
10. What Women Want
11. The Princess Diaries
12. Americas Sweethearts
13. American Pie
14. The World Is Not Enough
15. Big Trouble
16. Fast and Furious
17. Kiss Of The Dragon
18. The One
19. Rush Hour

Figure 7.15. The distribution of Movies on the basis of visual disturbance and *Average shot length.* Notice that action movies appear to have large motion content and short average shot length. *Non-action* movies, on the other hand, show opposite characteristics.

Our next step is to make classes within each group. This is done by analyzing the key frames. Using the intensity distribution we label movies as *comedy, horror* and *drama/other.* "Dracula", "Sleepy Hollow" and "The Others" were classified as horror movies. "What Lies Beneath", which is actually a *horror/drama* movie was also labelled as a *horror* movie. Movies that are neither comedy nor horror including "Ali", "Jackpot", "Hannibal" and "What Women Want" were also labelled correctly. There is a misclassification of the movie "Mandolin" which was marked as a comedy although it is a drama according to its official website. The only cue used here is the intensity images of key frames. We expect that by incorporating the further information, such as the audio, a better classification with more classes will be possible. We sort action movies on the basis of the number of shots showing fire/explosions. Our algorithm detected that the movie "The World Is Not Enough" contains more explosions/gunfire than the other movies, and therefore may be violent and unsuitable for young children, whereas "Rush Hour" contains the least explosion shots.

7.5 Related Work

There have been many studies on indexing and retrieval for image databases. [Vailaya et al., 2001; Schweitzer, 2001; Liu et al., 2001], are some of them. A large portion of research in this field uses content extraction and matching. Features such as edges, shape, texture and GLCM (gray level consistency matrix) are extracted for all images in the database and indexed on the basis of similarity. Although these techniques work well for single images, they cannot be applied directly to video databases. The reason is that in the audio-visual data the content changes with time. Even though videos are collections of still images, meaning is derived from the change in these images over time, which cannot be ignored in the indexing and retrieval task.

The Informedia Project [Informedia,] at Carnegie Mellon University is one of the earliest works in this area. It has spearheaded the effort to segment and automatically generate a database of news broadcasts every night. The overall system relies on multiple cues, such as video, speech, close-captioned text. A large amount of work has also been reported in structuring videos, resulting in several interactive tools to provide navigation capabilities to the viewers. Virage [Hampapur et al., 1997], VideoZoom [Smith, 1999; Smith and Kanade, 1997; DeMenthon et al., 2000], are some examples. [Yeung et al., 1998], were the first ones to propose a graphical representation of video data by constructing a *Scene Transition Graph* (*STG*). The *STG* is then split into several sub-graphs using *complete-link* method of hierarchical clustering. Each subgraph satisfies a similarity constraint based on color, and represents a scene. [Hanjalic et al., 1999], use a similar approach of shot clustering using graph and find *logical story units.*

Content-based video indexing also constitutes a significant portion of the work in this area. [Chang et al., 1998], have developed an interactive system for video retrieval. Several attributes of video such as color, texture, shape and motion are computed for each video in the database. The user provides a set of parameters for attributes of video to look for. These parameters are compared with those in the database using a weighted distance formula for the retrieval. A similar approach has also been reported by [Deng and Manjunath, 1997].

The use of Hidden Markov Models has been very popular in the research community for video categorization and retrieval. [Naphade and Huang, 2001], have proposed a probabilistic framework for video indexing and retrieval. Low-level features are mapped to high-level semantics as probabilistic multimedia objects called *multijects.* A Bayesian belief network, called *multinet,* is developed to perform the semantic indexing using Hidden Markov Models. Some other examples that make use of probabilistic approaches are [Wolf, 1997; Dimitrova et al., 2000; Boreczky and Wilcox., 1997]. [Haering et al., 1999], have also suggested a semantic framework for video indexing and detection of *events.* They have presented an example of *hunt* detection in videos.

A large amount of research work on video categorization has also been done in the compressed domain using MPEG-1 and MPEG-2. The work in this area utilizes extractable features from compressed video and audio. The compressed information may not be very precise, however, it avoids the overhead of computing features in the pixel domain. [Kobla et al., 1997], have used the DCT coefficients, macroblock and motion vector information of MPEG videos for indexing and retrieval. Their proposed method is based on *query by example.* The methods proposed by [Yeo and Liu, ; Patel and Sethi, 1997] are other examples of work on compressed video data. [Lu et al., 2001], have applied the HMM approach in the compressed domain and promising results have been presented. Recently, MPEG-7 has focused on video indexing using embedded semantic descriptors, [Benitez et al., 2002]. However, at the time of this writing, the standardization of MPEG-7 is still in progress and content-to-semantic interpretation for retrieval of videos is still an open question for the research community.

7.6 Conclusion

In our approach, we exploited domain knowledge and used *film grammar* for video segmentation. We were able to distinguish between the shots of host and guests by analyzing the shot transitions. We also studied the cinematic principles used by the movie directors and mapped low-level features, such as the intensity histogram, to high-level semantics, such as the movie genre. Thus, we have provided an automatic method of video content annotation which is crucial for efficient media access.

References

Arijon, D. (1976). *Grammar of the Film Language*. Hasting House Publishers, NY.

Benitez, A. B., Rising, H., Jrgensen, C., Leonardi, R., Bugatti, A., Hasida, K., Mehrotra, R., Tekalp, A. M., Ekin, A., and Walker, T. (2002). Semantics of Multimedia in MPEG-7. In *IEEE International Conference on Image Processing*.

Boreczky, J. S. and Wilcox., L. D. (1997). A hidden Markov model framework for video segmentation using audio and image features. In *IEEE International Conference on Acoustics, Speech and Signal Processing*.

Chang, S. F., Chen, W., Horace, H., Sundaram, H., and Zhong, D. (1998). A fully automated content based video search engine supporting spatio-temporal queries. *IEEE Transaction on Circuits and Systems for Video Technology*, pages 602–615.

DeMenthon, D., Latecki, L. J., Rosenfeld, A., and Vuilleumier-Stuckelberg, M. (2000). Relevance ranking of video data using hidden Markov model distances and polygon simplification. In *Advances in Visual Information Systems, VISUAL 2000*, pages 49–61.

Deng, Y. and Manjunath, B. S. (1997). Content-based search of video using color, texture and motion. In *IEEE Intl. Conf. on Image Processing*, pages 534–537.

Dimitrova, N., Agnihotri, L., and Wei, G. (2000). Video classification based on HMM using text and faces. In *European Conference on Signal Processing*.

Haering, N. (1999). A framework for the design of event detections, (Ph.D. thesis). School of Computer Science, University of Central Florida.

Haering, N. C., Qian, R., and Sezan, M. (1999). A semantic event detection approach and its application to detecting hunts in wildlife video. *IEEE Transaction on Circuits and Systems for Video Technology*.

Hampapur, A., Gupta, A., Horowitz, B., Shu, C. F., Fuller, C., Bach, J., Gorkani, M., and Jain, R. (1997). Virage video engine. In *SPIE, Storage and Retrieval for Image and Video Databases*, volume 3022, pages 188–198.

Hanjalic, A., Lagendijk, R. L., and Biemond, J. (1999). Automated high-level movie segmentation for advanced video-retrieval systems. *IEEE Transaction on Circuits and Systems for Video Technology*, 9(4):580–588.

Informedia. *Informedia Project, Digital video library.* http:// www. informedia. cs.cmu.edu.

Jahne, B. (1991). *Spatio-tmporal Image Processing: Theory and Scientific Applications.* Springer Verlag.

Kjedlsen, R. and Kender, J. (1996). Finding skin in color images. In *International Conference on Face and Gesture Recognition.*

Kobla, V., Doermann, D., and Faloutsos, C. (1997). Videotrails: Representing and visualizing structure in video sequences. In *Proceedings of ACM Multimedia Conference*, pages 335–346.

Liu, Y., Emoto, H., Fujii, T., and Ozawa, S. (2001). A method for content-based similarity retrieval of images using two dimensional dp matching algorithm. In *11th International Conference on Image Analysis and Processing*, pages 236–241.

Lu, C., Drew, M. S., and Au, J. (2001). Classification of summarized videos using hidden Markov models on compressed chromaticity signatures. In *ACM International Conference on Multimedia.*

Lyman, P. and Varian, H. R. (2000). *School of Information Management and Systems at the University of California at Berkeley.* http:// www.sims. berkeley.edu/ research/ projects/ how-much-info/.

Naphade, M. R. and Huang, T. S. (2001). A probabilistic framework for semantic video indexing, filtering, and retrieval. *IEEE Transactions on Multimedia*, pages 141–151.

Patel, N. V. and Sethi, I. K. (1997). *The Handbook of Multimedia Information Management.* Prentice-Hall/PTR.

Perona, P. and Malik, J. (1990). Scale-space and edge detection using anisotropic diffusion. *IEEE Transactions on Pattern Analysis and Machine Intelligence*, 12(7):629–639.

Reynertson, A. F. (1970). *The Work of the Film Director.* Hasting House Publishers, NY.

Rilla, W. (1970). *A-Z of movie making, A Studio Book.* The Viking Press, NY.

Schweitzer, H. (2001). Template matching approach to content based image indexing by low dimensional euclidean embedding. In *Eight IEEE International Conference on Computer Vision*, pages 566–571.

Smith, J. R. (1999). Videozoom spatio-temporal video browser. *IEEE Transactions on Multimedia*, 1(2):157–171.

Smith, M. A. and Kanade, T. (1997). Video skimming and characterization through the combination of image and language understanding techniques.

Vailaya, A., Figueiredo, M., Jain, A. K., and Zhang, H.-J. (2001). Image classification for content-based indexing. *IEEE Transactions on Image Processing*, 10(1):117–130.

Vasconcelos, N. and Lippman, A. (1997). Towards semantically meaningful feature spaces for the characterization of video content. In *IEEE International Conference on Image Processing.*

Wolf, W. (1997). Hidden Markov model parsing of video programs. In *International Conference on Acoustics, Speech and Signal Processing*, pages 2609–2611.

Yeo, B. L. and Liu, B. Rapid scene change detection on compressed video. 5:533–544.

Yeung, M. M., Yeo, B.-L., and Liu, B. (1998). Segmentation of video by clustering and graph analysis. *Computer Vision and Image Understanding*, 71(1).

Zettl, H. (1990). *Sight Sound Motion: Applied Media Aesthetics.* Wadsworth Publishing Company, second edition.

Chapter 8

UNDERSTANDING THE SEMANTICS OF MEDIA

Malcolm Slaney, Dulce Ponceleon and James Kaufman
IBM Almaden Research Center
San Jose, California
malcolm@ieee.org

Abstract It is difficult to understand a multimedia signal without being able to say something about its semantic content or its meaning. This chapter describes two algorithms that help bridge the semantic understanding gap that we have with multimedia. In both cases we represent the semantic content of a multimedia signal as a point in a high-dimensional space. In the first case, we represent the sentences of a video as a time-varying semantic signal. We look for discontinuities in this signal, of different sizes in a one-dimensional scale space, as an indication of a topic change. By sorting these changes, we can create a hierarchical segmentation of the video based on its semantic content. The same formalism can be used to think about color information and we consider the different media's temporal correlation properties. In the second half of this chapter we describe an approach that connects sounds to semantics. We call this semantic-audio retrieval; the goal is to find a (non-speech) audio signal that fits a query, or to describe a (non-speech) audio signal using the appropriate words. We make this connection by building and clustering high-dimensional vector descriptions of the audio signal and its corresponding semantic description. We then build models that link the two spaces, so that a query in one space can be mapped into a model that describes the probability of correspondence for points in the opposing space.

Keywords: Segmentation, semantics, multimedia signal, high-dimensional space, video retrieval, audio analysis, latent semantic indexing (LSI), semantic content, clustering, non-speech audio signal, hierarchical segmentation, color information, topic change, sorting, temporal correlation, mixture of probability experts, MPESAR, SVD, scale space, acoustic space.

8.1 Semantic Understanding Problem

Due to the proliferation of personal cameras and inexpensive hard disk drives, we are drowning in media. Unfortunately, the tools we have to understand this media are very limited. In this chapter we describe tools that help us understand the meaning of our media. We will demonstrate that tools that analyze the semantic content are possible and this represents a high-level understanding of the media.

There are many systems which find camera shot boundaries—low-level events in the video where the camera changes to a new view of the scene [Srinivasan et al., 1999]. There are some tools, described below, which attempt to segment video at a higher level. But this level of analysis does not tell us much about the meaning represented in the media.

Only recently have researchers constructed higher-level understanding from multimedia signals. Aner [Aner and Kender, 2002] suggest an approach that finds the background in a video shot, and then clusters shots into physical scenes by noting shots with common backgrounds. This is one way to build up a higher-level representation of the video, but we argue that the most important information is in the words.

Retrieving media is a similarly hard problem. Systems such as IBM's QBIC system [Flickner et al., 1993] allow users to search for images based on the colors and images in an image. This is known as query-by-example, but most people don't think about their image requests in terms of colors or shapes. A better tool uses semantic information to retrieve objects based on the meaning in the media.

In the remainder of this section we will talk about specific approaches for segmentation and retrieval, and describe how our approaches differ. Section 8.1.3 describes the rest of this chapter.

8.1.1 Segmentation Literature

Our work extends previous work on text and video analysis and segmentation in several different ways.

Latent semantic indexing (LSI) has a long history, starting with Deerwester's paper [Deerwester et al., 1990], as a powerful means to summarize the semantic content of a document and measure the similarity of two documents. We use LSI because it allows us to quantify the position of a portion of the document in a multi-dimensional semantic space.

Hearst [Hearst, 1994] proposes to use the dips in a similarity measure of adjacent sentences in a document to identify topic changes. Her method is powerful because the size of the dip is a good indication of the relative amount of change in the document. We extend this idea using scale-space techniques to allow us to talk about similarity or dissimilarity over larger portions of the document.

Miller and her colleagues proposed Topic Islands [Miller et al., 1998], a visualization and segmentation algorithm based on a wavelet analysis of text documents. Their wavelets are localized in both time (document position) and frequency (spectral content) and allow them to find and visualize topic changes at many different scales. The localized nature of their wavelets makes it difficult to isolate and track segmentation boundaries through all scales. We propose to summarize the text with LSI and analyze the signal with smooth Gaussians, which are localized in time but preserve the long-term correlations of the semantic path.

Segmentation is a popular topic in the signal and image processing worlds. Witkin [Witkin, 1984] introduced scale-space ideas to the segmentation problem and Lyon [Lyon, 1984] extended Witkin's approach to multi-dimensional signals. A more theoretical discussion of the scale-space segmentation ideas was published by Leung [Leung et al., 2000]. The work described here extends the scale-space approach by using LSI as a basic feature and changing the distance metric to fit semantic data.

The key concept in our segmentation work is to think about a video signal's path through space, and detect jumps at multiple scales. The signal processing analysis proposed in this chapter is just one part of a complete system. We use a singular-value decomposition (SVD) to do the basic analysis, but more sophisticated techniques are also applicable. Any method which allows us to summarize the image and semantic content of the document can also be used in conjunction with the techniques described here.

8.1.2 Semantic Retrieval Literature

There are many multimedia retrieval systems that use a combination of words or examples to retrieve audio (and video) for users. Our algorithm, mixtures of probability experts for semantic-audio retrieval (MPESAR), is a more sophisticated model connecting words and media.

An effective way to find an image of the space shuttle is to enter the words "space shuttle jpg" into a text-based web search engine. The original Google system did not know about images, but, fortunately, many people created web pages with the phrase "space shuttle" and a JPEG image of the shuttle. The MPESAR work expands those search techniques by considering the acoustic and semantic similarity of sounds to allow users to retrieve sounds without running searches on the exact words used on the web page.

Barnard [Barnard and Forsyth, 2001] used a hierarchical clustering algorithm to build a model that combined words and image features to create a single hierarchical model that spanned both semantic and image features. He demonstrated the effectiveness of coupled clustering for an information-retrieval task and argued that the words written by a human annotator describing an image (e.g., "a rose") often provide information that complements the obvious information in the image (it is red).

MPESAR improves on three aspects of Barnard's approaches. First, the semantic and image features do not have the same probability distributions. Barnard's algorithm assumes that image features can be described by a multinomial distribution, while a Gaussian is probably more appropriate. Second, and perhaps most important, there is nothing in Barnard's algorithm that guarantees that the features used to build each stage of the model include both semantic and image features. Thus, the algorithm is free to build a model that completely ignores the image features and clusters the 'documents' based on only semantic features. Third, MPESAR interpolates between models. Previous work assigned each document to a single cluster and used a single model (winner-take-all) to map to the opposite domain. On the other hand, MPESAR calculates the probability that each cluster generates the query and then calculates a weighted average of models based on the cluster probabilities.

The MPESAR algorithm is appropriate for mapping one type of media to another. We illustrate the idea here using audio and semantic documents because audio retrieval is a simpler problem.

8.1.3 Overview

In this chapter, we describe semantic tools for understanding media[1]. The key to these tools is a representation of the media's content based on the words contained in the media, or words describing the media. In this work we use mathematical tools to represent a set of words as a point in a vector space. We will then use this vector representation of the semantic content to allow us to create a hierarchical table of contents for a multimedia signals, or to build a query-by-semantics system.

Our description of the semantic tools is structured as follows. In Section 8.2 of this chapter, we will describe some common tools and mathematics we use to analyze multimedia signals. Section 8.3 describes an algorithm for hierarchical segmentation that uses the color, acous-

[1]This chapter combines material first published elsewhere [Slaney et al., 2001; Slaney, 2002]

tic, and semantic information in the signal. Section 8.4 describes our semantic-retrieval algorithm, which is applied to audio retrieval.

8.2 Analysis Tools

We use two types of transformations to reduce raw text and video signals into meaningful spaces where we can find edges or events.

The SVD provides a principled way to reduce the dimensionality of a signal in a manner which is optimum, in a least-squared sense. In the next sub-sections, we describe how we apply thc SVD to color and semantic information. The SVD transformation allows us to summarize different kinds of video data and combine the results into a common representation (Section 8.3.5).

8.2.1 SVD Principles

We express both semantic and video data as vector-valued functions of time, $\vec{x}(t)$. We collect data from an entire video and put the data into a matrix, $\mathbf{X}$, where the columns of $\mathbf{X}$ represent the signal at different times. Using an SVD, we rewrite the matrix $\mathbf{X}$ in terms of three matrices, $\mathbf{U}$, $\mathbf{S}$ and $\mathbf{V}$, such that

$$\mathbf{X} = \mathbf{USV}^T. \tag{8.1}$$

The columns of the $\mathbf{U}$ and $\mathbf{V}$ matrices are orthonormal; $\mathbf{S}$ is a diagonal matrix. The values of $\mathbf{S}$ along the diagonal are ordered such that

$$S_{11} >= S_{22} >= S_{33} >= ... >= S_{nn} \tag{8.2}$$

where n is the minimum of the number of rows or columns of $\mathbf{X}$.

The SVD allows us to generate approximations of the original data. If the first k diagonal terms of $\mathbf{S}$ are retained, and the rest are set to zero, then the rank k approximation to $\mathbf{X}$, or $\mathbf{X}^k$, is the best possible approximation to X (in the least squares sense):

$$|\mathbf{X} - \mathbf{X}^k| = \min_{rank(\mathbf{Y}) \leq k} |\mathbf{X} - \mathbf{Y}| \geq |\mathbf{X} - \mathbf{X}_{k+1}|. \tag{8.3}$$

The first equality in equation 8.3 says that $\mathbf{X}^k$ is the best approximation in all k-dimensional subspaces. The second inequality states that, as we add more terms, and thus increase the size of the subspace, the approximation will not deteriorate (it typically improves).

Typically the first singular values are large; they then decay until a noise floor is reached. We want to keep the dimensions that are highly significant, while setting the dimensions that are dominated by noise to zero.

The columns of the $\mathbf{U}$ matrix are an ordered set of vectors that approximate the column space of the original data. In our case, each column of the $\mathbf{X}$ matrix is the value of our function at a different point in time. As we use more terms of $\mathbf{S}$, the columns of $\mathbf{U}$ provide a better and better approximation to the cloud of data that forms from $\vec{x}(t)$.

Given the left-singular vectors $\mathbf{U}$ and our original data $\mathbf{X}$, we project our data into the optimal k-dimensional subspace by multiplying

$$\mathbf{Y}^k = (\mathbf{U}^k)^T \mathbf{X} \tag{8.4}$$

where $\mathbf{U}^k$ contains only the first k columns of $\mathbf{U}$, and $\mathbf{Y}^k = \vec{x^k}(t)$ is a k-dimensional function of time. We compute a new SVD and a new $\mathbf{U}$ matrix for each video, essentially creating movie-dependent subspaces with all the same advantages of speaker-dependent speech recognition.

We use the SVD to reduce the dimensionality of both our audio and image video data. The reduced representation is nearly as accurate as the original data, but is more meaningful (the noise dimensions have been dropped) and is easier to work with (the dimensionality is significantly lower).

8.2.2 Color Space

Color changes provide a useful metric for finding the boundary between shots in a video [Srinivasan et al., 1999]. We can represent the color information by collecting a histogram of the colors within each frame and noting the temporal positions in the video where the histogram indicates large frame-to-frame differences.

We collected color information by using 512 histogram bins. We converted the three red, green, and blue intensities— each of which range in value from 0 to 255— to a single histogram bin by finding the log, in base 2, of the intensity value, and then packing the three colors into a 9-bit number using floor() to convert to an integer:

$$Bin = 64\,\text{floor}(\log_2(R)) + 8\,\text{floor}(\log_2(G)) + \text{floor}(\log_2(B)) \tag{8.5}$$

We chose this logarithmic scaling because it equalizes the counts in the different bins for our test videos.

The color histogram of the video frames converts the original video images into a 512-dimensional signal that is sampled at 29.97 Hz. The order of the dimensions is arbitrary and meaningless; the SVD will produce the same subspace regardless of how the rows or columns of the $\mathbf{X}$ matrix are arranged.

8.2.3 Word Space

Latent semantic indexing (LSI), a popular technique for information retrieval [Dumais, 1991], uses an SVD in direct analogy to the color analysis described above. As we did with the color data, we start analyzing the audio data by collecting a histogram of the words in a transcript of the video.

Normally, in information retrieval, each document is one of a large collection of electronically-formatted documents from which we want to retrieve the best match. In our case we want to study only a single document, so we consider portions of that document—sentences. The sentences of a document define a semantic space; each sentence, in general, represents a specific point in the semantic space.

Two difficult problems associated with semantic information retrieval are posed by synonyms and polysemy. Often, two or more words have the same meaning—synonyms. For information retrieval, we want to be able to use any synonym to retrieve the same information. Conversely, many words have multiple meanings—polysemy. For example, apple in a story about a grocery store is likely to have a different meaning from Apple in a story about a computer store.

The SVD allows us to capture both relationships. Words that are frequently used in the same section of text are given similar counts in the histogram. The SVD is sensitive to this correlation, in that one of the singular vectors points in the combined direction. Furthermore, words such as apple show up in two different types of documents, representing the two types of stories and will thus contribute to two different directions in the semantic space.

Changes in semantic space are based on angles, rather than on distance. A simple "sentence" such as "Yes!" has the same semantic content as "Yes, yes!" Yet the second sentence contains twice as many words, and, in semantic space, it will have a vector magnitude that is twice as large. Instead of using a Euclidean metric, we describe the similarity of two points in semantic space by the angle between the two vectors. We usually compute this value by finding the cosine of the angle between the two vectors,

$$cos(\phi) = (\nu_1 \cdot \nu_2)/(|\nu_1||\nu_2|). \tag{8.6}$$

8.3 Segmenting Video

Browsing videotapes of image and sound (hereafter referred to as "videos") is difficult. Often, there is an hour or more of material, and there is no roadmap to help viewers find their way through the medium.

It would be tremendously helpful to have an automated way to create a hierarchical table of contents that listed major topic changes at the highest level, with subsegments down to individual shots. DVDs provide the chapter indices; we would like to find the position of the sub-chapter boundaries. Realization of such an automated analysis requires the development of algorithms which can detect changes in the video or semantic content of a video as a function of time. We propose a technology that performs this indexing task by combining the two major sources of data—images and words—from the video into one unified representation.

With regard to the words in the sound track of a video, the information-retrieval world has used, with great success, statistical techniques to model the meaning, or semantic content, of a document. These techniques, such as LSI, allow us to cluster related documents, or to pose a question and find the document that most closely resembles the query. We can apply the same techniques within a document or, in the present case, the transcript of a video. These techniques allow us to describe the semantic path of a video's transcript as a signal, from the initial sentence to the conclusions. Thinking about this signal in a scale space allows us to find the semantic discontinuities in the audio signal and to create a semantic table of contents for a video.

Our technique is analogous to one that detects edges in an image. Instead of trying to find similar regions of the video, called segments, we think of the audio–visual content as a signal and look for "large" changes in this signal or peaks in its derivative. The location of these changes are edges; they represent the entries in a table of contents.

8.3.1 Temporal Properties of Video

The techniques we describe in this chapter allow us to characterize the temporal properties of both the audio and image data in the video. The color information in the image signal and the semantic information in the audio signal provide different information about the content.

Color provides robust evidence for a shot change in a video signal. An easy way to convert the color data into a signal that indicates scene changes is to compute each frame's color histogram and to note the frame-by-frame differences [Srinivasan et al., 1999]. In general, however, we do not expect the colors of the images to tell us anything about the global structure of the video. The color balance in a video does not typically change systematically over the length of the film. Thus, over the long term, the video's overall color often does not tell us much about the overall structure of the video.

Random words from a transcript, on the other hand, do not reveal much about the low-level features of the video. Given just a few words from the audio signal, it is difficult to define the current topic. But the words indicate a lot about the overall structure of the story. A documentary script may, for instance, progress through topic 1, then topic 2, and finally topic 3.

We describe any time point in the video by its position in an color–semantic vector space. We represent the color and the semantic information in the video as two separate vectors as a function of time. We concatenate these two vectors to create a single vector that encodes the color and the semantic data. Using scale-space techniques we can then talk about the changes that the color–semantic vector undergoes as the video unwinds over time. We label as segment boundaries large jumps in the combined color–semantic vector. "Large jumps" are defined by a scale-space algorithm that we describe in Section 8.3.4.

8.3.2 Segmentation Overview

This chapter proposes a unified representation for the audio–visual information in a video. We use this representation to compare and contrast the temporal properties of the audio and images in a video. We form a hierarchical segmentation with this representation and compare the hierarchical segmentation to other forms of segmentation. By unifying the representations we have a simpler description of the video's content and can more easily compare the temporal information content in the different signals.

As we have explained, we combine two well-known techniques to find the edges or boundaries in a video. We reduce the dimensionality of the data and put them all into the same format. The SVD and its application to color and word data were described in Section 8.2. We describe the test material we use to illustrate our algorithm in Section 8.3.3.

Scale-space techniques give us a way to analyze temporal regions of the video that span a time range from a few seconds to tens of minutes. Properties of scale spaces and their application to segmentation are described in Section 8.3.4.

In Section 8.3.5, we describe our algorithm, which combines these two approaches.

We discuss several temporal properties of video, and present simple segmentation results, in Section 8.3.6. Our representation of video allows us to measure and compare the temporal properties of the color and words. We perform a hierarchical segmentation of the video, automatically creating a table of contents for the video.

We conclude in Section 8.3.7 with some observations about this representation.

8.3.3 Test Material

We evaluated our algorithm using the transcript from two different videos.

The shortest test was the manual transcript of a 30 minute CNN Headline News television show [Linguistic Data Consortium, 1997]. This transcript is cleaner than those typically obtained from closed-captioned data or automatic speech recognition.

We also looked at the words and images from a longer documentary video, "21st Century Jet," about the making of the Boeing 777 airplane [PBS Home Video, 1995]. We analyzed the color information from the first hour of this video, and the words from all six hours.

In these two cases we have relatively clean transcripts and the ends of sentences are marked with periods. We can also use automatic speech recognition (ASR) to provide a transcript of the audio, but sentence boundaries are not reliably provided by ASR systems. In that case, we divide the text arbitrarily into 20 word groups or "sentences." We believe that a statistical technique such as LSI will fail gracefully in the event of word errors. For the remainder of this chapter we will use the word "sentence" to indicate a block of text, whether ended by a period or found by counting words.

8.3.4 Scale Space

Witkin [Witkin, 1984] introduced the idea of using scale-space segmentation to find the boundaries in a signal. In scale space, we analyze a signal with many different kernels that vary in the size of the temporal neighborhood that is included in the analysis at each point in time. If the original signal is $s(t)$, then the scale-space representation of this signal is given by

$$s_\sigma(t) = \int s(\tau)g(\sigma, t-\tau)d\tau, \tag{8.7}$$

where $g(\sigma, t-\tau)$ is a Gaussian kernel with a variance of σ. With σ approaching zero, $s_\sigma(t)$ is nearly equal to $s(t)$. For larger values of σ, the resulting signal, $s_\sigma(t)$, is smoother because the kernel is a low-pass filter. We have transformed a one-dimensional signal into a two-dimensional image that is a function of t and σ.

An important feature of scale space is that the resulting image is a continuous function of the scale parameter, σ. Because the location of a local maximum in scale space is well behaved [Babaud et al., 1986], we can start with a peak in the signal at the largest scale and trace it back to the exact point at zero scale where it originates. The range of scales over which the peak exists is a measure of how important this peak is to the signal.

In scale-space segmentation, we look for changes in the signal over time. We do so by calculating the derivative of the signal with respect to time and then finding the local maximum of this derivative. Because the derivative and the scale-space filter are linear, we can exchange their order. Thus, the properties of the local maximum described previously also apply to the signal's derivative.

Lyon [Lyon, 1984] extended the idea of scale-space segmentation to multi-dimensional signals, and used it to segment a speech signal. The basic idea remains the same: He filtered the signal using a Gaussian kernel with a range of scales. By performing the smoothing independently on each dimension, he traced with the new signal a smoother path through his 92-dimensional space. To segment the signal, he looked for the local peaks in the magnitude of the vector derivative.

Cepstral analysis transforms each vocal sound into a point in a high-dimensional space. This transformation makes it easy to recognize each sound (good for automatic speech recognition) and to perform low-level segmentation of the sound (as demonstrated by Lyon). Unfortunately, the cepstral coefficients contain little information about high-level structures. Thus, we consider the image and the semantic content of the video.

Combining LSI analysis with scale-space segmentation is straightforward. This process is illustrated in Figure 8.1. We describe the scale-space process as applied to semantic content. The analysis of the acoustic and color data is identical to the semantic information.

The semantic data is first grouped into a time sequence of sentences, s_i. From these groups, we create a histogram of word frequencies, $\vec{H}(s_i)$, a vector function of sentence number s_i. LSI/SVD analysis of the full histogram produces a k-dimensional representation, $\vec{H}_k(s_i) = X^k$ of the document's semantic path (where the dimensionality k is much less than the original histogram.) In this work[2] we arbitrarily set $k = 10$.

We use a low-pass filter on each dimension of the reduced histogram data $\vec{H}_k(s_i)$, replacing s in equation 8.7 with each component of $\vec{H}_k(s_i) =$

[2]Information retrieval systems often use 100–300 dimensions to distill thousands of documents, but those collections cover a larger number of topics than we see in a single document.

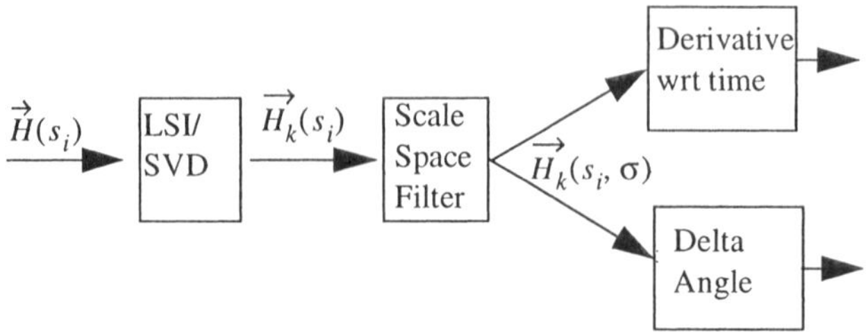

Figure 8.1. The LSI-SS algorithm. The top path shows the derivative based on euclidean distance. The bottom path shows the proper distance metric for LSI based on angle. See Section 8.3.4 for definitions.

$[H_1(s_i)H_2(s_i)...H_k(s_i)]^T$ to find a low-pass filtered version of the semantic path. This replacement gives $\vec{H_k}(s_i, \sigma)$, a k-dimensional vector function of sentence number and scale.

We are interested in detecting edges in acoustic, color and semantic scale spaces. An important property of the scale-space segmentation is that the length of a boundary in scale space is a measure of the importance of that boundary. It is useful to think about a point representing the document's local content wandering through the space in a pseudo-random walk. Each portion of the video is a slightly different point in space, and we are looking for large jumps in the topic space. As we increase the scale, thus lowering the cutoff frequency of a low-pass filter, the point moves more sluggishly. It eventually moves to a new topic, but small variations in topic do not move the point much. Thus, the boundaries that are left at the largest scales mark the biggest topic changes within the document.

The distance metric in Witkin's original scale-space work [Witkin, 1984] was based on Euclidean distance. When we use LSI as input to a scale-space analysis, our distance metric is based on angle. The dot product of adjacent (filtered and normalized) semantic points gives us the cosine of the angle between the two points. We convert this value into a distance metric by subtracting the cosine from one.

When we use LSI within a document, we must choose the appropriate block size. Placing the entire document into a single histogram gives us little information that we can use to segment the document. On the other hand, one-word chunks are too small; we would have no way to link single-word subdocuments. The power of LSI is available for segments that comprise a small chunk of text, where words that occur in close proximity are linked together by the histogram data.

Choosing the proper segment size is straightforward during the segmentation phase, since projecting onto a subspace is a linear operation.

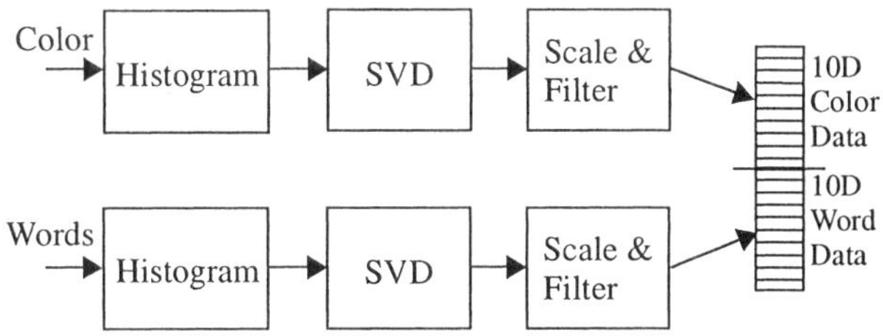

Figure 8.2. Combining color, words and scale space analysis. The result is a 20-dimensional vector function of time and scale.

Thus, even if we start with single-word histograms, the projection of the (weighted) sum of the histograms is the same as the (weighted) sum of the projections of the histograms.

The story is not so simple with the SVD calculation. For this study, we chose a single sentence as the basic unit of analysis, based on the fact that one sentence contains one subject. It is possible that larger subdocuments, or documents keyed by other parameters of a video, such as color information, might be more meaningful. The results of the temporal studies, described in Section 8.3.6.0, suggest that the optimal segment size is four to eight sentences, or a paragraph.

8.3.5 Combined Image and Audio Data

Our system for hierarchical segmentation of video combines the audio (semantic) and image (color) information into a single unified representation, and then uses scale-space segmentation on the combined signal (Figure 8.2).

Our algorithm starts by analyzing a video, using whatever audio and image features are available. For this chapter, we concentrated on the color and the semantic histograms. We perform an SVD on each feature, gaining noise tolerance and the ability to handle synonyms and polysemy (see Section 8.2.3).

The SVD, for either the color or the words, is performed in two steps. We build a model by collecting all the features of the signal into a matrix and then computing that matrix's SVD to find the k left-singular vectors that best span the feature space. We use the model by projecting the same data onto these k-best vectors to reduce the dimensionality of the signal. The semantic information typically starts with more than 1,000 dimensions; the color information has 512 dimensions. For the examples described in this chapter, we reduced all signals to individual 10-dimensional spaces.

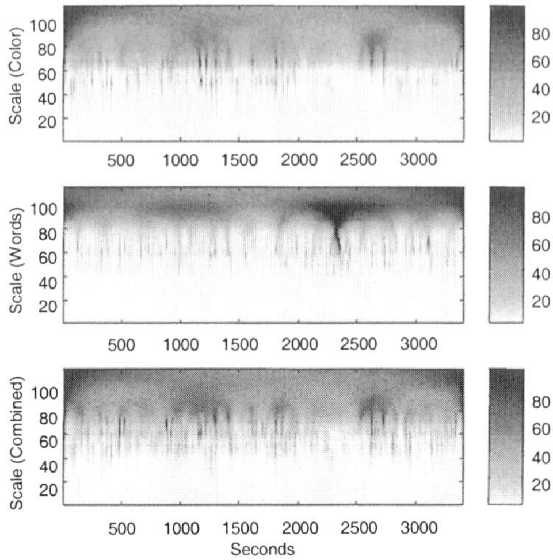

Figure 8.3. These three plots show the derivatives of the scale space representations for the colors (top), words (middle) and combined (bottom) spaces of the Boeing 777 video. Many details are lost because the 102089 frames are collapsed into only a few inches on this page.

The challenge when combining information in this manner is to not allow one source of information to overwhelm the others. The final steps before combining the independent signals are scaling and filtering. Scaling confers similar power on two independent sources of data. Typically, color histograms have larger values, since the number of pixels in an image tends to be much greater than the number of words in a semantic segment. Without scaling, the color signal is hundreds of times larger than the word signal; the combined signal makes large jumps at every color change, whereas semantic discontinuities have little effect.

To avoid this problem and to normalize the data, we balance the color and the semantic vectors such that both had an average vector magnitude of 1. Other choices are possible; for example, one might decide that the semantic signal contains more information about content changes than does the image signal and thus should have a larger magnitude. Plots showing the derivative of the color, word and the combined scale spaces are shown in Figure 8.3.

As we will discuss in Section 8.3.6.0, each signal has a natural frequency content, which we can filter to select a scale of interest. Thus, it might be appropriate to high-pass filter the color information to minimize the effects of changes over time scales greater than 10 seconds, while low-pass filtering the semantic information to preserve the information over scales greater than 10 seconds. We did not do this kind of filtering for the results presented in this chapter.

We combined the audio and visual data by aligning and concatenating the individual vectors. Alignment (and resampling) is important because the audio and image data have different natural sampling rates. Typically, the color data are available at the frame rate, 29.97 Hz, whereas the word information is available only at each sentence boundary—which occurred every 8 seconds, on average, in the Boeing 777 video that we studied.

We marked manually the start of each sentence in the video's audio channel. The marking was approximate, delineating the beginning of each sentence within a couple of seconds. We then created a new 10-dimensional vector by replicating each sentence's SVD-reduced representation at all the appropriate frame times. Then, based on the approximate sentence delineations, we smoothed the semantic vector with a 2-second rectangular averaging filter.

We concatenate the video and semantic vectors at each frame time, turning two 10-dimensional signals, sampled at 29.97 Hz, into a single 20-dimensional vector. We then can use these data as input to the scale-space algorithm.

8.3.6 Hierarchical Segmentation Results

We evaluated our approach with two studies. First, we studied the temporal properties of videos and text, by characterizing the temporal autocorrelation of the color and semantic information in news and documentary videos (Section 8.3.6.0). Second, to quantify the results of our segmentation algorithm, we performed scale-space hierarchical segmentation on two multimedia signals and compared the results to several types of segmentations (Section 8.3.6.0).

Temporal Results. There are many ways to characterize the temporal information in a signal. The autocorrelation analysis we describe in this section tells us the minimum and maximum interesting temporal scales for the audio and image data. This information is important in the design and characterization of a segmentation algorithm.

Autocorrelation. We investigated the temporal information in the signals by computing the autocorrelation of our representations:

$$R_{xx}(\tau) = \int_{-\infty}^{\infty} x'(t)x'(t+\tau)dt, \tag{8.8}$$

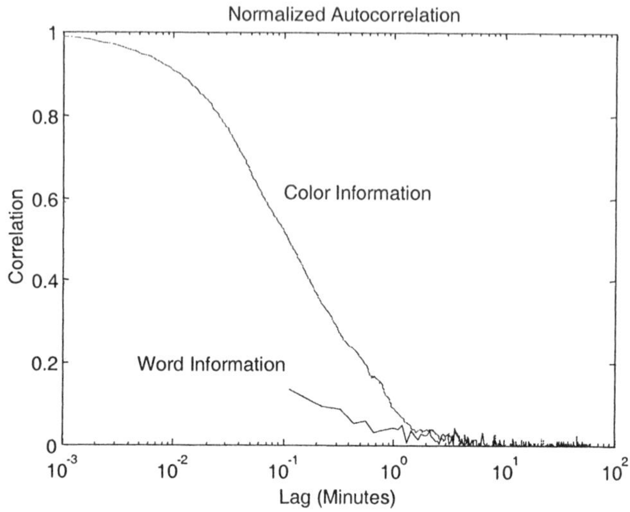

Figure 8.4. Color and word autocorrelations for the Boeing 777 video.

where x' is the original signal with the mean subtracted. There are six one-hour videos in the Boeing 777 documentary. The short length makes it difficult to estimate very long autocorrelation lags (more than 30 minutes). We computed the autocorrelation individually for each hour of video, then averaged the results across all videos to obtain a more robust estimate of the autocorrelation.

For both the image and the semantic data we used the reduced-dimensionality signals. We assumed that each dimension is independent and summed the autocorrelation over the first four dimensions to find the average correlation. The results of this analysis are shown in Figure 8.4 for both the image and the semantic signals.

The correlation for the color data is high until about 1/10 minute, when it falls rapidly to zero. This behavior makes sense, since the average shot length in this video, as computed by YesVideo (see Section 8.3.6.0), is 8 seconds.

Grouped Autocorrelation. At first, we were surprised by the semantic-signal results: There was little correlation at even the smallest time scale. We postulated that individual sentences have little in common with one another, but that groups of consecutive sentences might show more similarity. Usually, the same words are not repeated from one sentence to the next, and neighboring sentences should be nearly orthogonal.

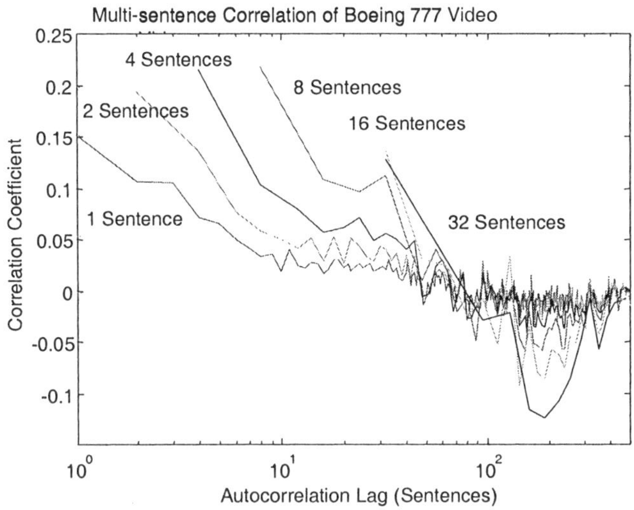

Figure 8.5. Grouping 4–8 sentences produces a larger semantic autocorrelation (data from the Boeing 777 video). This peak corresponds to 29–57 seconds of the original video.

By grouping sentences—averaging several points in semantic space—we formed a more robust estimate of the exact location of a given portion of a transcript or document in semantic space. In Figure 8.5, we show the results that we obtained by grouping sentences of the Boeing 777 video. In the line marked "8 sentences," we grouped (averaged) the reduced-dimensionality representation of eight contiguous sentences, and computed the correlations between that group and other groups of eight, non overlapping, sentences.

Figure 8.5 shows that, indeed, the correlation starts small when we consider individual sentences, and gradually grows to a maximum for groups of between four and eight sentences, and then falls again as the group size increases. Evidently, grouping four to eight sentences allows us to estimate reliably a single point in semantic space. The correlation reaches a minimum at approximately 200 sentences.

Interestingly, in two documents we saw a strong anti-correlation around 200 sentences [Slaney et al., 2001]. This is interesting because it indicates that the topic has moved from one side of the semantic space to the opposite side in the course of 200 sentences.

Segmentation Results. We evaluated our hierarchical representation's ability to segment the 30-minute Headline News television show and the first hour of the Boeing 777 documentary. We describe qualitative results and a quantitative metric, and show how our results compare to those obtained with automatic shot-boundary and manual topical segmentations.

Most videos are not organized in a perfect hierarchy. In text, the introduction often presents a number of ideas, which are then explored in subsequent sections; a graceful transition is used between ideas. The lack of hierarchy is much more apparent in a news show, the structure of which may be somewhat hierarchical, but is designed to be watched in a linear fashion. For example, the viewer is teased with information about an upcoming weather segment, and the "top of the news" is repeated at various stages through the broadcast.

We illustrate our hierarchical segmentation algorithm by showing intermediate results using just the semantic information from the Headline News video. The results of hierarchical segmentations are compared with the ground truth. The LDC [Linguistic Data Consortium, 1997] provided story boundaries for this video, but we estimated the high-level structure based on our familiarity with this news program. The timing and other meta information were removed from the transcript before analysis. We found 257 sentences in this broadcast transcript; which after the removal of stop words, contained 1032 distinct words.

Intermediate Results. Our segmentation algorithm measured the changes in a signal over time as a function of the scale size. A scale-space segmentation algorithm produced a boundary map showing the edges in the signal, as shown in Figure 8.6. At the smallest scale there were many possible boundaries; at the largest scale, with a long smoothing window, only a small number of edges remained.

Due to the local peculiarities of the data, the boundary deviated from its true location as we moved to large windows. We traced the boundary back to its true location (at zero scale) and drew the straightened boundary map shown at the bottom of Figure 8.6. For any one boundary, indicated by its vertical lines, strength is represented by line height, and is a measure of how significant this topic change is to the document.

Qualitative Measure. The classic measures for the evaluation of text-retrieval performance [Allan et al., 1998] do not extend easily to a system that has hierarchical structure. Instead, we evaluated our results by examining a plot that compared headings and the scale-space segmentation strength. The scale-space analysis produced a large number of possible segmentations; for each study, we plotted only twice the number of boundaries indicated by the ground truth.

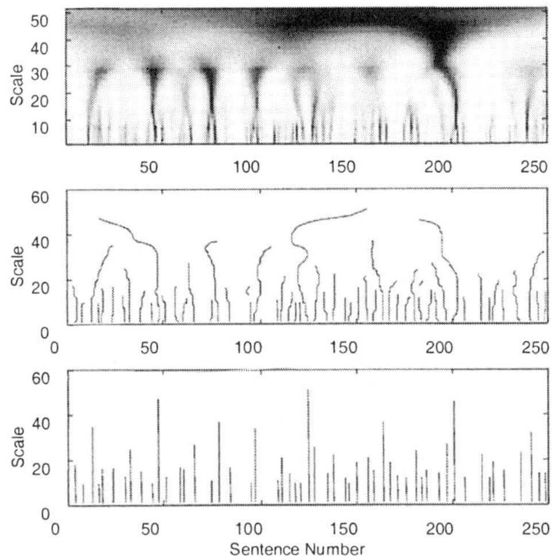

Figure 8.6. Representations of the semantic information in the Headline News video in scale space. The top image shows the cosine of the angular change of the semantic trajectory with different amounts of low-pass filtering. The middle plot shows the peaks of the scale-space derivative for the tomography chapter. The bottom plot shows the peaks traced back to their original starting point. These peaks represent topic boundaries.

Our results of calculating the hierarchical segmentations of the Headline News are shown in Figure 8.7. On the right, the major (left most text) and the minor (right most text) headings are shown. The left side of the plot shows the strength of the boundary. The "Weather," "Tech Trends" and "Lifestyles" sections are indicated within a few sentences, yet there are large peaks at other locations in the transcript. Interestingly, there is a large boundary near sentence 46, which neatly divides the softer news stories at the start of this broadcast from the political stories that follow.

We measured the degree of agreement between two segmentations by looking at both ends of a fixed window passed over two sets of segmentation data. Figure 8.8 summarizes the process for one set of data labeled with ground truth and for another set labeled "experimental." The segmentation, at this point, is successful if both ends of the window fall within the same segment or if each is in a different segment. The segmentation is wrong if, for example, the ground-truth window falls entirely within one segment and the experimental window covers two or more segmentation boundaries. We move the window over the entire document and calculate the fraction of correct windows.

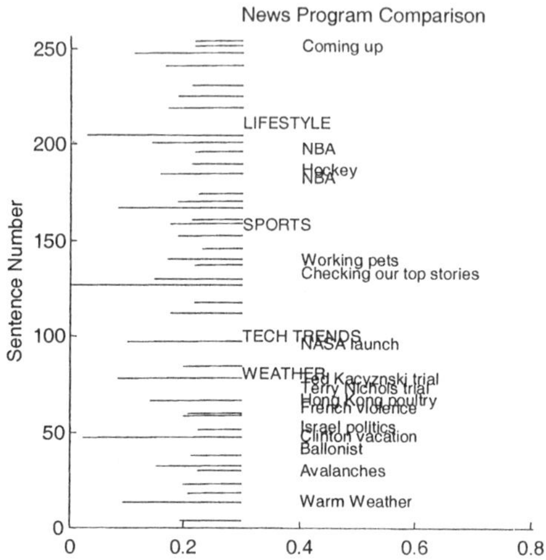

Figure 8.7. A comparison of ground truth (right) and the size of boundaries for the Headline News video as determined by scale-space segmentation. The major headings are in all capitals, and the sub-headings are in upper and lower case.

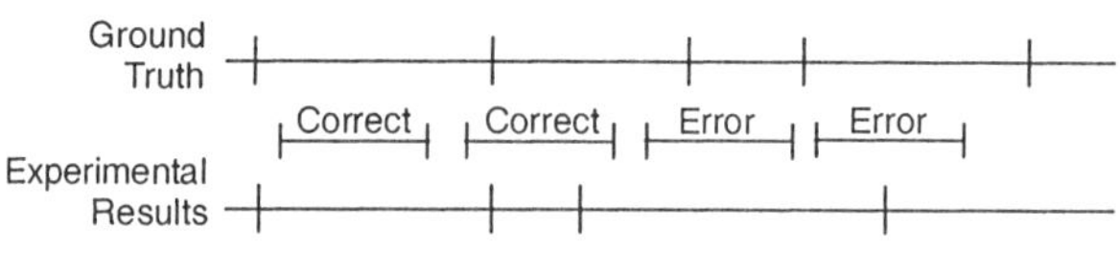

Figure 8.8. We evaluate accuracy by measuring whether the ends of a fixed-size window fall in the same or different segments.

Our quantitative measure of segmentation performance was suggested by Lafferty and refined by Doddington [Allan et al., 1998] An especially important property of Lafferty's measure for semantic segmentations is that small offsets in the segmentation lower the performance metric, but do not cause complete misses. We used a fixed window size that was 50 percent of the length of the average segment calculated using the ground-truth segmentation.

Lafferty's segmentation metric has several properties that were reflected in our data. Assume that the probability that any particular frame is a boundary is independent and is fixed at $p = 1/(2N)$ where N is the window length or half the average segment length. If we measure the accuracy of an experimental result with just one (large) segment, then Lafferty's measure is asymptotically equal to

$$(1 - p)^N \approx 0.606. \tag{8.9}$$

Conversely, if we measure the accuracy of a segmentation that puts a boundary at every time step, then Lafferty's measure is equal to

$$1 - (1 - p)^N \approx 0.394. \tag{8.10}$$

Finally, if we compare two random segmentations, each with the same probability of a boundary, Lafferty's measure indicates that the segmentation accuracy is

$$(1 - p)^{2N} + [1 - (1 - P)^N]^2 \approx 0.523. \tag{8.11}$$

Shot Boundary Segmentation. We used the segmentation produced by a state-of-the-art commercial product, designed by YesVideo [YesVideo, Inc., 2002], as our shot-boundary ground truth. They reported that, on a database of professionally produced wedding videos, their segmenter had an overall precision of 93% and a recall of 91%. For their test set, most of the errors, both false positives and false negatives, were due to uncompensated camera motion.

We performed quantitative tests on the first hour of the Boeing 777 video. This video had 102,089 frames. The semantic analysis found 1314 distinct words in 537 sentences. There were shot boundaries on average every 242 frames. The standard deviation of the Gaussian blur used in scale-space filtering is $\sigma = 1.1^{s-1}$, where s is the scale number.

To evaluate our combined representation, we show the results here using only the color data, only the word data, and the combination. We evaluate Lafferty's measure for the segmentation boundaries predicted at each scale, effectively assuming that a single scale would produce the best segmentation. Assuming all segmentation boundaries are at the

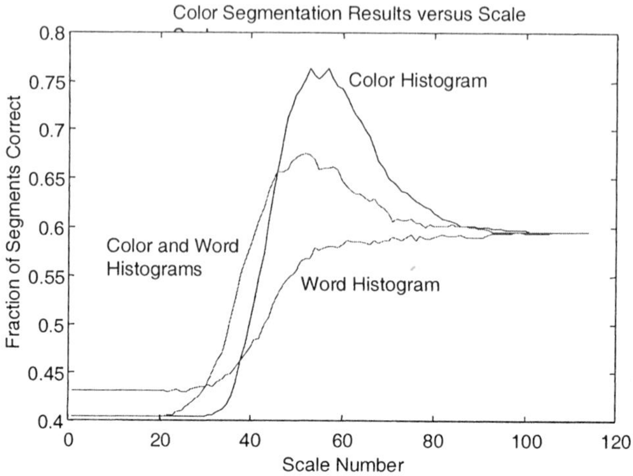

Figure 8.9. This figure shows the accuracy of the scale-space segmentation algorithm, at any one scale, at finding shot boundaries. Video was the first hour of the Boeing 777 video, compared to ground truth from YesVideos segmenter [YesVideo, Inc., 2002].

same scale is not the best solution; instead, the information from the scale-space segmentation metric should be used as input to a higher-level model of video transitions, as suggested by Srinivasan [Srinivasan et al., 1999].

Figure 8.9 shows how segmentation accuracy varied with scale, comparing the segmentation at each scale to the YesVideo results. At small scales the probability, as predicted by equation 8.10, was 40%. At large scales, only one or two boundaries were found, and, as predicted by equation 8.9, the accuracy was 60%. At the middle scale, the segmentation accuracy was 77%—well above that of random segmentations (52%). As expected, the semantic signal does not predict the color boundaries. Adding the semantic information to the color information does reduce the highest accuracy at any one scale to 67%.

Semantic Segmentation. We also compared our algorithm's semantic segmentations to those of humans. Two of the authors of this chapter and a colleague segmented the transcript of the first hour of the Boeing 777 video. There was a wide range in what these three readers described as a segment; they chose to segment the text with 29, 37, and 122 segment boundaries. They found it difficult to produce a hierarchical segmentation of the text. The video was designed to be watched in one sitting; it transitions smoothly from topic to topic, weaving a single story.

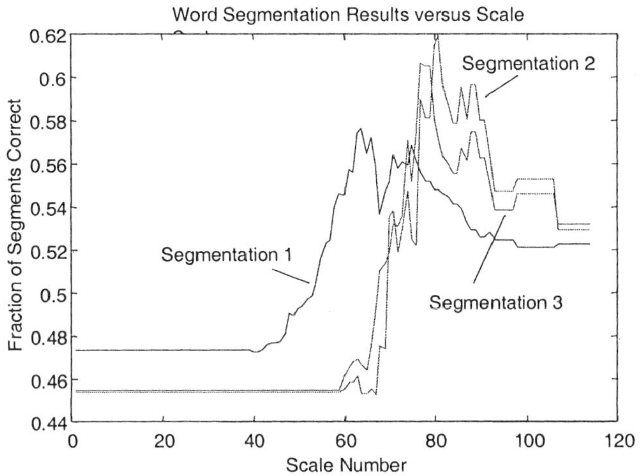

Figure 8.10. Manual segmentation versus scale, tested with Lafferty's measure (all three manual segmentations). Source data from the first hour of the Boeing 777 video.

Figure 8.10 shows how the scale-space segmentation compares, across scale, to the manual segmentations. As expected, the best scale is larger than that shown for the color segmentations in Figure 8.9. The scale-space segmentation algorithm matches each of the humans' segmentations equally well. Perhaps most surprisingly, the color information is a good predictor of the semantic boundaries. This correlation may indicate that the color signal carries information regarding content changes that is richer than we assumed.

8.3.7 Segmentation Conclusions

We have demonstrated a new framework for combining into a unified representation and for segmenting information from multiple types of information from a video. We used the SVD to reduce the dimensionality of each signal. Then, we applied scale-space segmentation to find edges in the signals that corresponded to large changes. We demonstrated how these ideas apply to words and to color information from a video.

These techniques are an important piece of a complete system. The system we have described does not have the domain knowledge to know that, for example when it is considering a videotape of a news broadcast, the phrase "coming up after the break" is a pointer to a future story and is not a new story in its own right. Systems that include domain knowledge about specific types of video content [Dharanipragada et al., 2000] show how this knowledge is incorporated.

We have described the natural-frequency content of information in a video. Autocorrelation analysis showed that the color information was correlated for about 0.1 minute, whereas the semantic content showed significant correlation for hundreds of sentences (tens of minutes). These results suggest the smallest meaningful unit of semantic information is about 8 sentences.

We described our hierarchical segmentation results by comparing them to conventional segmentations. Qualitatively, the automatic segmentation has many similarities to a manual segmentation. It is hard to evaluate the quantitative results, but we were surprised by the amount of information that was available in the color information for topical segmentation.

The methods we described here are equally useful with other information from a video. This includes speaker identification features, musical key, speech/music indicators, and even audio emotion. These techniques do not give any assistance with professional video production techniques, such as L-cuts, which change the audio topic and the camera shot at different times.

8.4 Semantic Retrieval

The previous section described an algorithm which used the semantic (and perhaps also the color information) to segment media. In this section, we build links between the media and the semantic content.

This section describes a method of connecting sounds to words, and words to sounds. Given a description of a sound, the system finds the audio signals that best fit the words. Thus, a user might make a request with the description "the sound of a galloping horse," and the system responds by presenting recordings of a horse running on different surfaces, and possibly of musical pieces that sound like a horse galloping. Conversely, given a sound recording, the system describes the sound or the environment in which the recording was made. Thus, given a recording made outdoors, the system says confidently that the recording was made at a horse farm where several dogs reside.

A system that has these functions, called MPESAR (mixtures of probability experts for semantic–audio retrieval), learns the connections between a semantic space and an acoustic space. Semantic space maps words into a high-dimensional probabilistic space. Acoustic space describes sounds by a multidimensional vector. In general, the connection between these two spaces will be many to many. Horse sounds, for example, might include footsteps and neighs.

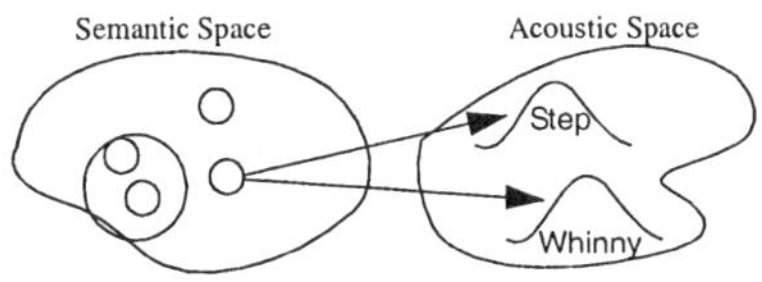

Figure 8.11. MPESAR models all of semantic space with overlapping multinomial clusters, each portion in the semantic model is linked to equivalent sound documents in acoustic space with a GMM.

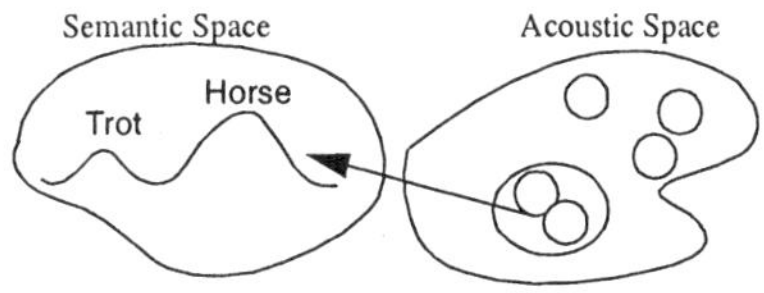

Figure 8.12. MPESAR describes with words an audio query by partitioning the audio space with a set of acoustic models and then linking each cluster of audio files (or documents) to a probability model in semantic space.

Figure 8.11 shows one half of MPESAR: how to retrieve sounds from words. Annotations that describe sounds are clustered and represented with multinomial models. The sound files, or acoustic documents, that correspond to each node in the semantic space are modeled with Gaussian mixture models (GMMs). Given a semantic request, MPESAR identifies the portion of the semantic space that best fits the request, and then measures the likelihood that each sound in the database fits the GMM linked to this portion of the semantic space. The most likely sounds are returned to satisfy the user's semantic request.

Figure 8.12 shows the other half of MPESAR: how to generate words to describe a sound. MPESAR analyzes the collection of sounds and builds models for arbitrary sounds. This approach gives us a multidimensional representation of any sound, and a distance metric that permits agglomerative clustering in the acoustic space. Given an acoustic request, MPESAR identifies the portion of the acoustic space that best fits the request. Each portion of the acoustic space has an associated multinomial word model, and from this model MPESAR generates words to describe the query sound.

In general, sounds that are close in acoustic space might correspond to many different points in semantic space, and vice versa. Thus, MPESAR builds two completely separate sets of models: one connecting audio to semantic space and the other connecting semantic to audio space.

8.4.1 The Algorithm

Mixture of Probability Experts. MPESAR uses a mixture of experts approach [Waterhouse, 1997] to link semantic and audio spaces. A mixture of experts approach uses a different expert for different regions of an input space. Thus, one expert might be responsible for horse sounds while another is responsible for bird sounds.

Mathematically, a mixture of probability experts for semantic to audio retrieval is summarized by the following equation

$$P(a|q) = \sum_{c} P(c|q)P(a|c) \tag{8.12}$$

Here $P(c|q)$ represents the probability that a semantic query (q) matches a cluster (c). The probability that a particular portion of acoustic space is associated with an expert or cluster (c) is given by $P(a|c)$. To find the overall probability of a point in audio space given the query, $P(a|q)$, we sum over all possible clusters, essentially interpolating the different expert's opinions to arrive at the final probability estimate.

We want to calculate the probability of a cluster given a query. We group semantic documents into clusters and then estimate $P(q|c)$. Using Bayes' rule: $P(c|q) = P(c)P(q|c)/P(q)$. The $P(c)$ and $P(q|c)$ terms are calculated using clustering algorithms described in Sections 8.4.1.0 and 8.4.1.0 Since the query is given, we can ignore the $P(q)$ term. The same formalism is used for the audio to semantic problem.

Semantic Features. MPESAR uses multinomial models to represent and cluster a collection of semantic documents. The likelihood that a document matches a given multinomial model is described by $L = \prod p_i^{n_i}$, where p_i is the probability that word i occurs in this type of document, and n_i is the number of times that word i is found in this document. The set of probabilities, p_i, is different for different types of documents. Thus, a model for documents about cows will have a relatively high probability for containing "cow" and "moo," whereas a model for documents that describe birds with have a high probability of containing "feather." These multinomial models accomplish the same task as LSI—convert a bag of words into a multinomial vector—but have a more principled theory.

A semantic document contains the text used to describe an audio clip. MPESAR uses the PORTER stemmer to remove common suffixes from the words, and deletes common words on the SMART list before further processing [Porter, 1980]. In effect, a 705-dimensional vector (the multinomial coefficients) describes a point in semantic space, and MPESAR partitions the space into overlapping clusters of regions.

Smoothing is used in statistical language modeling to compensate for a paucity of data. It is called smoothing because the probability associated with likely events is reduced and distributed to events that were not seen in the training data. The most successful methods [Chen and Goodman, 1996] use a back-off method, where data from simpler language models are used to set the probability of rare events. MPESAR uses a unigram word model, so the back-off model suggests a uniform low probability for all words.

Acoustic Features. Sound is difficult to analyze because it is dynamic. The sound of a horse galloping is constantly changing at time scales in the hundreds of milliseconds; a hoofstep is followed by silence, and then by another hoofstep. Yet we would like a means to transform the sound of a galloping horse into a single point in an acoustic space. This section describes acoustic features that allow us to describe each sound as a single point in acoustic space, and to cluster related sounds.

Conventional acoustic features for speech recognition and for sound identification use a short-term spectral slice to characterize the sound at 10-ms intervals. A combination of signal-processing and machine-learning calculations endeavors to capture the sound of a horse as a point in auditory space.

MFCC (mel-frequency cepstral coefficient) is a popular technique in the automatic speech recognition community to analyze speech sounds. Based on auditory perception, the MFCC representation captures the overall spectral shape of a sound, while throwing away the pitch information. This allows MFCC to distinguish one vowel from another or one instrument from another; while ignoring the melody. We use MFCC to reduce an audio signal from it's original representation (22kHz sampling rate) to a 13-dimensional vector sampled 100 times a second.

MFCC has been used in earlier music segmentation work [Foote, 1999], but it is not the ideal representation. In this work it allows us to capture the overall timbral qualities of the sound, but ignore the detailed pitch information. The dimensionality reduction is similar to that performed by an SVD, but takes into account specialized knowledge about audio and dimensions that are safe to ignore. The MFCC representation does not give us any high-level information about rhythm or other musical properties of the signal. Eventually we hope that more sophisticated acoustic features will allow us to segment musical accompaniment at phrase or beat boundaries, or even to detect mood changes in movie scores.

The process of converting a waveform into a point in acoustic space is shown in Figure 8.13. The MFCC algorithm [Quatieri, 2002] decomposes each signal into broad spectral channels and compresses the loudness of the signal. RASTA filtering [Quatieri, 2002] is used on the MFCC coefficients to remove long-term spectral characteristics that often occur due to the different recording environments. Then seven frames of data—

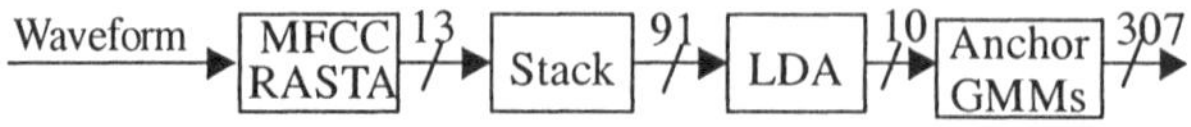

Figure 8.13. The acoustic signal processing chain. Arrows are marked with the signals dimensionality. All but the last are sampled at 100Hz. The final output is sampled once per sound.

three before the current frame, the current frame, and the three frames following the current frame—are stacked together. Finally, linear discriminant analysis (LDA) [Quatieri, 2002] uses the intra- and inter-class scatter matrices for a hand-labeled set of classes to project the data onto the optimum dimensions for linear separability.

The long-term temporal characteristics of each sound are captured using a GMM. One of the Gaussians might capture the start of the footstep, a second captures the steady-state portion, a third captures the footstep's decay, and, finally, a fourth captures the silence between footsteps. The GMM measures the probability that a vector sequence fits a probabilistic model learned from the training sounds. Unlike hidden Markov models (HMMs), a GMM ignores temporal order.

MPESAR converts the MFCC-RASTA-LDA plus GMM recognition system into an auditory space by using model likelihood scores to measure the closeness of a sound to pre-trained acoustic models. The negative log-likelihood that a sound fits a model is a measure of the distance of the new sound from the test model.

Acoustic to Semantic Lookup. Given representations of acoustic and semantic spaces, we can now build models to link the two spaces together. The overall algorithm for both acoustic to semantic and semantic to acoustic lookup is shown in Figure 8.14.

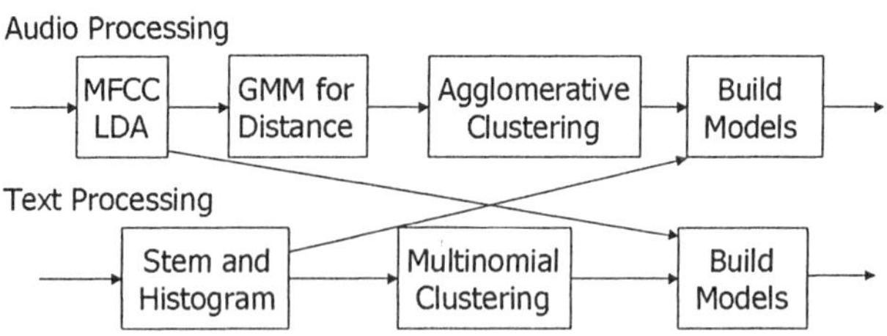

Figure 8.14. A schematic showing the process of building the MPESAR models. The top line shows the construction of the audio to semantic model and the bottom line shows the construction of the semantic to audio model.

Acoustic space is clustered into regions using agglomerative clustering [Jain and Dubes, 1988]. We compute the distance between each pair of training sounds $[L(model\ a|sound\ b) + L(model\ b|sound\ a)]/2$ where $L(modela|soundb)$ represents the likelihood that sound b is generated by model a. At each step, agglomerative clustering grows another layer of a hierarchical model by merging the two remaining clusters that have between them the smallest distance. MPESAR uses "complete" linkage, which uses the maximum distance between the points that form the two clusters, to decide which clusters should be combined. While agglomerative clustering generates a hierarchy, MPESAR only uses the information about which sounds are clustered. Leaves at the bottom of the tree are considered clusters containing a single document.

Each acoustic cluster is composed of a number of audio tracks and their associated descriptive text. A new 10-element GMM with diagonal covariance models all the sounds in this cluster and estimates the probability density for acoustic frames in this cluster, $P(a|c)$. Given a new sound, MPESAR uses this model to estimate the probability that a new sound belongs to this cluster. The text associated with each acoustic sample in the cluster is used to estimate the semantic model associated with this cluster. This is written as a simple multinomial model; there is not enough text in this study to form a richer model.

Given a new waveform, MPESAR queries all acoustic GMMs to find the probability that each possible cluster generated this query. Each cluster comes with an associated semantic model. MPESAR uses a weighted average of all the semantic models, based on cluster probabilities, to estimate the semantic model that describes the test sound. The words that describe the test sound are entries in the semantic multinomial model with the highest probabilities.

Semantic to Acoustic Lookup. A similar procedure is used for semantic to acoustic lookup. A document's point in semantic space is described by the coefficients of a unigram multinomial model. Semantic space is clustered into regions using a multinomial clustering algorithm, which uses an iterative expectation-maximization algorithm [Nigam et al., 1998] to group documents with similar (multidimensional) models. In this work, we assign each document to its own cluster, and then split the entire corpus into a number of arbitrary-sized clusters (32, 64, 128 and 256 clusters for the corpus).

Each text cluster is composed of a number of text documents and their associated audio tracks. All the text associated with each cluster is used to form a unigram multinomial model of the text documents. All of the audio associated with a cluster is used to form a 10-element

GMM to describe the link to audio space. (Note there are three sets of GMMs used in this work: the GMMs used to compute the distances as part of audio clustering, the GMMs used to model each audio cluster, and the GMMs used to model the sounds associated with each semantic cluster.)

Given a text query, MPESAR finds the probability that each semantic cluster generated the query. Then the acoustic models are averaged (weighted by the cluster probabilities) to find the probability that any one sound fits the query.

8.4.2 Testing

This section describes several tests performed using the algorithms described above.

Data. The animal sounds from two sets of sound effect CDs were used as training and testing material. Seven CDs from the BBC Sound Effects Library (#6, 12, 30, 34, 35, 37, 38) contained 261 separate tracks and 390 minutes of animal sounds. Two CDs from the General 6000 Sound Effect library (all tracks from CD6003 and tracks 18 to 40 of CD6023) totaled 122 tracks and 110 minutes of animal sounds.

The concatenated name of the CD (e.g., "Horses I") and track description (e.g., "One horse eating hay and moving around") forms a semantic label for each track. The audio from the CD track and the liner notes form a pair of acoustic and semantic documents used to train the MPESAR system.

The system training and testing described in this chapter were performed on distinct sets of data. 80% of the tracks (307) from both sets of CDs were randomly assigned as training data in the procedure shown in Figure 8.15. The remaining 20% of the tracks (93) were reserved for testing. Mixing the data obtained from the two sets of CDs is important for several reasons. First, the acoustic environments of the two data sets are different; RASTA reduces these effects. Second, the words and description are different because the sounds are labeled by different organizations with different needs. For example, the BBC describes the sound of a cat's vocalization as miaow and the General Sound Effects CD uses meow. Finally, the two sets of audio data do not contain the same sounds: There are many sounds in the General set which are not represented in the BBC training set.

Acoustic Feature Reduction and Language Smoothing. The audio-feature reduction using LDA was computed using portions of the audio data from both sets of CDs. We chose ten broad classes of dis-

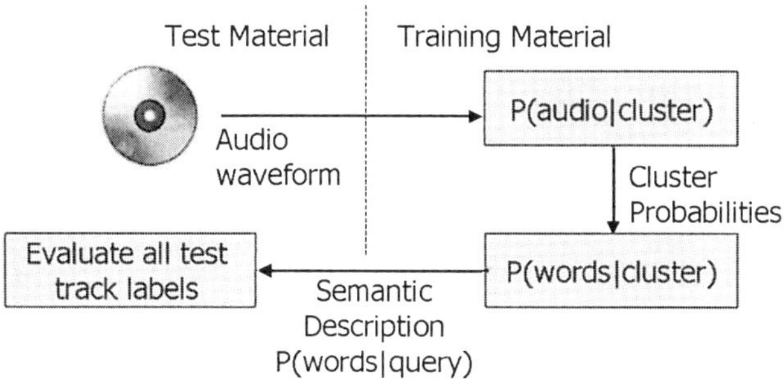

Figure 8.15. A schematic of the audio to semantic testing procedure.

tinct sound types (baboon, bird, cat, cattle, dog, fowl, goat, horse, lion, pig, sheep). The stacked features from only those audio tracks that fit these classes were used as input to the LDA algorithm. This computation produced a matrix that reduced the 91-dimensional data to the 10-dimensional subspace that best discriminates between these 10 classes. This dimensionality reduction was fixed for all experiments.

A simple test was used to set the amount of smoothing in the language models. Without smoothing, the semantic lookup results were poor because many of the General sounds were labeled with the word "animal," which was seldom used in the BBC labels. The results here were generated using a back-off method that added a small constant probability ($1/N_w$, where N_w is the number of words in the vocabulary) to each word model.

Labeling Tests. Figure 8.15 shows the test procedure for the acoustic-to-semantic task (a similar procedure is used to test semantic-to-audio labeling.) Audio from each test track is applied as an acoustic query to the system. The MPESAR system calculates the probability of each cluster given this acoustic query. These cluster probabilities are used to weight the semantic models associated with each cluster. The result is a multinomial probability distribution that represents the probabilities that each word in the dictionary describes the acoustic test track. The likelihoods that each test-track description fit the query's semantic description were sorted and the rank of the true test label was recorded.

Figures 8.16 and 8.17 show histograms of the true test ranks for both directions of the MPESAR algorithm. Figure 8.16 shows the acoustic-to-semantic results and the median rank of the true result over all the test tracks is 17.5. Figure 8.17 shows the semantic-acoustic results and the median rank of the true result for this direction is 9. At this point

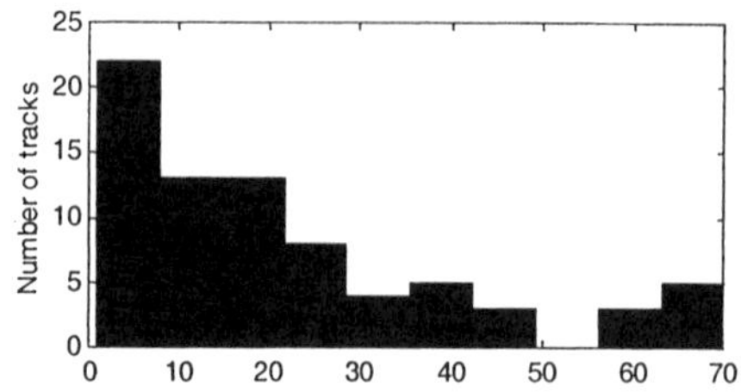

Figure 8.16. Histogram of true label ranks based on likelihoods from audio-to-semantic tests.

we do not understand the difference in performance between these two directions.

8.4.3 Retrieval Conclusions

This chapter described a system that uses a mixture of probability experts to learn the connection between an audio and a semantic space, and the reverse. It describes the conversion of sound and text into acoustic and semantic spaces and the process of creating the mixture of probability experts. The system was tested using commercial sound-effect CDs and is effective at labeling acoustic queries with the most appropriate words, and for finding sounds that fit a semantic query.

There are several improvements to this system that are worth pursuing. First, an algorithm that integrates the clustering and the MPE training will improve the system's models. Second, a richer acoustic description, perhaps replacing the GMMs with hidden Markov models, will provide more discrimination power. Finally, larger training sets will improve the system's knowledge.

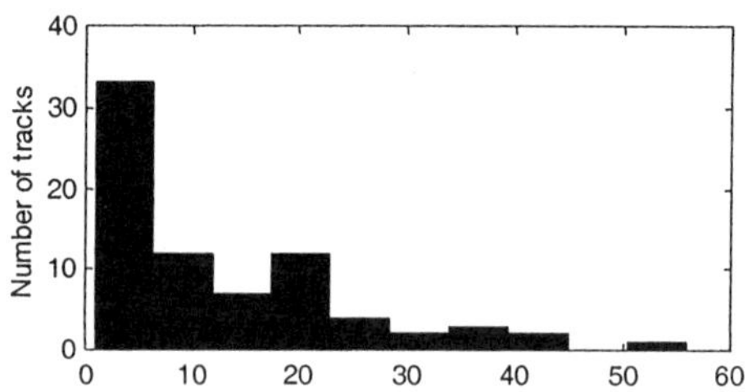

Figure 8.17. Histogram of true label ranks based on likelihoods from semantic-to-audio tests.

8.5 Acknowledgements

We appreciate the assistance that we received from Byron Dom, Arnon Amir, Myron Flickner, John Fisher, Clemens Drews and Michele Covell. Ian Nabney's NETLAB software was used to calculate the GMMs; Roger Jang provided the clustering code.

References

Allan, J., Carbonell, J., Doddington, G., Yamron, J., and Yang, Y. (1998). Topic detection and tracking pilot study: Final report.

Aner, A. and Kender, J. R. (2002). Video summaries through mosaic-based shot and scene clustering. In *ECCV*, pages 388–402.

Babaud, J., Witkin, A. P., Baudin, M., and Duda, R. O. (1986). Uniqueness of the gaussian kernel for scale-space filtering. *IEEE Transactions on Pattern Analysis and Machine Intelligence*, PAMI-8(1):26–33.

Barnard, K. and Forsyth, D. (2001). Learning the semantics of words and pictures. In *Proceedings of the 2001 International Conference on Computer Vision*, volume 2, pages 408–415.

Chen, S. F. and Goodman, J. (1996). An empirical study of smoothing techniques for language modeling. In Joshi, A. and Palmer, M., editors, *Proceedings of the Thirty-Fourth Annual Meeting of the Association for Computational Linguistics*, pages 310–318, San Francisco. Morgan Kaufmann Publishers.

Deerwester, S. C., Dumais, S. T., Landauer, T. K., Furnas, G. W., and Harshman, R. A. (1990). Indexing by latent semantic analysis. *Journal of the American Society of Information Science*, 41(6):391–407.

Dharanipragada, S., Franz, M., McCarley, J. S., Papineni, K., S.Roukos, T.Ward, and Zhu, W.-J. (2000). Statistical models for topic segmentation. In *Proc. of ICSLP-2000.*

Dumais, S. T. (1991). Improving the retrieval of information from external sources. *Behavior Research Methods, Instruments, & Computers*, 23:229–236.

Flickner, M., Sawhney, H., Niblack, W., Ashley, J., Huang, Q., Dom, B., Gorkani, M., Hafner, J., Lee, D., Petkovic, D., Steele, D., and Yanker, P. (1993). The qbic project: Query images by content using color, texture and shape. In *SPIE Storage and Retrieval of Image and Video Database*, pages 173–181.

Foote, J. (1999). Visualizing music and audio using self-similarity. In *Proceedings of the seventh ACM international conference on Multimedia (Part 1)*, pages 77–80. ACM Press.

Hearst, M. (1994). Multi-paragraph segmentation of expository text. In *32nd. Annual Meeting of the Association for Computational Linguis-*

tics, pages 9–16, New Mexico State University, Las Cruces, New Mexico.

Jain, A. K. and Dubes, R. C. (1988). *Algorithms for Clustering Data.* Prentice-Hall.

Leung, Y., Zhang, J. S., and Xu, Z. B. (2000). Clustering by scale-space filtering. In *IEEE Transactions on PAMI*, volume Vol. 22(12), pages 1396–1410.

Linguistic Data Consortium (1997). 1997 english broadcast news speech (hub-4).

Lyon, R. F. (1984). Speech recognition in scale space. In *Proc. of 1984 ICASSP*, pages 29.3.1–4.

Miller, N. E., Wong, P. C., Brewster, M., and Foote, H. (1998). TOPIC ISLANDS - A wavelet-based text visualization system. In Ebert, D., Hagen, H., and Rushmeier, H., editors, *IEEE Visualization '98*, pages 189–196.

Nigam, K., McCallum, A. K., Thrun, S., and Mitchell, T. M. (1998). Learning to classify text from labeled and unlabeled documents. In *Proceedings of AAAI-98, 15th Conference of the American Association for Artificial Intelligence*, pages 792–799, Madison, US. AAAI Press, Menlo Park, US.

PBS Home Video (1995). 21st century jet: The building of the 777.

Porter, M. F. (1980). An algorithm for suffix stripping. *Program*, 14(3):130–137.

Quatieri, T. F. (2002). *Discrete-Time Speech Signal Processing: Principles and Practice.* Prentice-Hall.

Slaney, M. (2002). Mixtures of probability experts for audio retrieval and indexing. In *Proceedings 2002 IEEE International Conference on Multimedia and Expo*, volume 1, pages 345–348.

Slaney, M., Ponceleon, D., and Kaufman, J. (2001). Multimedia edges: finding hierarchy in all dimensions. In *Proceedings of the ninth ACM international conference on Multimedia*, pages 29–40. ACM Press.

Srinivasan, S., Ponceleona, D., Amir, A., and Petkovic, D. (1999). 'what is in that video anyway?' in search of better browsing. In *Proceedings IEEE International Conference on Multimedia Computing and Systems*, pages 388–393.

Waterhouse, S. (1997). Classification and regression using mixtures of experts.

Witkin, A. P. (1984). Scale-space filtering: A new approach to multi-scale description. In *Proceedings of ICASSP*, pages 39A.1.1–39A.1.4.

YesVideo, Inc. (2002).

Chapter 9

STATISTICAL TECHNIQUES FOR VIDEO ANALYSIS AND SEARCHING

John R. Smith, Ching-Yung Lin, Milind Naphade, Apostol (Paul) Natsev and Belle Tseng

IBM T.J. Watson Research Center
30 Saw Mill River Road
Hawthorne, NY 10532 USA
jrsmith@watson.ibm.com

Abstract With the growing amounts of digital video data, effective methods for video indexing are becoming increasingly important. In this chapter, we investigate statistical techniques for video analysis and searching. In particular, we examine a novel method for multimedia semantic indexing using model vectors. Model vectors provide a semantic signature for multimedia documents by capturing the detection of concepts broadly across a lexicon using a set of independent binary classifiers. We also examine a new method for querying video databases using interactive search fusion in which the user interactively builds a query by interactively choosing target modalities and descriptors and by selecting from various combining and score aggregation functions to fuse results of individual searches.

Keywords: Image and video databases, content-based retrieval (CBR), model-based retrieval (MBR), multimedia indexing, MPEG-7, statistical analysis, video search, video analysis, concept detection, descriptors, fusion of query results, semantic indexing, model vectors, semantic signature, interactive search, interactive queries, user interaction, query building, TREC Video Track, normalization of scores, model vectors.

Introduction

The growing amount of multimedia data in the form of video, images, graphics, speech and text is driving the need for more effective methods for indexing, searching, categorizing and organizing this information. Recent advances in content analysis, feature extraction and classification are improving capabilities for effectively searching and filtering multimedia documents. However, a significant gap remains between the low-level feature descriptions that can be automatically extracted from the multimedia content, such as colors, textures, shapes, motions, and so forth, and the semantic descriptions of objects, events, scenes, people and concepts that are meaningful to users of multimedia systems. In this chapter, we investigate statistical techniques for video analysis and retrieval based on model vector indexing and interactive search fusion, respectively.

Related work. The problem of multimedia semantic indexing is being addressed in a number of ways relying on manual, semi-automatic, or fully-automatic methods. The use of manual annotation tools allows humans to manually ascribe labels to multimedia documents. However, the manual cataloging is time consuming and often subjective, leading to incomplete and inconsistent annotations. Recently, semi-automatic methods have been developed for speeding up the annotation process, such as those based on active learning [Naphade et al., 2002b]. Fully-automatic approaches have also been recently investigated based on statistical modeling of the low-level audio-visual features. The statistical modeling approach is useful for allowing searching based on the fixed set of labels, however, the indexing is limited to the labels in the lexicon [Naphade and Smith, 2003].

Recent research has explored new methods for extracting rich feature descriptors, classifying content and detecting concepts using statistical models, extracting and indexing speech information, and so forth. While research continues on these directions to develop more effective and efficient techniques, the challenge remains to integrate this information together to effectively answer user queries. There are a number of approaches for video database access, which include search methods based on the above extracted information, as well as techniques for browsing, clustering, visualization, and so forth. Each approach provides an important capability. For example, content-based retrieval (CBR) allows searching and matching based on perceptual similarity of video content.

On the other hand, model-based retrieval (MBR) allows searching based on automatically extracted labels and detection results [Naphade et al., 2002a]. New hybrid approaches, such as model vectors, allow similarity searching based on semantic models [Smith et al., 2003b]. Text-based retrieval (TBR) applies to textual forms of information related to the video, which includes transcripts, embedded text, speech, metadata, and so on. Furthermore, video retrieval using speech techniques can leverage important information that often cannot be extracted or detected in the visual aspects of the video. Given these varied approaches, there is a great need to develop a solution for integrating these different search methods of data sources given their complementary nature to bring the maximum resources to bear on satisfying a user's information need from a video database [Smith et al., 2003a].

Model vectors. Given that multimedia indexing systems are improving capabilities for automatically detecting concepts in multimedia documents, new techniques are needed that leverage these classifiers to extract more meaningful semantic descriptors for searching, classifying and clustering. Furthermore, new methods are needed that take into account uncertainty or reliability as well as the relevance of any labels assigned to multimedia documents in order to provide an effective index.

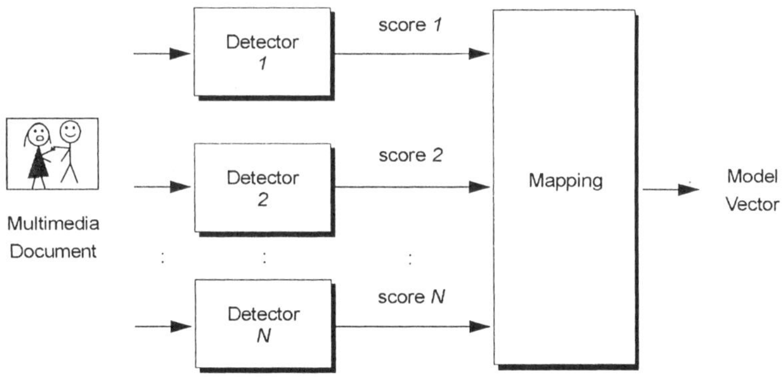

Figure 9.1. Overview of model vector extraction from multimedia documents using a set of N semantic concept detectors.

Figure 9.1 shows how the model vector representation aggregates the results of applying a series of independent binary classifiers to the multimedia documents. The advantage of the model vector representation is that it captures the concept labeling broadly across an entire lexicon. It also provides a compact representation that captures the uncertainty of the labels. The model vector approach also leverages the knowledge captured by each specific binary detector by allowing independent and possibly specialized classification techniques to be leveraged for each concept. For example, Support Vector Machines could be used for detecting "indoors" and Gaussian Mixture Models used for detecting "music". The real-valued multidimensional nature of model vectors also allows efficient indexing in a metric space using straightforward computation of model vector distances. This allows development of effective systems for similarity searching, relevance feedback-based searching, classification, clustering and filtering that operate at a semantic-level based on automatic analysis of multimedia content.

Search Fusion. We propose an interactive search fusion approach for video database retrieval that provides users with resources for building queries of video databases sequentially using multiple individual search tools. For example, the search fusion method allows the user to use form a query using CBR, MBR, text-retrieval, and cluster navigation. In some cases, the user can compile together these resources for forming a query. For example, consider a simple case in which the user wants video clips of 'Thomas Jefferson' and issues the query "retrieve video clips that look like given example clip of 'Thomas Jefferson' and have detection of 'face"' This query involves both CBR ("looks like") and MBR ("face detection").

Although we might want to assume that the retrieval system should effectively retrieve matching clips using only the CBR search, in practice, CBR is not sufficient for retrieving matches based on semantics. However, the addition of the MBR search when combined with the CBR can improve retrieval effectiveness. In other cases, the user can build the query interactively based on the intermediate results of the searches. For example, consider user wanting to retrieve "scenes showing gardens". The user can issue a MBR search for "scenes classified as landscape." Then, the user can select some of the best examples of scenes of "gardens" issue a second CBR search for similar scenes and fuse with the result of the "landscape" model. In this chapter, we develop a solution for supporting this type of search fusion problem by providing controls for fusing multiple searches, which involves selecting from normalization and combination methods and aggregation functions. This allows the user the greatest flexibility and power for composing and expressing complex queries of video databases.

Outline. In this chapter, we present the model vector approach for indexing multimedia documents and investigate different strategies for computing model vectors. The outline of the chapter is as follows: in Section 9.1, we examine methods for extracting model vectors from multimedia based on statistical analysis. In Section 9.1.4, we examine different methods for matching model vectors. In Section 9.2, we describe the basic video database search functions for content-based retrieval (CBR), model-based retrieval (MBR), and text-based retrieval (TBR). In Section 9.2.4, we describe the approach for multi-example searches. In Section 9.2.5, we describe the fusion methods including the normalization and combination methods and aggregation functions. in

Section 9.3, we evaluate the model vector approach in video retrieval experiments and evaluate queries of a large video database using search fusion methods.

9.1 Model Vectors

The generation of model vectors involves two stages of processing: (1) *a priori* learning of detectors (as shown in Figure 9.2) and (2) concept detection and score mapping to produce model vectors (as shown in Figure 9.1). The output of the detectors is transformed in a mapping process to produce the model vectors. The model vectors provide an aggregate scoring of the multimedia documents in relation to the concepts of the lexicon. While the model vector method can work with any set of concept detectors, we describe the concept learning process in order to better understand how information about confidence, reliability, relevance and correlation can be used.

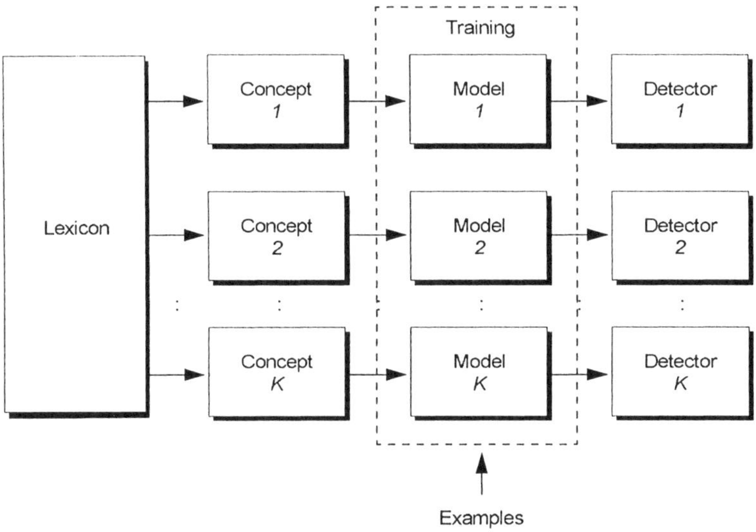

Figure 9.2. Overview of learning of N semantic concept detectors from a lexicon using training examples.

The scores that are output from the detectors are concatenated to produce the model vectors as follows: let $\mathcal{C}$ give a lexicon of K concepts, where l_m is the label of the m-th concept. Let $\mathcal{D}$ give the set of K concept detectors, where d_m is the detector corresponding to the concept l_m in the lexicon $\mathcal{C}$. Let $c_m[j]$ give the confidence of detection of concept l_m by detector d_m in video j, where $c_m \in [0, 1]$, and $c_m = 1$ gives highest value of confidence of detection of l_m. Then,

$$\mathbf{m}[j] = [c_1[j], c_2[j], \ldots, c_K[j]] \tag{9.1}$$

is the k-dimensional model vector for video j.

9.1.1 Concept Learning

The concept learning process uses ground-truth labeled examples as training data for building statistical models for detecting semantic concepts. We construct a set of N binary detectors, each corresponding to a unique concept in a fixed lexicon. For example, considering the size $N = 3$ lexicon: $\mathcal{C} = \{$"face", "indoors", "music"$\}$, the following three binary detectors are constructed: $D_1 =$ "face", $D_2 =$ "indoors", $D_3 =$ "music". The detectors may take any number of forms including Support Vector Machines (SVM), Gaussian Mixture Models (GMM), Hidden Markov Models (HMM), Neural Nets, Bayes Nets, Linear Discriminant Models, and so on. For creating model vectors, we have primarily investigated using SVMs for visual concepts.

Features. The statistical models can use a variety of information related to the multimedia document for modeling the semantic concepts of the lexicon. We used the following descriptors: color histograms (166-bin HSV color space), edge histogram (8 angles and 8 magnitudes, wavelet texture (spatial-frequency energy of 12 bands), color correlogram (8 radii depths in 166-color HSV color space), co-occurrence texture (24 orientations), motion vector histogram (6 bins), visual perception texture (coarseness, contrast and directionality).

Lexicon. Since the detectors are used as a basis for constructing model vectors, the design of the semantic concept lexicon is an important aspect of the model vector approach. Fixed lexicons and classification schemes can be used as the model vector basis. For example, MPEG-7 provides a number of Classification Schemes (CS), such as the Genre CS, which provides a hierarchy of genre categories for classifying multimedia content. Alternatively, more extensive classification systems can be used such as the Library of Congress Thesaurus of Graphical Material (TGM), which provides a set of categories for cataloging photographs and other types of graphical documents. In this chapter, we adopt a lexicon of $N = 7$ concepts that comprises the visual concepts from the TREC-2002 video retrieval concept detection benchmark [Adams et al., 2002] as follows:

$$\begin{aligned}\mathcal{C} = & \{\text{“indoors”}, \text{“outdoors”}, \text{“cityscape”}, \text{“landscape”}, \\ & \text{“face”}, \text{“people”}, \text{“text overlay”}\}. \end{aligned} \tag{9.2}$$

9.1.2 Concept Detection

Once the N detectors are constructed, multimedia documents are analyzed, classified and scored using by each detector. We base the scoring on the confidence of detection of each concept. Additionally, we allow the incorporation of detector correlations in the mapping process (see Sec 9.1.3.0) and detector reliability and concept relevance score in the matching process (see Sec 9.1.4).

Scoring. For each of the N detectors a confidence score $s_n \in [0 \ldots 1]$ is produced for each multimedia document that measures the degree of certainty of detection of concept c_n. We base the confidence score on proximity to the decision boundary for each detector, where a high confidence score is given for documents far from the decision boundary and a low score is given when close to the boundary. A relevance score indicates how relevant the concept is to the multimedia document. For example, if a "face" is only partly depicted or does not comprise a significant part of the multimedia document, then a low relevance score may be determined. Additionally, a reliability score indicates how reliable the detector is for detecting its respective concept. For example, if D_1 was trained using only a few examples of "faces", then a low reliability score may be determined. Alternatively, the reliability score can be based on the classification accuracy on a validation data set.

Detector Score Normalization. In some cases, it is advantageous to normalize the confidence scores to adjust weighting of each of the detectors. We consider a normalization strategy based on Gaussian assumption of each detector scoring as follows: let μ_n and σ_n given the mean and standard deviation of confidence scores $c_n[j]$ over the collection of J documents, respectively, then

$$w_n[j] = \frac{c_n[j] - \mu_n}{\sigma_n}. \tag{9.3}$$

Alternatively, we consider normalization based on range values as follows: let $\min_n$ and $\max_n$ give the minimum and maximum of values of $c_n[j]$ for the collection, respectively, then

$$w_n[j] = \frac{c_n[j] - \min_n}{max_n - \min_n}. \tag{9.4}$$

9.1.3 Model Vector Construction

Once the concepts are detected, the scores are mapped to produce the model vectors.

Mapping. The confidence scores c_n corresponding to the detectors D_n are mapped to produce the model vectors. In general, the mapping involves a transformation of the N confidence scores to a K dimensional vector space, where typically, $K \leq N$. By basing the transformation on techniques such as Principal Components Analysis (PCA) or Fisher Discriminant Analysis (FDA), the mapping can reduce dimensionality. However, in the simplest case, a one-to-one mapping of N confidence scores is achieved by concatenating the confidence scores C_n to build an $K = N$ dimensional vector. For example, considering the three detectors above, D_1 = "face", D_2 = "indoors", D_3 = "music', the $K = 3$ dimensional model vector $\mathbf{m} = \{c_1, c_2, c_3\}$ is produced by concatenating the confidence scores. As a result, the model vector $\mathbf{m}$ provides an aggregate scoring of the document with respect to the concepts "face", "indoors", and "music".

Model Vector Normalization. In addition to normalization based the detector scoring, we consider normalization of the model vectors $\mathbf{m}$ in the K-dimensional model vector space $\mathcal{R}^k$ as follows, where r indicates the norm:

$$m'_k = \frac{m_k}{(\sum_{j=1}^{K} |m_j|^r)^{1/r}}. \tag{9.5}$$

Values of $r > 1$ have the effect of emphasizing dominant detector values or high-energy dimensions.

Figure 9.3 illustrates examples of video key-frames mapped to a model vector space. In this example, we assume a fixed $K = 4$ size lexicon as follows: $\mathcal{C} = \{$"cityscape", "face", "car", "sky"$\}$. Figure 9.3(a) shows the scattering of images along the dimensions $k = 1$ ("cityscape") and $k = 2$ ("face"). The upper-right quadrant identifies those key-frames that have high confidence in both detection of "cityscape" and "face". The scores themselves are not thresholded or used to determine the actual presence of the concepts, although the scores are correlated with detection. Instead, the scores are used to provide a basis for semantic organization of the key-frames. Similarly, Figure 9.3(b) shows the scattering of images along the dimensions $k = 3$ ("car") and $k = 4$ ("sky").

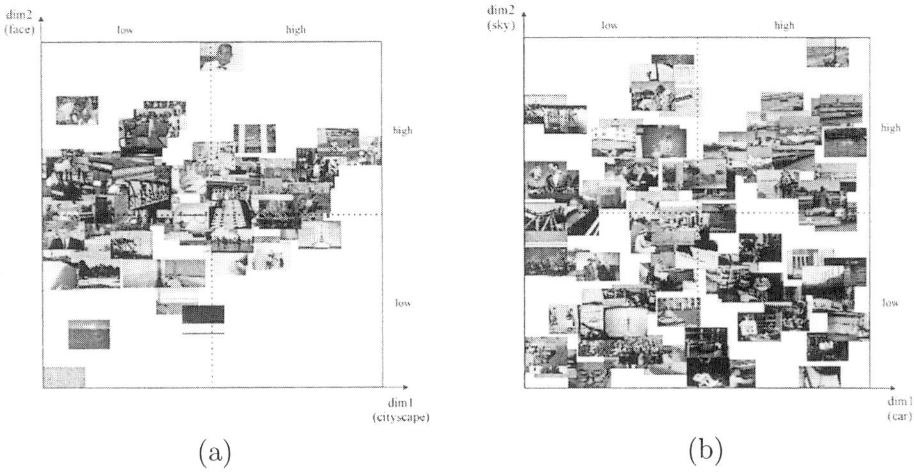

Figure 9.3. Example plots of sub-spaces for model vectors: (a) l_1 = "cityscape" and l_2 = "face", (b) l_1 = "car" and l_2 = "sky".

9.1.4 Model Vector-Based Retrieval

The model vector-based retrieval approach for video is illustrated in Figure 9.4. The system works by classifying and scoring each video (or scene, shot, key-frame, etc.), using a set of concept detectors. The detectors correspond to a fixed set of concepts in a lexicon. For example, the lexical entities may refer to concepts such as "face", "indoors", "sky", "people", and so forth. The resulting detection scores are then mapped to a fixed multi-dimensional vector space. Since the target videos are similarly mapped to that space, a query of the video database can be carried out by searching the multi-dimensional space and identifying nearest neighbors.

While the model vector-based retrieval approach provides a powerful approach for automatically indexing video content based on semantics, some challenges need to be overcome, such as the case when underlying concept detectors are of varying quality. Currently, the state-of-the-art allows reasonably good concept detection performance for some simple concepts, such as "face", "people", "indoors", "outdoors", etc. (as demonstrated in the NIST TREC-2002 video retrieval benchmark, see [Adams et al., 2002]. However, challenges remain for developing large numbers of good performing video concept detectors due to the paucity of training data and the complexity of concepts. However, we show that model vector-based retrieval can be tuned based on the quality of the detectors by incorporating detector validity scores in the model vector distance metric. This allows the variable weighting of the detection scores based on the relative qualities of the underlying detectors. For example, a highly accurate "face" detector can be given higher weighting than a less accurate "cityscape" detector.

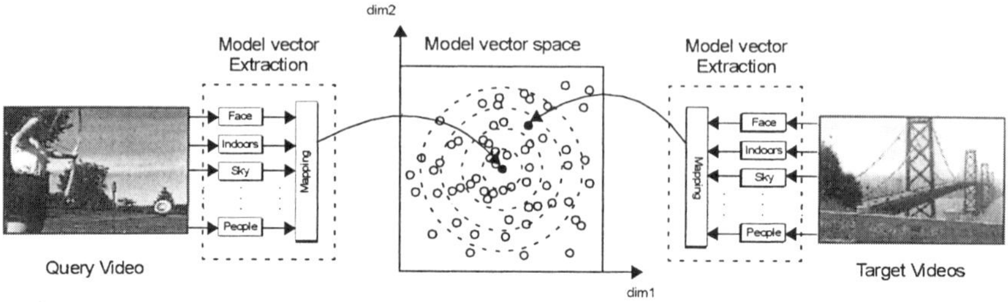

Figure 9.4. Model vector-based retrieval of video allows semantics-based similarity searching by mapping content to multi-dimensional vector representation that captures the confidence of detection of fixed set of concepts from a lexicon.

Model vector-based retrieval is enabled by indexing the vectors in a multi-dimensional metric space. The distance between model vectors indicates the similarity of the videos with respect to the detector confidence scores. The Euclidean distance metric (Eq 9.6) can be used to measure the similarity of the model vectors. However, since all dimensions are equally weighted using Euclidean distance, equal emphasis is given to both good and poorly performing detectors. Alternatively, we explore a validity weighted distance metric (Eq 9.7), that weighs the model vector dimensions using an indicator score of validity of each underlying detector.

9.1.5 Metrics

- **Model vector distance**: the dissimilarity of model vectors is measured using a Euclidean distance metric as follows: let $\mathbf{m}^q$ and $\mathbf{m}^t$ be K-dimensional query and target model vectors, respectively, then distance D is given by

$$D = \frac{1}{K} \sum_{k=1}^{K} (m_k^q - m_k^t)^2. \tag{9.6}$$

- **Validity weighted distance**: given a validity indicator score v_k for each detector d_k, the validity-weighted model vector distance D_v is given by:

$$D_v = \frac{1}{K} \sum_{k=1}^{K} v_k (m_k^q - m_k^t)^2. \tag{9.7}$$

We study the case in which the validity scores v_k are based on the number of training examples used to build each underlying detector. Potentially, other criteria can be used such as the detection accuracy on a test collection. While the number of training examples does not accurately capture the specific characteristics of the models used for detection, in general, poorly trained detectors are likely to give poor performance, whereas, detectors trained with sufficient examples have potential to give good performance. Let n_k give the number of examples used to train detector d_k. Then, the validity score for detector d_k is given as follows:

$$v_k = \frac{n_k}{\sum_{i=1}^{K} n_i} \tag{9.8}$$

9.1.6 Lexicon

In order to evaluate the performance of validity weighting for model vector-based retrieval, we explore two lexicons of size $K = 7$ and size $K = 29$, respectively. Table 9.1 gives the first lexicon $\mathcal{C}_1$, where n_k gives the number of training examples for concept l_k, and v_k gives the validity score using Eq 9.8. Table 9.2 gives the second lexicon $\mathcal{C}_2$.

Concept (l_k)	n_k	v_k
Cityscape	106	0.0159
Face	1108	0.1666
Indoors	1351	0.2031
Landscape	82	0.0123
Outdoors	1668	0.2508
People	1836	0.2760
Text Overlay	500	0.0752

Table 9.1. Lexicon $\mathcal{C}_1$: seven visual concepts.

9.2 Video Search Fusion

Video database systems typically provide a number of facilities for searching based on feature descriptors, models, concept detectors, clusters, speech transcript, associated text, and so on. We classify these techniques broadly into three basic search functions: content-based retrieval (CBR), model-based retrieval (MBR), and text-based retrieval (TBR).

Concept (l_k)	n_k	v_k
Airplane	100	0.0126
Animal	55	0.0070
Beach	24	0.0030
Boat	79	0.0100
Bridge	9	0.0011
Building	918	0.1161
Car	237	0.0300
Cartoon	194	0.0245
Cloud	247	0.0312
Desert	10	0.0013
Factory Setting	317	0.0401
Flag	1	0.0001
Flower	144	0.0182
Graphics	202	0.0255
Greenery	954	0.1206
Horse	28	0.0035
House Setting	279	0.0353
Land	230	0.0291
Man-Made Setting	5	0.0006
Mountain	205	0.0259
Office Setting	222	0.0281
Road	387	0.0489
Sky	1778	0.2248
Smoke	67	0.0085
Tractor	59	0.0075
Train	155	0.0196
Transportation	63	0.0080
Tree	572	0.0723
Water Body	369	0.0466

Table 9.2. Lexicon $\mathcal{C}_2$: 29 visual concepts with varying support from training examples.

9.2.1 Content-Based Retrieval (CBR)

Content-based retrieval (CBR) is an important technique for indexing video content. While CBR is not a robust surrogate for indexing based on semantics of image content (scenes, objects, events, and so forth), CBR has an important role in searching. For one, CBR complements traditional querying by allowing "looks like" searches, which can be useful for pruning or re-ordering result sets based on visual appearance. Since CBR requires example images or video clips, CBR be only typically be used to initiate the query when the user provides the example(s), or within an interactive query in which the user selects from the retrieved

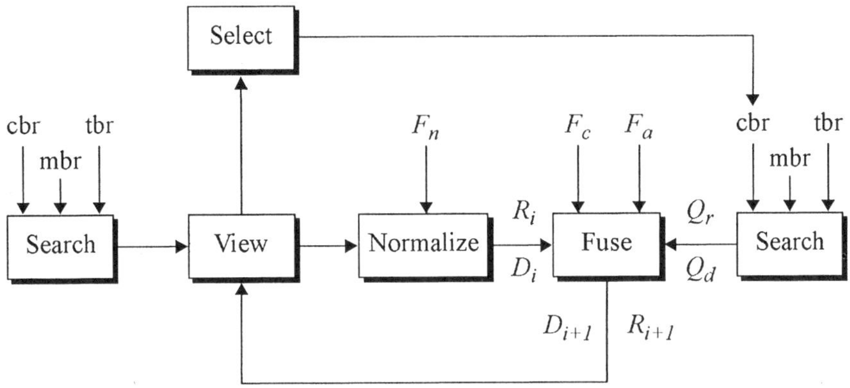

Figure 9.5. Overview of search fusion framework for video databases.

results to search the database again, as shown in Figure 9.5 [Smith et al., 2002]. CBR produces a ranked, scored results list in which the similarity is based on distance in feature space.

9.2.2 Model-Based Retrieval (MBR)

Model-based retrieval (MBR) allows the user to retrieve matches based on the concept labels produced by statistical models, concept detectors, or other types of classifiers. Since both supervised and unsupervised techniques are used, MBR applies for labels assigned from a lexicon with some confidence as well as clusters in which the labels do not necessarily have a specific meaning. In MBR, the user enters the query by typing label text, or the user selects from an inverted list of label terms. Since a confidence score is associated with each automatically assigned label, MBR ranks the matches using a distance $\mathcal{D}$ derived from confidence $\mathcal{C}$ using $\mathcal{D} = 1 - \mathcal{C}$. As shown in Figure 9.5, MBR applies equally well in manual and interactive searches, since it can be used to initiate query, or can be applied at intermediate stage to fuse with prior search results.

9.2.3 Text-Based Retrieval (TBR)

Text-based retrieval (TBR) applies to various forms of textual data associated with video, which includes speech recognition results, transcript, closed captions, extracted embedded text, and metadata. In some cases, TBR is scored and results are ranked. For example, similarity of words is often used to allow fuzzy matching. In other cases, crisp matching of search text with indexed text, the matches are retrieved but not scored and ranked. As in the case for MBR, TBR applies equally well in manual and interactive searches as shown in Figure 9.5.

9.2.4 Multi-Example Search

Multi-example search allows the user to provide or select multiple examples from a results list and issue a query that is executed as a sequence of independent searches using each of the selected items. The user can also select a descriptor for matching and an aggregation function for combining and re-scoring the results from the multiple searches. Consider for each search k of K independent searches the scored result $S_k(n)$ for each item n, then the final scored result $Q_d(n)$ for each item with id $= n$ is obtained using a choice of the following fusion functions:

- Average: Provides "and" semantics. This can be useful in searches such as "retrieve matches similar to item "A" *and* item "B".

$$Q_d(n) = \frac{1}{K} \sum_k (S_k(n)) \tag{9.9}$$

- Minimum: Provides "or" semantics. This can be useful in searches such as "retrieve items that are similar to item "A" or item "B".

$$Q_d(n) = \min_k (S_k(n)) \tag{9.10}$$

- Maximum:

$$Q_d(n) = \max_k (S_k(n)) \tag{9.11}$$

- Sum: Provides "and" semantics.

$$Q_d(n) = \sum_k (S_k(n)) \tag{9.12}$$

- Product: Provides "and" semantics and better favors those items that have low scoring matches compared to "average".

$$Q_d(n) = \prod_k (S_k(n)) \tag{9.13}$$

9.2.5 Search Fusion

The search fusion methods provide a way for normalizing, combining and re-scoring results lists using aggregation functions through successive operations. Normalization is important for fusion since individual searches may use different scoring mechanisms for retrieving and ranking matches. On the other hand, the combination methods determine how multiple results lists are combined from the set point of view. The aggregation functions determine how the scores from multiple results lists are aggregated to re-score the combined results.

Normalization. The normalization methods provide a user with controls to manipulate the scores of a results list. Given a score $D_k(n)$ for each item with id $= n$ in results set k, the normalization methods produce the score $D_{i+1}(n) = \mathcal{F}_z(D_i(n))$ for each item n as follows:

- Invert: Re-ranks the results list from bottom to top. Provides "not" semantics. This can be useful for searches such as "retrieve matches that are *not* cityscapes."

$$D_{i+1}(n) = 1 - D_i(n) \tag{9.14}$$

- Studentize: Normalizes the scores around the mean and standard deviation. This can be useful before combining results lists.

$$D_{i+1}(n) = \frac{D_i(n) - \mu_i}{\sigma_i}, \tag{9.15}$$

  where μ_i gives the mean and σ_i the standard deviation, respectively, over the scores $D_i(n)$ for results list i.

- Range normalize: Normalizes the scores within the range $0 \ldots 1$.

$$D_{i+1}(n) = \frac{D_i(n) - \min(D_i(n))}{\max(D_i(n)) - \min(D_i(n))} \tag{9.16}$$

Combination Methods. Consider results list R_k for query k and results list Q_r for current user-issued search, then the combination function $R_{i+1} = \mathcal{F}_c(R_i, Q_r)$ combines the results lists by performing set operations on list membership. We explored the following combination methods:

- Intersection: retains only those items present in both results lists.

$$R_{i+1} = R_i \cap Q_r \tag{9.17}$$

- Union: retains items present in either results list.

$$R_{i+1} = R_i \cup Q_r \tag{9.18}$$

Aggregation Functions. Consider scored results list R_k for query k, where $D_k(n)$ gives the score of item with id $= n$ and $Q_d(n)$ the scored result for each item n in the current user-issued search, then the aggregation function re-scores the items using the function $D_{i+1}(n) = \mathcal{F}_a(D_i(n), Q_d(n))$. We explored the following aggregation functions:

- Average: takes the average of scores of prior results list and current user-search. Provides "and" semantics. This can be useful for searches such as "retrieve items that are indoors *and* contain faces."

$$D_{i+1}(n) = \frac{1}{2}(D_i(n) + Q_d(n)) \tag{9.19}$$

- Minimum: retains lowest score from prior results list and current user-issued search. Provides "or" semantics. This can be useful in searches such as "retrieve items that are outdoors *or* have music."

$$D_{i+1}(n) = \min(D_i(n), Q_d(n)) \tag{9.20}$$

- Maximum: retains highest score from prior results list and current user-issued search.

$$D_{i+1}(n) = \max(D_i(n), Q_d(n)) \tag{9.21}$$

- Sum: takes the sum of scores of prior results list and current user-search. Provides "and" semantics.

$$D_{i+1}(n) = D_i(n) + Q_d(n) \tag{9.22}$$

- Product: takes the product of scores of prior results list and current user-search. Provides "and" semantics and better favors those matches that have low scores compared to "average".

$$D_{i+1}(n) = D_i(n) \times Q_d(n) \tag{9.23}$$

- A: retains scores from prior results list. This can be useful in conjunction with "intersection" to prune a results list, as in searches such as "retrieve matches of beach scenes but retain only those showing faces."

$$D_{i+1}(n) = D_i(n) \tag{9.24}$$

- B: retains scores from current user-issued search. This can be useful in searches similar to those above but exchanges the arguments.

$$D_{i+1}(n) = Q_d(n) \tag{9.25}$$

9.2.6 Manual *vs.* Interactive Fusion

The video retrieval system allows the CBR, MBR, TBR, and other methods of searching and browsing. The following pedagogical queries show how the facilities of the video retrieval system can be used for querying for "beach" scenes in the case of manual and interactive searches.

Manual Search Fusion. Consider user looking for items showing a beach scene. For manual search operations the user can issue a query with the following sequence of searches, which corresponds to the following query statement ((((beach color) AND (sky model)) AND water model) OR (beach text)):

1 Search for images with color similar to example query images of beach scenes,

2 Combine results with model = "sky" using "average" aggregation function (Eq 9.20),

3 Combine with model = "water" using "product" aggregation function (Eq 9.23),

4 Combine with text = "beach" using "minimum" aggregation function (Eq 9.20).

Interactive Search Fusion. On the other hand, for interactive search operations the user can issue the following sequence of searches in which the user views the results at each stage, which corresponds to the following query statement: ((beach text) AND (beach color) AND (sky model) AND (water model)):

1 Search for text = "beach",

2 Select results that best depict beach scenes. Search based similar color. Combine with previous results using "product" aggregation function (Eq 9.23),

3 Combine with model = "sky" using "average" aggregation function (Eq 9.20),

4 Combine with model = "water" using "product" aggregation function (Eq 9.23).

9.3 Experiments

We evaluate the model vector approach in terms of retrieval effectiveness of querying of a video retrieval testbed as follows:

9.3.1 Experimental Testbed

We use the video retrieval testbed from the NIST TREC-2002 video retrieval benchmark[1]. We developed the $N = 7$ visual semantic concept detectors using a training corpus of 9,495 video clips [Adams et al., 2002]. In the experiments below, we evaluate the retrieval effectiveness using a separate retrieval corpus of 2,249 video clips.

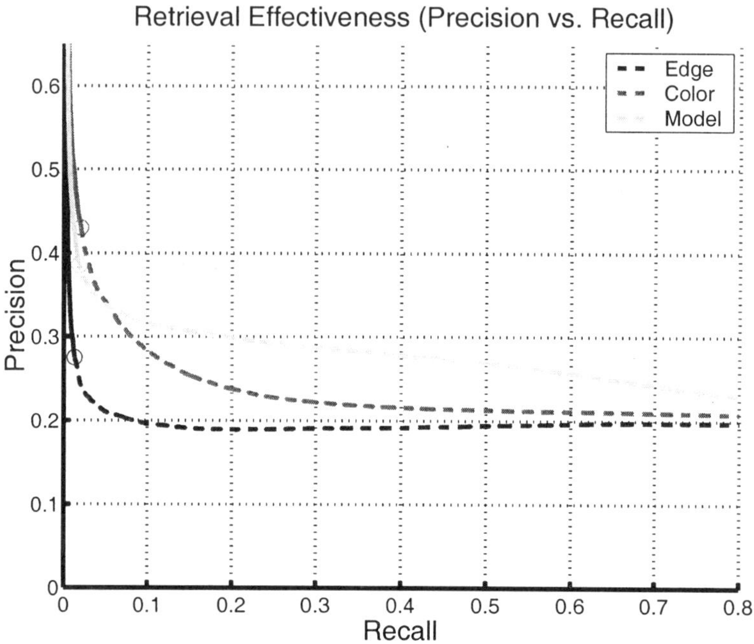

Figure 9.6. Retrieval effectiveness for query topic = "sky" using similarity search using color histograms ("color") histograms, edge histograms ("edge") and model vectors ("model').

9.3.2 Experiment 1: MBR *vs* CBR

The first experiment compares model vector to content-based retrieval. The queries are conducted for four topics, where the ground-truth has been labeled manually as follows, where number of matches is indicated in parentheses: "building" (230), "car" (114), "sky" (427), "trees" (142). In the retrieval experiments, for each topic, each relevant video clip is used in turn to query the database. Overall, the following three descriptors are compared: $K = 7$ dimensional model vectors, 166-dimensional

[1] http://www-nlpir.nist.gov/projects/t2002v/t2002v.html (TREC-2002 Video Track)

color histograms and 64-dimensional edge histograms. The average retrieval effectiveness is computed over the searches using each method for each topic. The results are shown in Figure 9.6 (retrieval effectiveness plot for "sky" topic") and Table 9.3 (mean average precision for all topics).

Figure 9.6 plots the average precision *vs.* average recall over the sequence of searches for the "sky" topic. The 7-dimensional model vectors ("model") provide significantly higher retrieval effectiveness compared to 166-dimensional color histograms ("color") and 64-dimensional edge histograms ("edge"). The R-value, which measures the precision at the number of matches, which is 427 for "sky" topic, is 25% higher for model vectors compared to best of color and edge. The three-point score, which gives average precision at recall of 0.2, 0.5 and 0.8, is 20% higher for model vectors.

Topic	# Searches	Color	Edge	Model	Gain
Building	230	0.14	0.13	0.14	1.4%
Car	114	0.13	0.10	0.17	24.1%
Sky	427	0.25	0.20	0.29	15.1%
Trees	142	0.11	0.11	0.15	36.1%

Table 9.3. Mean average precision computed over number of searches for four topics ("building", "car", "sky", "trees") using three methods: color histograms ("color"), edge histograms ("edge") and model vectors ("model').

Table 9.3 gives the mean average precision (MAP), which corresponds to the weighted area under the precision *vs.* recall curve, for the four topics comparing the three search methods. The smallest margin of improvement is shown for the "building" topic, which has MAP that is 1.4% higher for model vectors than best of color and edge. Other topics, "car", "sky" and "building" show significant improvement in MAP score, giving between 15.1% and 36.1% increase in MAP score.

9.3.3 Experiment 2: Model Vector Norms

Figure 9.7 plots the precision *vs.* recall over the sequence of searches for the "sky" topic using different model vector normalization methods as follows: "Reg" corresponds to the as-is vector of confidence scores, "Student" corresponds to studentization normalization using Eq 9.3, "Range" corresponds to range normalization using Eq 9.4 and "L2" corresponds to L2 normalization using Eq 9.5. The highest average retrieval effectiveness is given by range normalization.

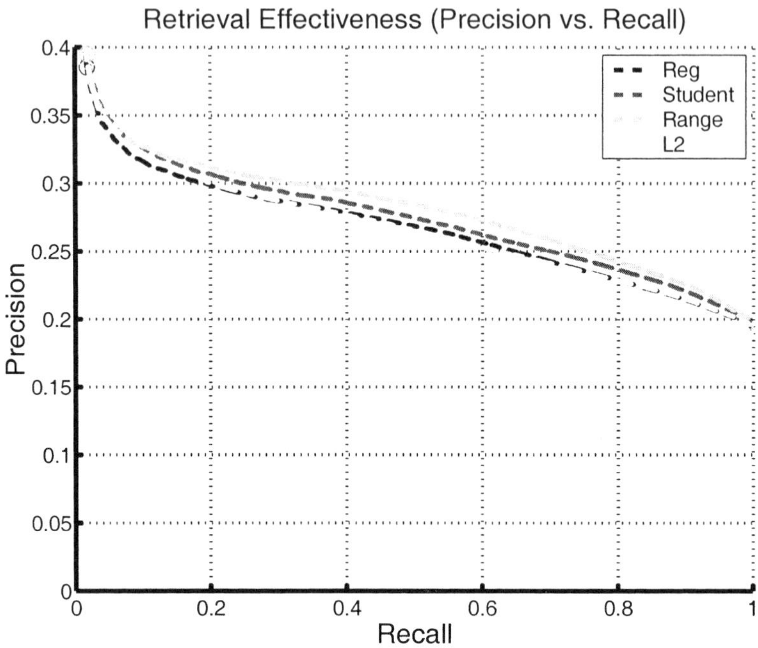

Figure 9.7. Retrieval effectiveness for query topic = "sky" using different model vector normalization methods.

9.3.4 Experiment 3: Validity Weighting

The first experiment evaluates the relative retrieval effectiveness of model vector-based retrieval using the two different lexicons using validity weighting and without validity weighting. Figure 9.8 plots the mean average precision (MAP) *vs.* cut-off over the sequence of searches for the "sky" query topic. The results show that in cases of both lexicons, the validity weighting improves the MAP.

9.3.5 Experiment 4: MBR *vs.* CBR

The second experiment compares the model vector-based retrieval performance to color-based retrieval (using 166-dimensional color histograms) and random retrieval as a base-line.

Figure 9.9 plots the MAP for "sky". The results show that in the case of lexicon $\mathcal{C}_2$, validity weighting gives the best performance compared to unweighted model vectors, color-based retrieval, and random baseline. Figure 9.10 plots the MAP for the "building" query topic. The results show that in the case of lexicon $\mathcal{C}_1$, model vector-based retrieval using validity weighting gives better retrieval performance than without va-

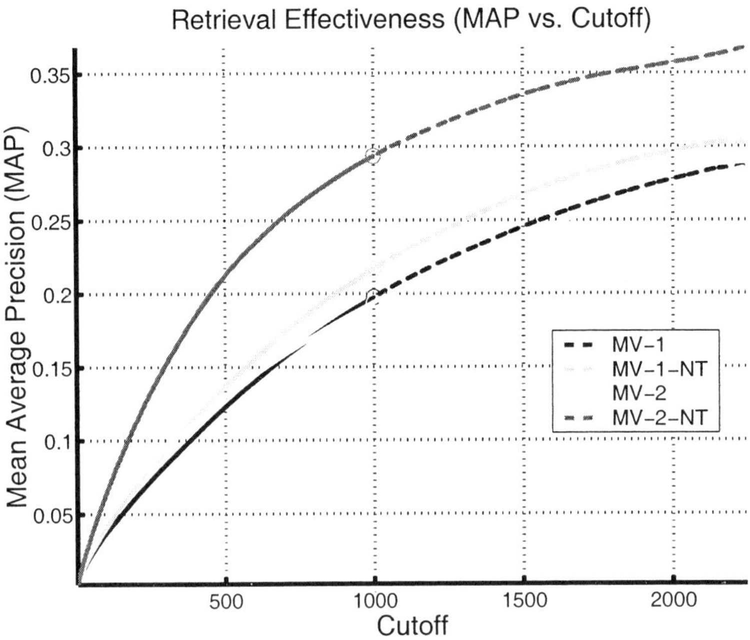

Figure 9.8. Retrieval effectiveness for query topic = "sky" using similarity searching of model vectors: MV-1 = model vectors and MV-1-NT = validity weighted matching for $\mathcal{C}_1$, MV-2 = model vectors and MV-2-NT = validity weighted matching for $\mathcal{C}_2$.

lidity weighting. Both give higher MAP than color-based retrieval and random baseline.

9.3.6 Manual Search

We explore queries that fuse search results for CBR and MBR searches. In general, we adopt the following strategy for manually composing the queries:

1 Select example content and issue a CBR search;

2 Select model and issue MBR search; models;

3 Select fusion methods and fuse CBR and MBR results lists.

For example, the following sequence of operations was executed for query of video clips of "Garden scenes":

1 Select example image of "garden scene" and perform CBR search using color correlograms;

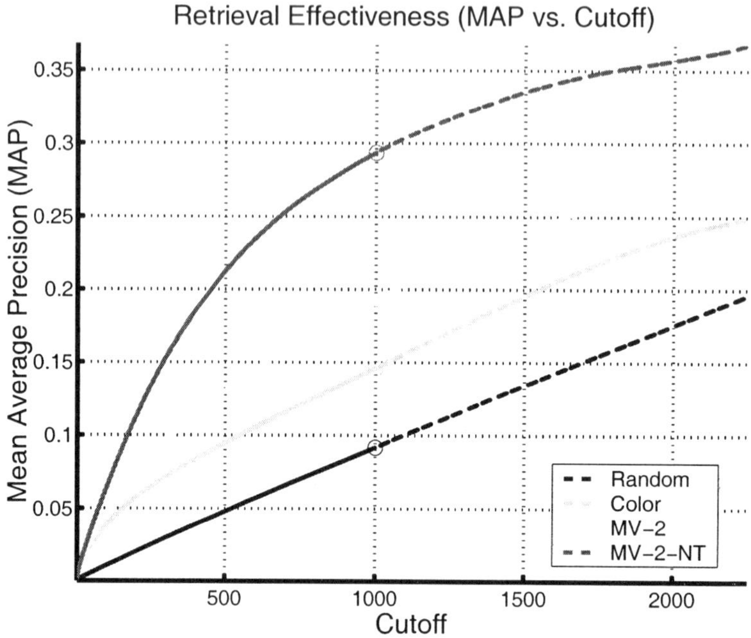

Figure 9.9. Retrieval effectiveness for query topic = "sky" using random baseline ("random"), color histograms ("color"), model vectors ("MV-2"), and validity-weighted model vectors ("MV-2-NT").

2 Issue MBR using "Landscape" model and fuse with CBR result using "intersection" combining method (Eq 9.17) and "product" aggregation function (Eq 9.23).

The results of the CBR search return 16 matches in the top 100 video clips in the results list. After issuing the MBR search and combining with the "Landscape" results, 28 matches are retrieved in the top 100 including 16 in the top 26. Figure 9.11 shows the results of the fused searches using CBR and MBR for the "garden" query.

9.3.7 Interactive Search

We explored interactive search using CBR and MBR to examine whether search performance increases with interactivity. Consider the expanded search for gardens, which uses the initial search steps above, then an additional interactive step is applied as follows:

1 Select top 10 example results from manual search;

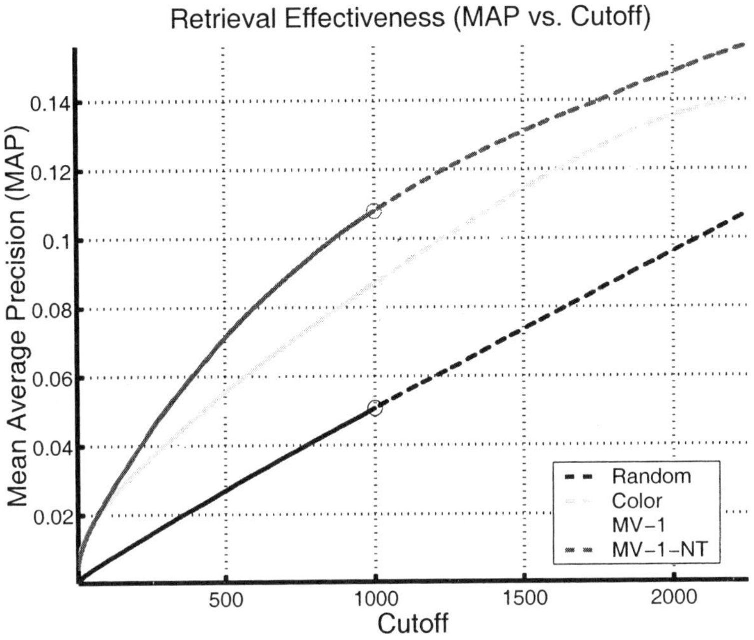

Figure 9.10. Retrieval effectiveness for query topic = "building" using model vectors ("MV-1") and validity-weighted model vectors ("MV-1-NT").

2 Perform multi-example CBR using color correlograms and combine with manual search results using "intersection" combining method (Eq 9.17) and "average" aggregation function (Eq 9.19).

The results give 32 matches in the top 100 and 16 matches in the top 21, which shows that the additional interactivity further improves the retrieval performance.

9.4 Summary

We presented a novel approach for video database retrieval based on interactive search fusion. The system supports basic search methods for content-based, model-based, and text-based retrieval. The system provides additional facilities for user to fuse search results using the different basic search methods and control the normalization of scores, combination of results lists, and aggregation of scores of multiple results lists. We explore the application to both manual and interactive queries. We demonstrated the results applied to a large video database, which show improvements in retrieval performance using the proposed search fusion methods.

Figure 9.11. Results of query for "gardens" that fuses CBR and MBR.

References

Adams, W., Amir, A., Dorai, C., Ghosal, S., Iyengar, G., Jaimes, A., Lang, C., C.-Y.Lin, Naphade, M. R., Natsev, A., Neti, C., Nock, H. J., Permuter, H., Singh, R., Smith, J. R., Srinivasan, S., Tseng, B. L., Varadaraju, A., and Zhang, D. (2002). IBM Research TREC-2002 Video Retrieval System. In *Proc. Text Retrieval Conference (TREC)*, Gaithersburg, MD.

Naphade, M., Basu, S., Smith, J. R., Lin, C., and Tseng, B. (2002a). Modeling semantic concepts to support query by keywords in video. In *IEEE Proc. Int. Conf. Image Processing (ICIP)*, Rochester, NY.

Naphade, M. R., Ling, C.-Y., Smith, J. R., Tseng, B., and Basu, S. (2002b). Learning to annotate video databases. In *IS&T/SPIE Sym-*

posium on Electronic Imaging: Science and Technology - Storage & Retrieval for Image and Video Databases, San Jose, CA.

Naphade, M. R. and Smith, J. R. (2003). The role of classifiers in video indexing. In *IS&T/SPIE Symposium on Electronic Imaging: Science and Technology - Storage & Retrieval for Image and Video Databases*, San Jose, CA.

Smith, J. R., Basu, S., Lin, C., Naphade, M., and Tseng, B. (2002). Interactive content-based retrieval of video. In *IEEE Proc. Int. Conf. Image Processing (ICIP)*, Rochester, NY.

Smith, J. R., Jaimes, A., Lin, C.-Y., Naphade, M., Natsev, A., and Tseng, B. (2003a). Interactive search fusion methods for video database retrieval. In *IEEE Proc. Int. Conf. Image Processing (ICIP)*, Barcelona, Spain. To appear.

Smith, J. R., Naphade, M., and Natsev, A. (2003b). Multimedia semantic indexing using model vectors. In *IEEE Intl. Conf. on Multimedia and Expo (ICME)*, Baltimore, MD. To appear.

Chapter 10

UNSUPERVISED MINING OF STATISTICAL TEMPORAL STRUCTURES IN VIDEO

Lexing Xie, Shih-Fu Chang
Department of Electrical Engineering, Columbia University, New York, NY
{xlx,sfchang}@ee.columbia.edu

Ajay Divakaran and Huifang Sun
Mitsubishi Electric Research Labs, Cambridge, MA
{ajayd,hsun}@merl.com

Abstract In this chapter we present algorithms for unsupervised mining of structures in video using multi-scale statistical models. Video structure are repetitive segments in a video stream with consistent statistical characteristics. Such structures can often be interpreted in relation to distinctive semantics, particularly in structured domains like sports. While much work in the literature explores the link between the observations and the semantics using supervised learning, we propose unsupervised structure mining algorithms that aim at alleviating the burden of labelling and training, as well as providing a scalable solution for generalizing video indexing techniques to heterogeneous content collections such as surveillance and consumer video. Existing unsupervised video structuring work primarily uses clustering techniques, while the rich statistical characteristics in the temporal dimension at different granularities remain unexplored. Automatically identifying structures from an unknown domain poses significant challenges when domain knowledge is not explicitly present to assist algorithm design, model selection, and feature selection. In this work we model multi-level statistical structures with hierarchical hidden Markov models based on a multi-level Markov dependency assumption. The parameters of the model are efficiently estimated using the EM algorithm. We have also developed a model structure learning algorithm that uses stochastic sampling techniques to find the optimal model structure, and a feature selection algorithm that automatically finds compact relevant feature sets using hybrid wrapper-filter methods. When tested on sports videos, the unsupervised learning scheme achieves very promising results: (1) The automatically selected feature set for soccer and baseball videos matches sets that are manually selected with domain knowledge. (2) The system automatically discovers high-level structures that match the semantic events in the video. (3) The system achieves better accuracy in detecting semantic events in unlabelled soccer videos than a competing supervised approach designed and trained with domain knowledge.

Keywords: Multimedia mining, structure discovery, unsupervised learning, video indexing, statistical learning, model selection, automatic feature selection, hierarchical hidden Markov model (HHMM), hidden Markov model (HMM), Markov chain Monte-Carlo (MCMC), dynamic Bayesian network (DBN), Bayesian Information Criteria (BIC), maximum likelihood (ML), expectation maximization (EM).

Introduction

In this chapter, we present algorithms for jointly discovering statistical structures, using the appropriate model complexity, and finding informative low-level features from video in an unsupervised setting. These techniques address the challenges of automatically mining salient structures and patterns that exist in video streams from many practical domains. Effective solutions to video indexing require detection and recognition of *structure* and *event* in the video, where *structure* represents the syntactic level composition of the video content, and *event* represents the occurrences of certain semantic concepts. In specific domains, high-level syntactic structures may correspond well to distinctive semantic events. Our focus is on temporal structures, which is defined as the repetitive segments in a time sequence that possess consistent deterministic or statistical characteristics. This definition is general to various domains, and it is applicable at multiple levels of abstraction. At the lowest level for example, structure can be the frequent triples of symbols in a DNA sequence, or the repeating color schemes in a video; at the mid-level, the seasonal trends in web traffics, or the canonical camera movements in films; and at a higher level, the genetic functional regions in DNA sequences, or the game-specific temporal state transitions in sports video. Automatic detection of structures will help locate semantic events from low-level observations, and facilitate summarization and navigation of the content.

The structure discovery problem. The problem of identifying structure consists of two parts: finding a description of the structure (a.k.a *the model*), and locating segments that match the description. There are many successful cases where these two tasks are performed in separate steps. The former is usually referred to as *training*, while the latter, *classification* or *segmentation*. Among various possible models, hidden Markov model (HMM) [Rabiner, 1989] is a discrete state-space stochastic model with efficient learning algorithms that work well for temporally correlated data streams. HMM has been successfully applied to many different domains such as speech recognition, handwriting recognition, motion analysis, or genome sequence analysis. For video analysis in particular, different genres in TV programs have been distinguished with HMMs trained for each genre in [Wang et al., 2000], and the high-level structure of soccer games (e.g. play versus break) was also delineated with a pool of HMMs trained for each category in [Xie et al., 2002b].

The structure detection methods above fall in the conventional category of supervised learning - the algorithm designers manually identify important structures, collect labelled data for training, and apply supervised learning tools to learn the classifiers. This methodology works for domain-specific problems at a small scale, yet it cannot be readily extended to diverse new domains at a large scale. In this chapter, we propose a new paradigm that uses fully unsupervised statistical techniques and aims at automatic discovery of salient structures and simultaneously recognizing such structures in unlabelled data without prior domain knowledge. Domain knowledge, if available, can be used to assign semantic meanings to the discovered structures in a post-processing stage. Although unsupervised clustering techniques date back to several decades ago [Jain et al., 1999], most of the data sets were treated as independent samples, while the temporal correlation between samples were largely unexplored. Classical time series analysis techniques has been widely used in many domains such as financial data and web stat analysis [Iyengar et al., 1999], where the problem of identifying seasonality reduces to the problem of parameter estimation with a known order ARMA model, where the order is determined with prior statistical tests. Yet this model does not readily adapt to domains with dynamically changing model characteristics, as is often the case with video. New statistical methods such as Monte Carlo sampling have also appeared in genome sequence analysis [Lawrence et al., 1993], where unknown short motifs were recovered by finding the best alignment among all protein sequences using Gibbs sampling techniques on a multinomial model, yet independence among amino acids in adjacent positions is still assumed. Only a few instances have been explored for video. Clustering techniques are used on the key frames of shots [Yeung and Yeo, 1996] or the principal components of color histogram of image frames [Sahouria and Zakhor, 1999], to detect the story units or scenes in the video, yet the temporal dependency of video has not been fully explored. In the independent work in [Clarkson and Pentland, 1999; Naphade and Huang, 2002], several left-to-right HMMs were concatenated to identify temporally evolving events in ambulatory videos captured by wearable devices or in films. In the former, the resulting clusters correspond to different locations such as the lab or a restaurant; while in the latter, some of the clusters correspond to recurrent events such as explosion.

Unsupervised learning of statistical structures also involve automatic selection of features extracted from the audio-visual stream. The computational front end in many real-world scenarios extracts a large pool of observations (i.e. features) from the stream, and in the absence of expert knowledge, picking a subset of relevant and compact features becomes a bottleneck. Automatically identifying informative features, if done, will improve both the learning quality and computation efficiency. Prior work in feature selection for supervised learning mainly divides into *filter* and *wrapper* methods according to whether or not the classifier is in-the-loop [Koller and Sahami, 1996]. Many directions of existing work address the supervised learning scenario, and evaluate the fitness of a feature with regard to its information gain against training labels (*filter*) or the quality of learned classifiers (*wrapper*). For unsupervised learning on spatial data (i.e. assuming temporally adjacent samples are independent), [Xing and Karp, 2001] developed a method that iterated between cluster assignment and filter/wrapper methods under the scenario that the number of clusters is known; [Dy and Brodley, 2000] used scatter separability and maximum likelihood (ML) criteria to evaluate fitness of features. To the best of our knowledge, no prior work has been reported for our particular problem of interest: unsupervised learning on temporally dependent sequences with unknown cluster size.

Characteristics of Video Structure. Our main attention in this chapter is on the particular domain of video (i.e. audio-visual streams), where the structures have the following properties from our observations: (1) Video structure is in a discrete state-space, since we humans understand video in terms of concepts, and we assume there exist a small set of concepts in a given domain; (2) The features, i.e. observations from data,s are stochastic, as segments of video seldom have exactly the same raw features even if they are conceptually similar; (3) The sequence is highly correlated in time, since the videos are sampled at a rate much higher than that of the changes in the scene.

In this chapter, several terms are used without explicit distinction in referring to the *video structures*, despite the differences in their original meanings: by *structure* we emphasize the statistical characteristics in raw features. Given specific domains, such statistic structures often correspond to *events*, which represent occurrences of objects, or changes of the objects or the current scene.

In particular, we will focus on *dense* structures in this chapter. By *dense* we refer to the cases where constituent structures can be modelled as a common parametric class, and representing their alternation would be sufficient for describing the whole data stream. In this case, there is no need for an explicit *background* class, which may or may not be of the same parametric form, to delineate *sparse* events from the majority of the background.

Based on the observations above, we model stochastic observations in a temporally correlated discrete state space, and adopt a few weak assumptions to facilitate efficient computation. We assume that within each *event*, states are discrete and Markov, and observations are associated with states under a fixed parametric form, usually Gaussian. Such assumptions are justified based on the satisfactory results from previous work using supervised HMM to classify video events or genre [Wang et al., 2000; Xie et al., 2002b]. We also model the transitions of events as a Markov chain at a higher level; this simplification will enable efficient computation at a minor cost of modelling power.

Our approach. In this chapter, we model the temporal dependencies in video and the generic structure of events in a unified statistical framework. Adopting the multi-level Markov dependency assumptions above for computational efficiency in modelling temporally structures, we model the recurring events in each video as HMMs, and the higher-level transitions between these events as another level of Markov chain. This hierarchy of HMMs forms a Hierarchical Hidden Markov Model (HHMM); its hidden state inference and parameter estimation can be efficiently learned in $O(T)$ using the expectation-maximization (EM) algorithm. This framework is general in that it is scalable to events of different complexity; yet it is also flexible in that prior domain knowledge can be incorporated in terms of state connectivity, number of levels of Markov chains, and the time scale of the states.

We have also developed algorithms to address model selection and feature selection problems that are necessary in unsupervised settings when domain knowledge is not used. Bayesian learning techniques are used to learn the model complexity automatically, where the search over model space is done with reverse-jump Markov chain Monte Carlo, and Bayesian Information Criteria (BIC) is used as model posterior. We use an iterative filter-wrapper methods for feature selection, where the wrapper step partitions the feature pool into consistent groups that agree with each other with a mutual information gain criterion, and the filter step eliminates redundant dimensions in each group by finding an approximate Markov blanket, and finally the resulting groups are ranked with modified BIC with respect to their *a posteriori* fitness. The approach is elegant in that maximum likelihood parameter estimation, model and feature selection, structure decoding, and content segmentation are done in a single unified process.

Evaluation on real video data shows very promising results. We tested the algorithm on multiple sports videos, and our unsupervised approach automatically discovers the high-level structures, namely, *plays* and *breaks* in soccer and baseball. The feature selection method also automatically discovered a compact relevant feature set, which matched the features manually selected using domain knowledge. The new unsupervised method discovers the statistical descriptions of high-level structure from unlabelled video, yet it achieves even slightly higher accuracy (75.7% and 75.2% for unsupervised vs. 75.0% for supervised, Section 10.5.1) when compared to our previous results using supervised classification with domain knowledge and similar HMM models. We have also compared the proposed HHMM model with left-to-right models with single entry/exit states as in [Clarkson and Pentland, 1999; Naphade and Huang, 2002], and the average accuracy of the HHMM is 2.3% better than that of the constrained models. So the additional hierarchical structure imposed by HHMM over a more constrained model introduces more modelling power on our test domain.

The rest of this chapter is organized as follows: Section 10.1 presents the structure and semantics of the HHMM model; Section 10.2 presents the inference and parameter learning algorithms for HHMM; Section 10.3 presents algorithms for learning HHMM strucutre; Section 10.4 presents our feature selection algorithm for unsupervised learning over temporal sequences; Section 10.5 evaluates the results of learning with HHMM on sports video data; Section 10.6 summarizes the work and discusses open issues.

10.1 Hierarchical hidden Markov models

Based on the two-level Markov setup described above, we use a two-level hierarchical hidden Markov model to model structures in video. In this model, the higher-level structure elements usually correspond to semantic events, while the lower-level states represent variations that can occur within the same event, and these lower-level states in turn produce the observations, i.e., measurements taken from the raw video, with mixture-of-Gaussian distribution. Note the HHMM model is a special case of Dynamic Bayesian Network (DBN); also note the model can be easily extended to more than two levels, and feature distribution is not constrained to mixture-of-Gaussians. In the sections that follow, we will present algorithms that address the inference, parameter learning, and structure learning problems for general D-level HHMMs.

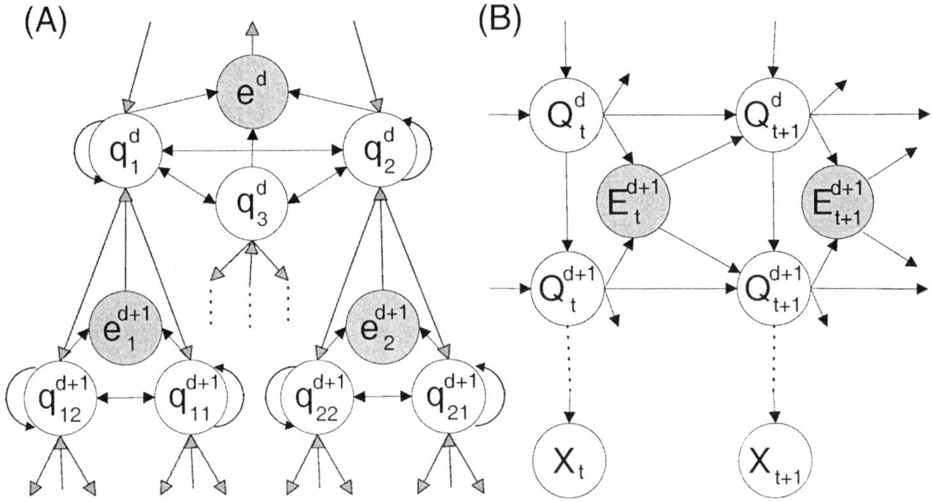

Figure 10.1. Graphical HHMM representation at level d and $d+1$ (A) Tree-structured representation; (B) DBN representations, with observations X_t drawn at the bottom. Uppercase letters denote the states as random variables in time t, lowercase letters denote the state-space of HHMM, i.e. values these random variables can take in any time slice. Shaded nodes are auxiliary *exit* nodes that turn on the transition at a higher level - a state at level d is not allowed to change unless the exiting states in the levels below are *on* ($E^{d+1} = 1$).

10.1.1 Structure of HHMM

Hierarchical hidden Markov modeling was first introduced in [Fine et al., 1998] as a natural generalization to HMM with a hierarchical control structure. As shown in Figure 10.1(A), every higher-level state symbol corresponds to a stream of symbols produced by a lower-level sub-HMM; a transition at the high level model is invoked only when the lower-level model enters an *exit* state (shaded nodes in Figure 10.1(A)); observations are only produced at the lowest level states.

This bottom-up structure is general in that it includes several other hierarchical schemes as special cases. Examples include the stacking of left-right HMMs [Clarkson and Pentland, 1999; Naphade and Huang, 2002], where across-level transitions can only happen at the first or the last state of a lower-level model; or the discrete counterpart of the jump Markov model [Doucet and Andrieu, 2001] with top-down(rather than bottom-up) control structure, where the level-transition probabilities are identical for each state that belongs to the same *parent* state at a higher level.

Prior applications of HHMM falls into three categories: (1) Supervised learning where manually segmented training data is available, hence each sub-HMM is learned separately on the segmented sub-sequences, and cross-level transitions are learned using the transition statistics across the subsequences. Examples include extron/intron recognition in DNA sequences [Hu et al., 2000], action recognition [Ivanov and Bobick, 2000]; more examples summarized in [Murphy, 2001] fall into this category. (2) Unsupervised learning, where segmented data at any level are not available for training, and parameters of different levels are jointly learned; (3) A mixture of the above, where state labels at the high level is given (with or without sub-model boundary), yet parameters still need to be estimated across several levels. Few instances of (2) can be found in the literature, while examples of (3), as a combination of (1) and (2), abound: the celebrated application of speech recognition systems with word-level annotation [The HTK Team, 2000], text parsing and handwriting recognition [Fine et al., 1998].

10.1.2 Complexity of Inferencing and Learning with HHMM

Fine et. al. have shown that multi-level hidden state inference with HHMM can be done in $O(T^3)$ by looping over all possible lengths of subsequences generated by each Markov model at each level, where T is the sequence length [Fine et al., 1998]. This algorithm is not optimal, however, an $O(T)$ algorithm has later been shown in [Murphy and Paskin, 2001] with an equivalent DBN representation by unrolling the multi-level states in time (Figure 10.1(B)). In this DBN representation, the hidden states Q_t^d at each level $d = 1, \dots D$, the observation sequence X_t, and the auxiliary *level-exiting* variables E_t^d completely specify the state of the model at time t. Note E_t^d can be turned on only if all lower levels of $E_T^{d+1:D}$ are on. The inference scheme used in [Murphy and Paskin, 2001] is the generic junction tree algorithm for DBNs, and the empirical complexity is $O(DT \cdot |Q|^{1.5D})$,[1] where D is the number of levels in the hierarchy, and $|Q|$ is the maximum number of distinct discrete values of any variable $Q_t^d, d = 1, \dots, D$.

For simplicity, we use a generalized forward-backward algorithm for hidden state inference, and a generalized EM algorithm for parameter

[1]More accurately, $O(DT \cdot |Q|^{\lceil 1.5D \rceil} 2^{\lceil 0.5D \rceil})$

estimation based on the forward-backward iterations. The algorithms is outlined in Section 10.2, and details can be found in [Xie et al., 2002a]. Note the complexity of this algorithm is $O(DT \cdot |Q|^{2D})$, with a similar running time as [Murphy and Paskin, 2001] for small D and modest Q.

10.2 Learning HHMM parameters with EM

In this section, we define notations to represent the states and parameter set of an HHMM, followed by a brief overview on deriving the EM algorithm for HHMMs. Details of the forward-backward algorithm for multi-level hidden state inference, and the EM update algorithms for parameter estimation are found in [Xie et al., 2002a]. The scope of the EM algorithm is the basic parameter estimation; we will assume that the size of the model is given, and the model is learned over a pre-defined feature set. These two assumptions are relaxed using the proposed model selection algorithms described in Section 10.3, and feature selection criteria in Section 10.4.

10.2.1 Representing an HHMM

Denote the maximum state-space size of any sub-HMM as N, we use the *bar notation* (Equation10.1) to write the entire configuration of the hierarchical states from the top (level 1) to the bottom (level D) with a N-ary D-digit integer, with the lowest-level states at the least significant digit:

$$k^{(D)} = q_{1:D} = \overline{(q_1 q_2 \cdots q_D)} = \sum_{i=1}^{D} q_i \cdot N^{D-i} \tag{10.1}$$

Here $1 \leq q_i \leq N; i = 1, \ldots, D$. We drop the superscript of k where there is no confusion, the whole parameter set Θ of an HHMM then consists of (1) Markov chain parameters λ^d in level d indexed by the state configuration $k^{(d-1)}$, i.e., transition probabilities A_k^d, prior probabilities π_k^d, and exiting probabilities from the current level e_k^d; (2) emission parameters B that specify the distribution of observations conditioned on the state configuration, i.e., the means μ_k and covariances σ_k when emission distributions are Gaussian.

$$\begin{aligned} \Theta &= (\bigcup_{d=1}^{D} \{\lambda^d\}) \bigcup \{B\} \\ &= (\bigcup_{d=1}^{D} \bigcup_{i=1}^{N^{d-1}} \{A_i^d, \pi_i^d, e_i^d\}) \bigcup (\bigcup_{i=1}^{N^D} \{\mu_i, \sigma_i\}) \end{aligned} \tag{10.2}$$

10.2.2 Overview of the EM algorithm

Denote Θ the old parameter set, $\hat{\Theta}$ the new (updated) parameter set, then maximizing the data likelihood L is equivalent to iteratively maximizing the expected value of the complete-data log-likelihood function $\Omega(\cdot, \Theta)$ as in Equation (10.3), for the observation sequence $X_{1:T}$ and the D-level hidden state sequence $Q_{1:T}$, according to the general EM presented in [Dempster et al., 1977]. Here we adopt the Matlab-like notation to write a temporal sequence of length T as $(\cdot)_{1:T}$, and its element at time t is simply $(\cdot)_t$.

$$\begin{aligned}\Omega(\hat{\Theta}, \Theta) &= E[\log(P(Q_{1:T}, X_{1:T}|\hat{\Theta}))|X_{1:T}, \Theta] \qquad (10.3)\\ &= \sum_{Q_{1:T}} P(Q_{1:T}|X_{1:T}, \Theta) \log(P(Q_{1:T}, X_{1:T}|\hat{\Theta}))\\ &= L^{-1} \sum_{Q_{1:T}} P(Q_{1:T}, X_{1:T}|\Theta) \log(P(Q_{1:T}, X_{1:T}|\hat{\Theta})) \qquad (10.4)\end{aligned}$$

Generally speaking, the "E" step evaluates this expectation based on the current parameter set Θ, and the "M" step finds the value of $\hat{\Theta}$ that maximizes this expectation. Special care must be taken in choosing a proper hidden state space for the "M" step of (10.4) to have a closed-form solution. Since all the unknowns lie inside the $\log(\cdot)$, it can be easily seen that if the complete-data probability $P(Q_{1:T}, X_{1:T}|\hat{\Theta})$ takes the form of product-of-unknown-parameters, we would get summation-of-individual-parameters in $\Omega(\hat{\Theta}, \Theta)$, hence each unknown can be solved separately in maximization and a closed-form solution is possible.

10.3 Bayesian model adaptation

Parameter learning for HHMM using EM is known to converge to a local maximum of the data likelihood since EM is an hill-climbing algorithm, and it is also known that searching for a global maximum in the likelihood landscape is intractable. Moreover, this optimization for data likelihood is only carried out over a predefined model structure, and in order to enable the comparison and search over a set of model structures, we will need not only a new optimality criterion, but also an alternative search strategy since exhausting all model topologies is super-exponential in complexity.

In this work, we adopt randomized search strategies to address the intractability problem on the parameter and model structure space; and the optimality criterion is generalized to maximum posterior from maximum likelihood, thus incorporating *Bayesian* prior belief on the model structure. Specifically, we use a Markov chain Monte Carlo (MCMC) method to maximize Bayesian information criterion (BIC) [Schwarz, 1978], and the motivation and basics structure of this algorithm are presented in the following subsections.

We are aware that alternatives for structure learning exist, such as the deterministic parameter trimming algorithm with entropy prior [Brand, 1999], which ensures the monotonic increasing of model priors throughout the trimming process. But we would have to start with a sufficiently large model in order to apply this trimming algorithm, which is not preferable for computational complexity purposes, and also impossible if we do not know a bound of the model complexity beforehand.

10.3.1 Overview of MCMC

MCMC is a class of algorithms that can solve high-dimensional optimization problems, which is seeing much recent success in Bayesian learning of statistical models [Andrieu et al., 2003]. In general, MCMC for Bayesian learning iterates between two steps: (1) The proposal step gives a new model sampled from certain *proposal distributions*, which depend on the current model, and statistics of the data; (2) The decision step computes an acceptance probability α based on the *fitness* of the proposed new model using model posterior and proposal strategies, and then this proposal is *accepted* or *rejected* with probability α.

MCMC will converge to the global optimum *in probability* if certain constraints [Andrieu et al., 2003] are satisfied for the proposal distributions, yet the speed of convergence largely depends on the *goodness* of the proposals. In addition to parameters learning, model selection can also be addressed in the same framework with reverse-jump MCMC (RJ-MCMC) [Green, 1995], by constructing reversible moves between parameter spaces of different dimensions. In particular, [Andrieu et al., 2001] applied RJ-MCMC to the learning of radial basis function (RBF) neural networks by introducing birth-death and split-merge moves to the RBF kernels. This is similar to our case of learning a variable number of Gaussians in the feature space that correspond to the emission probabilities.

In this work, we deployed a MCMC scheme to learn the optimal state-space of an HHMM model. We use a mixture of the EM and MCMC algorithms, where the model parameters are updated using EM, and model structure learning uses MCMC. We choose this hybrid algorithm in place of full Monte Carlo update of the parameter set and the model, since MCMC update of parameters will take much longer than EM, and the convergence behavior does not seem to suffer in practice.

10.3.2 MCMC for HHMM

Model adaptation for HHMM involves moves similar to [Andrieu et al., 2003] since many changes in the state space involve changing the number of Gaussian kernels that associate states in the lowest level with observations. We included four general types of movement in the state-space, as can be illustrated form the tree-structured representation of the HHMM in figure 10.1(a): (1) *EM*, regular parameter update without changing the state space size. (2) *Split(d)*, to split a state at level d. This is done by randomly partitioning the direct children (when there are more than one) of a state at level d into two sets, assigning one set to its original parent, the other set to a newly generated parent state at level d; when split happens at the lowest level(i.e. $d = D$), we split the Gaussian kernel of the original observation probabilities by perturbing the mean. (3) *Merge(d)*, to merge two states at level d into one, by collapsing their children into one set and decreasing the number of nodes at level d by one. (4) *Swap(d)*, to swap the parents of two states at level d, whose parent nodes at level $d-1$ was not originally the same. This special new move is needed for HHMM, since its multi-level structure is non-homogeneous within the same size of overall state-space. Note we are not including birth/death moves for simplicity, since these moves can be reached with multiple moves of split/merge.

Model adaptation for HHMMs is choreographed as follows:

1 Initialize the model Θ_0 from data.

2 At iteration i, based on the current model Θ_i, compute a probability profile $P_{\Theta_i} = [p_{em}, p_{sp}(1:D), p_{me}(1:D), p_{sw}(1:D)]$ according to Equations (10.A.1)-(10.A.4) in the appendix, then propose a move among the types $\{$*EM, Split(d), Merge(d), Swap(d)*$|d = 1, \ldots, D\}$

3 Update the model structure and the parameter set by appropriate action on selected states and their children states, as described in the appendix;

4 Evaluate the acceptance ratio r_i for different types of moves according to Equations (10.A.7)–(10.A.11) in the appendix, this ratio takes into account model posterior, computed with BIC (Equation 10.5), and alignment terms that compensate for the fact that the spaces between which we are evaluating the ratio are of unequal size. Denote the acceptance probability $\alpha_i = min\{1, r_i\}$; we then sample $u \sim U(0,1)$, and accept the this move if $u \leq \alpha_i$, reject otherwise.

5 Stop if converged, otherwise goto step 2

BIC [Schwarz, 1978] is a measure of *a posteriori* model fitness, it is the major factor that determines whether or not a proposed move is accepted.

$$BIC = \log(P(x|\Theta)) \cdot \lambda - \frac{1}{2}|\Theta| \log(T) \tag{10.5}$$

Intuitively, BIC is a trade-off between data likelihood $P(X|\Theta)$ and model complexity $|\Theta| \cdot \log(T)$ with weighting factor λ. Larger models are penalized by the number of free parameters in the model $|\Theta|$; yet the influence of the model penalty decreases as the amount of training data T increases, since $\log(T)$ grows slower than $O(T)$. We empirically choose the weighting factor λ as $1/16$ in the simulations of this section as well as those in Section 10.4, in order for the change in data likelihood and that in model prior to be numerically comparable over one iteration.

10.4 Feature selection for unsupervised learning

Feature extraction schemes for audio-visual streams abound, and we are usually left with a large pool of diverse features without knowing which ones are actually relevant to the important events and structures in the data sequences. A few features can be selected manually if adequate domain knowledge exists. But very often such knowledge is not available in new domains, or the connection between high-level structures and low-level features is not obvious. In general, the task of feature selection is divided into two aspects — eliminating *irrelevant* features, and eliminating *redundant* ones. Irrelevant features usually disturb the classifier and degrade classification accuracy, while redundant features add to computational cost without bringing in new information. Furthermore, for unsupervised structure discovery, different subsets of features may relate to different events, and thus the events should be described with separate models rather than being modelled jointly.

Hence the scope of our problem, is to select a relevant and compact feature subset that fits the HHMM model assumption in unsupervised learning over temporally correlated data streams.

10.4.1 Feature selection algorithm

Denote the feature pool as $F = \{f_1, \ldots, f_D\}$, the data sequence as $X_F = X_F^{1:T}$, then the feature vector at time t is X_F^t. The feature selection algorithm proceeds through the following steps, as illustrated in figure 10.2:

1 (Let $i = 1$ to start with.) At the i-th round, produce a *reference set* $\tilde{F}_i \subseteq F$ at random, learn HHMM $\tilde{\Theta}_i$ on $\tilde{F}_i$ with model adaptation,

perform Viterbi decoding of $X_{\tilde{F}_i}$, and obtain the *reference state-sequence* $\tilde{Q}_i = \tilde{Q}^{1:T}_{\tilde{F}_i}$.

2 For each feature $f_d \in F \backslash \tilde{F}_i$, learn HHMM Θ_d, get the Viterbi state sequence Q_d, then compute the information gain (Section 10.4.2) of each feature on the Q_d with respect to the reference partition $\tilde{Q}_i$. We then find the subset $\hat{F}_i \subseteq (F \setminus \tilde{F}_i)$ with significantly large information gain, and form the consistent feature group as union of the *reference set* and the *relevance set*: $\bar{F}_i \triangleq \tilde{F}_i \cup \hat{F}_i$.

3 Use Markov blanket filtering of Section 10.4.3, eliminate redundant features within the set $\bar{F}_i$ whose Markov blanket exists. We are then left with a relevant and compact feature subset $F_i \subseteq \bar{F}_i$. Learn HHMM Θ_i again with model adaptation on X_{F_i}.

4 Eliminate the previous candidate set by setting $F = F \setminus \bar{F}_i$; go back to step 1 with $i = i + 1$ if F is non-empty.

5 For each feature-model combination $\{F_i, \Theta_i\}_i$, evaluate their *fitness* using the normalized BIC criterion in Section 10.4.4, rank the feature subsets, and interpret the meanings of the resulting clusters.

After the feature-model combinations are generated automatically, a human operator can look at the structures marked by these models, then come to a decision on whether a feature-model combination shall be kept based on the meaningfulness of the resulting structures, and the BIC criterion.

10.4.2 Evaluating information gain

Step 1 in Section 10.4.1 produces a reference labelling of the data sequence induced by the classifier learned over the reference feature set.

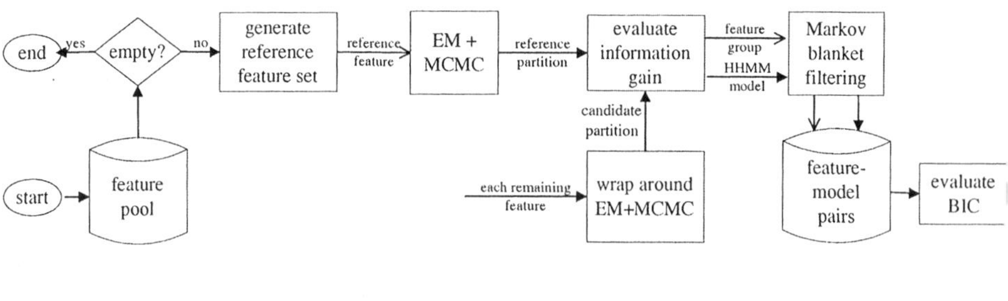

Figure 10.2. Feature selection algorithm overview

We want to find features that are *relevant* to this reference. One suitable measure to quantify the degree of *agreement* in each feature to the reference labelling, as used in [Xing and Karp, 2001], is the mutual information [Cover and Thomas, 1991], or the *information gain* achieved by the new partition induced with the candidate features over the reference partition.

A classifier Θ_F learned over a feature set F generates a partition, i.e. a label sequence Q_F, on the observations X_F, where there are at most N possible labels, we denote the label sequence as integers $Q_F^t \in \{1, \ldots, N\}$. We compute the probability of each label using the empirical portion, by counting the samples that bear label i over time $t = 1, \ldots, T$(Equation 10.6). We compute similarly the conditional probability of the reference labels $\tilde{Q}_i$ for the i-th iteration round given the new partition Q_f induced by a feature f (Equation 10.7), by counting over pairs of labels over time t. Then the information gain of feature f with respect to $\tilde{Q}_i$ is defined as the mutual information between $\tilde{Q}_i$ and Q_f(Equation 10.8).

$$
\begin{aligned}
P_{Q_f}(i) &= \frac{|\{t|Q_f^t = i, t = 1, \ldots, T\}|}{T}; && (10.6)\\
P_{\tilde{Q}_i|Q_f}(i \mid j) &= \frac{|\{t|(\tilde{Q}_i^t, Q_f^t) = (i,j), t = 1, \ldots, T\}|}{|\{t|Q_f^t = j, t = 1, \ldots, T\}|}; && (10.7)\\
I(Q_f; \tilde{Q}_i) &= H(P_{\tilde{Q}_i}) - \sum_j P_{Q_f} \cdot H(P_{\tilde{Q}_i|Q_f=j}) && (10.8)\\
&\text{where} \quad i, j = 1, \ldots, N
\end{aligned}
$$

Here $H(\cdot)$ is the entropy function. Intuitively, a larger information gain for candidate feature f suggests that the f-induced partition Q_f is more consistent with the reference partition $\tilde{Q}_i$. After computing the information gain $I(Q_f; \tilde{Q}_i)$ for each remaining feature $f_d \in F \setminus \tilde{F}_i$, we perform hierarchical agglomerative clustering on the information gain vector using a dendrogram [Jain et al., 1999], look at the top-most link that partitions all the features into two clusters, and pick features that lies in the upper cluster as the set with satisfactory consistency with the reference feature set.

10.4.3 Finding a Markov blanket

After wrapping information gain criterion around classifiers built over all feature candidates (step 2 in Section 10.4.1), we are left with a subset of features with consistency yet possible redundancy. The approach for identifying redundant features naturally relates to the conditional dependencies among the features. For this purpose, we need the notion of a Markov blanket [Koller and Sahami, 1996].

DEFINITION 10.1 *Let f be a feature subset, M_f be a set of random variables that does not contain f, we say M_f is the Markov blanket of f, if f is conditionally independent of all variables in $\{F \cup C\} \setminus \{M_f \cup f\}$ given M_f. [Koller and Sahami, 1996]*

Computationally, a feature f is redundant if the partition C of the data set is independent of f given its *Markov blanket* F_M. In prior work [Koller and Sahami, 1996; Xing and Karp, 2001], the Markov blanket is identified with the equivalent condition that the posterior probability distribution of the class given the feature set $\{M_f \cup f\}$ should be the same as that conditioned on the Markov blanket M_f only. i.e.

$$\Delta_f = D(\ P(C|M_f \cup f) \ || \ P(C|M_f)\ ) = 0 \tag{10.9}$$

where $D(P||Q) = \Sigma_x P(x)\log(P(x)/Q(x))$ is the Kullback-Leibler distance [Cover and Thomas, 1991] between two probability mass functions $P(x)$ and $Q(x)$.

For unsupervised learning over a temporal stream however, this criterion cannot be readily employed. This is because (1) the posterior distribution of a class depends not only on the current data sample, but also on adjacent samples; (2) we would have to condition the class label posterior over all dependent feature samples, and such conditioning quickly makes the estimation of the posterior intractable as the number of conditioned samples grows; (3) we will not have enough data to estimate these high-dimensional distributions by counting over feature-class tuples since the dimensionality is high. We therefore use an alternative necessary condition that the optimum state-sequence $C_{1:T}$ should not change conditioned on observing $M_f \cup f$ or M_f only.

Koller and Sahami have also proved that sequentially removing features one at a time with its Markov blanket identified will not cause divergence of the resulting set, since if we eliminate feature f and keep its Markov blanket M_f, f remains unnecessary in later stages when more features are eliminated. Additionally, as few if any features will have a Markov blanket of limited size in practice, we sequentially remove features that induce the least change in the state sequence given the change is small enough ($< 5\%$). Note this step is a filtering step in our HHMM learning setting, since we do not need to retrain the HHMMs for each candidate feature f and its Markov blanket M_f. Given the HHMM trained over the set $f \cup M_f$, the state sequence Q_{M_f}, decoded with the observation sequences in M_f only, is compared with the state sequence

$Q_{f \cup M_f}$ decoded using the whole observation sequence in $f \cup M_f$. If the difference between Q_{M_f} and $Q_{f \cup M_f}$ is *small enough*, then f is removed since M_f is found to be a Markov blanket of f.

10.4.4 Normalized BIC

Iterating over Section 10.4.2 and Section 10.4.3 results in disjoint small subsets of features $\{F_i\}$ that are compact and consistent with each other. The HHMM models $\{\Theta_i\}$ learned over these subsets are best-effort fits on the features, yet the $\{\Theta_i\}$s may not fit the multi-level Markov assumptions in Section 10.0.0.0.0.

There are two criteria proposed in prior work [Dy and Brodley, 2000], scatter separability and maximum likelihood (ML). Note the former is not suitable to temporal data since multi-dimensional Euclidean distance does not take into account temporal dependency, and it is non-trivial to define another proper distance measure for temporal data; while the latter is also known [Dy and Brodley, 2000] to be biased against higher-dimensional feature sets. We use a normalized BIC criterion (Equation 10.10) as the alternative to ML, which trades off normalized data likelihood $\tilde{L}$ with model complexity $|\Theta|$. Note the former has weighting factor λ in practice; the latter is modulated by the total number of samples $\log(T)$; and $\tilde{L}$ for HHMM is computed in the same forward-backward iterations, except all the emission probabilities $P(X|Q)$ are replaced with $P'_{X,Q} = P(X|Q)^{1/D}$, i.e. normalized with respect to data dimension D, under the *naive-Bayes* assumption that features are independent given the hidden states.

$$\widetilde{BIC} = \tilde{L} \cdot \lambda - \frac{1}{2}|\Theta| \log(T) \tag{10.10}$$

Initialization and convergence issues exist in the iterative partitioning of the feature pool. The strategy for producing the random *reference set* $\tilde{F}_i$ in step (1) affects the result of feature partition, as even producing the same $\tilde{F}_i$ in a different sequence may result in different final partitions. Moreover, the *expressiveness* of the resulting structures is also affected by the reference set. If the dimension of $\tilde{F}_i$ is too low for example, the algorithm tends to produce many small feature groups where features in the same group mostly *agree* with each other, and the learned model would not be able to identify potential complex structures that must be identified with features carrying complementary information, such as features from different modalities (audio and video). On the other hand, if $\tilde{F}_i$ is of very high dimension, then the information gain criterion will give a large feature group around $\tilde{F}_i$, thus mixing different event streams that would better be modelled separately, such as the activity of pedestrians and vehicles in a street surveillance video.

10.5 Experiments and Results

In this section, we report the tests of the proposed methods in automatically finding salient events, learning model structures, and identifying informative feature set in soccer and baseball videos. We have also experimented with variations in HHMM transition topology and found that the additional hierarchical structure imposed by HHMM over an ordinary HMM introduces more modelling power on our test domain.

Sports videos represent an interesting domain for testing the proposed techniques in automatic structure discovery. Two main factors contribute to this match between the video domain and the statistical technique: the distinct set of semantics in the sports domain exhibit strong correlations with audio-visual features; the well-established rules of games and production syntax in sports video programs impose strong temporal transition constraints. For example, in soccer videos, *plays* and *breaks* are recurrent events covering the entire time axis of the video data. In baseball videos, transitions among different perceptually distinctive mid-level events, such as pitching, batting, running, are semantically significant for the game.

Clip Name	Sport	Length	Resolution	Frame rate	Source
Korea	Soccer	25'00"	320×240	29.97	MPEG-7
Spain	Soccer	15'00"	352×288	25	MPEG-7
NY-AZ	Baseball	32'15"	320×240	29.97	TV program

Table 10.1. Sports video clips used in the experiment.

All our test videos are in MPEG-1 format, their profiles are listed in Table 10.1. For soccer videos, we have compared with our previous work using supervised methods on the same video streams [Xie et al., 2002b]. The evaluation basis for the structure discovery algorithms is for two semantic events, *play* and *break*, defined according to the rules of soccer. These two events are dense since they cover the whole time scale of the video, and distinguishing *break* from *play* will be useful for efficient browsing and summarization, since *break* takes up about 40% of the screen time, and viewers may browse through the game play by play, skipping all the breaks in between, or randomly access the break segments to find player responses or game announcements. For baseball videos, we conducted the learning without having labelled ground truth or manually identified features *a priori*, and an human observer (the first author) reported observations on the selected feature sets and the resulting structures afterwards. This is analogous to the actual application of structure discovery to an unknown domain, where evaluation and interpretation of the result is done after automatic discovery algorithms are applied.

It is difficult to define general evaluation criteria for automatic structure discovery results that are applicable across different domains, this is especially the case when domain-specific semantic labels are of interest. This difficulty lies in the gap between computational optimization and semantic meaning: the results of unsupervised learning are optimized with measures of statistical fitness, yet the link from statistical fitness to semantics needs a match between general domain characteristics and the computational assumptions imposed in the model. Despite this difficulty, our results have shown support for constrained domains such as sports. Effective statistic models built over statistically optimized feature sets have good correspondence with semantic events in the selected domain.

10.5.1 Parameter and structure learning

We first test the automatic model learning algorithms with a fixed feature set manually selected based on heuristics. The selected features, *dominant color ratio* and *motion intensity*, have been found effective in detecting soccer events in our prior work [Xu et al., 2001; Xie et al., 2002b]. Such features are uniformly sampled from the video stream every 0.1 second. Here we compare the learning accuracy of four different learning **schemes** against the ground truth.

1 Supervised HMM: This is developed in our prior work [Xie et al., 2002b]. One HMM per semantic event (i.e., play and break) is trained on manually defined chunks. For test video data with unknown event boundaries, the videos are first chopped into 3-second segments, where the data likelihood of each segment is evaluated with each of the trained HMMs. The final event boundaries are refined with a dynamic programming step taking into account the model likelihoods, the transition likelihoods between events, and the probability distribution of event durations.

2 Supervised HHMM: Individual HMMs at the bottom level of the hierarchy are learned separately, essentially using the models trained in scheme 1; across-level and top level transition statistics are also obtained from segmented data; and then segmentation is obtained by decoding the Viterbi path from the hierarchical model on the entire video stream.

3 Unsupervised HHMM without model adaptation: An HHMM is initialized with known size of state-space and random parameters; the EM algorithm is used to learn the model parameters; and segmentation is obtained from the Viterbi path of the final model.

4 Unsupervised HHMM with model adaptation: An HHMM is initialized with arbitrary size of state-space and random parameters; the EM and RJ-MCMC algorithms are used to learn the size and parameters of the model; the state sequence is obtained from the converged model with optimal size. Here we will report results separately for (a) model adaptation in the lowest level of HHMM only, and (b) full model adaptation across different levels as described in Section 10.3.

For supervised schemes 1 and 2, K-means clustering and Gaussian mixture fitting is used to randomly initialize the HMMs. For unsupervised schemes 3 and 4, as well as all full HHMM learning schemes in the sections that follow, the initial emission probabilities of the initial bottom-level HMMs are obtained with K-means and Gaussian fitting; and then the multi-level Markov chain parameters are estimated using a dynamic programming technique that groups the states into different levels by maximizing the number of within-level transitions, while minimizing inter-level transitions among the Gaussians. For schemes 1-3, the model size is set to six bottom-level states per event, corresponding to the optimal model size that schemes 4a converges to, i.e. six to eight bottom-level states per event. We run each algorithm for 15 times with random start, and compute the per-sample accuracy against manual labels. The median and semi-interquartile range [2] across multiple rounds are listed in Table 10.2.

Learning Scheme	Supervised?	Model type	Adaptation?		Accuracy	
			Bottom-level	High-levels	Median	SIQ
(1)	Y	HMM	N	N	75.5%	1.8%
(2)	Y	HHMM	N	N	75.0%	2.0%
(3)	N	HHMM	N	N	75.0%	1.2%
(4a)	N	HHMM	N	Y	75.7%	1.1%
(4b)	N	HHMM	Y	Y	75.2%	1.3%

Table 10.2. Evaluation of learning schemes (1)-(4) against ground truth using on clip *Korea*

[2]Semi-interquartile as a measure of the spread of the data, is defined as half of the distance between the 75th and 25th percentile, it is more robust to outliers than standard deviation.

Results show that the performance of the unsupervised learning scheme is comparable to the supervised learning, and sometimes it achieves even slightly better accuracy than the supervised learning counterpart. This is quite surprising since the unsupervised learning of HHMMs is not tuned to the particular ground-truth. The results maintain a consistent accuracy, as indicated by the low semi-interquartile range. Also note that the comparison basis using supervised learning is actually conservative since (1) unlike [Xie et al., 2002b], the HMMs are learning and evaluated on the same video clip and results reported for schemes 1 and 2 are actually training accuracies; (2) the models without structure adaptation are assigned the *a posteriori* optimal model size.

For the HHMM with full model adaptation (scheme 4b), the algorithm converges to two to four high-level states, and the evaluation is done by assigning each resulting cluster to the majority ground-truth label it corresponds to. We have observed that the resulting accuracy is still in the same range without knowing how many interesting structures there is to start with. And the reason for this performance match lies in the fact that the *additional* high level structures are actually a sub-cluster of *play* or *break*, they are generally of three to five states each, and two sub-clusters correspond to one *larger, true* cluster of play or break (refer to a three-cluster example in Section 10.5.2).

10.5.2 Feature selection

Based on the good performance of the model parameter and structure learning algorithm, we test the performance of the automatic feature selection method that iteratively *wraps* around, and *filters* (Section 10.4). We use the two test clips, Korea and Spain as profiled in Table 10.1. A nine-dimensional feature vector sampled at every 0.1 seconds is taken as the initial feature pool, including:

> Dominant Color Ratio (DCR), Motion Intensity (MI), the least-square estimates of camera translation (MX, MY), and five audio features - Volume, Spectral roll-off (SR), Low-band energy (LE), High-band energy (HE), and Zero-crossing rate (ZCR).

We run the feature selection method plus model learning algorithm on each video stream for five times, with a one or two-dimensional feature set as initial reference set in each iteration. After eliminating degenerate cases that only consist of one feature in the resulting set, we evaluate the feature-model pair that has the largest *Normalized BIC* value as described in Section 10.4.4.

For clip *Spain*, the selected feature set is {DCR, Volume} The model converges to two high-level states in the HHMM, each with five lower-level children states. Evaluation against the *play/break* labels showed a 74.8% accuracy. For clip *Korea*, the final selected feature set is {DCR, MX}, with three high-level states and {7, 3, 4} children states respectively. If we assign each of the three clusters to the semantic event that it agrees with for the most amount of times (which would be {*play, break, break*} respectively), per-sample accuracy would be 74.5%. The automatic selection of DCR and MX as the most relevant features is actually consistent with the manual selection of the two features DCR and MI in our prior work [Xie et al., 2002b; Xu et al., 2001]. MX is a feature that approximates the horizontal camera panning motion, which is the most dominant factor contributing to the overall motion intensity (MI) in soccer video, as the camera needs to track the ball movement in wide angle shorts, and wide angle shots are one major type of shot that is used to reveal overall game status [Xu et al., 2001].

The accuracies are comparable to their counterpart (scheme 4) in Section 10.5.1 without varying the feature set (75%). Yet the small discrepancy may due to (1) variability in RJ-MCMC (Section 10.3), for which convergence diagnostic is still an active area of research [Andrieu et al., 2003], and (2) possible inherent bias that may exist in the normalized BIC criterion (Equation 10.10), as we will need ways to further calibrate the criterion.

10.5.3 Testing on a different domain

We have also conducted a preliminary study on the baseball video clip described in Table 10.1. The same 9-dimensional feature pool as in Section 10.5.2 is extracted from the stream, also at 0.1 second per sample. The learning of models is carried out without having labelled ground truth or manually identified features *a priori*. Observations are reported based on the selected feature sets and the resulting structures of the test results. This is a standard process of applying structure discovery to an unknown domain, where automatic algorithms serve as a pre-filtering step, and evaluation and interpretation of the result can only be done afterwards.

HHMM learning with full model adaptation and feature selection is conducted, resulting in three consistent compact feature groups: (a) HE, LE, ZCR; (b) DCR, MX; (c) Volume, SR. It is interesting to see audio features fall into two separate groups, with visual features also in a individual group.

The BIC score for the second group, dominant color ratio and horizontal camera pan, is significantly higher than that of the other two. The HHMM model in (b) has two higher-level states, each has six and seven children states at the bottom level, respectively. Moreover, the resulting segments from the model learned with this feature set has consistent perceptual properties, with one cluster of segments mostly corresponding to pitching shots and other field shots when the game is in play, while the other cluster contains most of the cutaways shots, score boards and game breaks, respectively. It is not surprising that this result agrees with the intuition that the status of a game can mainly be inferred from visual information.

10.5.4 Comparing to HHMM with simplifying constraints

In order to investigate the *expressiveness* of the multi-level model structure, we compare unsupervised structure discovery performances of the HHMM with a similar model with constrains in the transitions each node can make.

The two model topologies being simulated are visualized in figure 10.3:

(a) The simplified HHMM where each bottom-level sub-HMM is a left-to-right model with skips, and cross level entering/exiting can only happen at the first/last node, respectively. Note that the right-most states serving as the single exit point from the bottom level eliminate the need for a special *exiting* state.

(b) The fully connected general 2-level HHMM model used in scheme 3, Section 10.5.1, a special case of the HHMM in figure 10.1. Note the dummy *exiting* cannot be omitted in this case.

Topology (a) is of interest because the left-to-right and single entry/exit point constraints enables the learning the model with the algorithms designed for ordinary HMMs by *collapsing* this model to an ordinary HMM. The *collapsing* can be done because unlike the general HHMM case (Section 10.1), there is no ambiguity in whether or not a cross-level has happened in the original model given the last state and the current state in the collapsed model, or equivalently, the *flattened* HMM transition matrix can be uniquely factored back to recover the multi-level transition structure. Note that here the trade-off for model generality is that parameter estimation of the flattened HMMs is of complexity $O(T|Q|^{2D})$, while HHMMs will need $O(DT|Q|^{2D})$, as analyzed in Section 10.1.2. With the total number of levels D typically a fixed small constant, this difference does not influence the scalability of the model to long sequences.

Topology (a) also contains models proposed in two prior publications as special cases: [Clarkson and Pentland, 1999] uses a left-to-right model

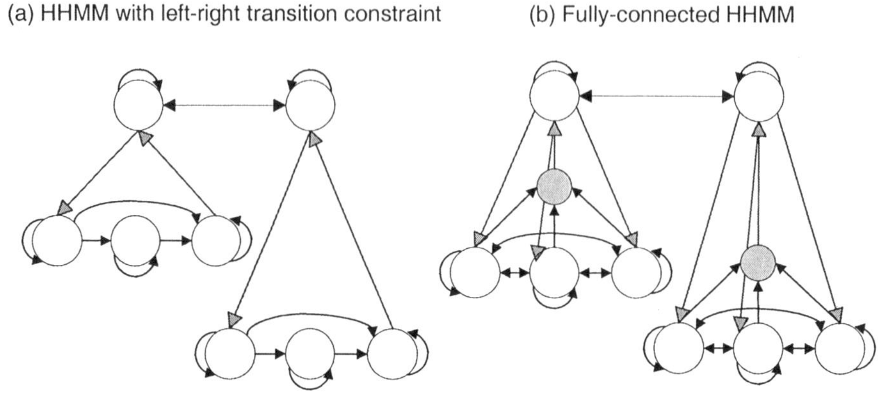

Figure 10.3. Comparison with HHMM with left-to-right transition constraints. Only 3 bottom-level states are drawn for the readability of this graph, models with 6-state sub-HMMs are simulated in the experiments.

without skip, and single entry/exit states; [Naphade and Huang, 2002] uses a left-to-right model without skip, single entry/exit states with one single high-level state, i.e. the probability of going to each sub-HMM is independent of which sub-HMM the model just came from, thus eliminating one more parameter from the model than [Clarkson and Pentland, 1999]. Both of the prior cases are learned with HMM learning algorithms.

This learning algorithm is tested on the soccer video clip *Korea*; it performs parameter estimation with fixed model structure of six states at the bottom level and two states at the top level, over the pre-defined features set of DCR and MI (Section 10.5.1). Results show that over 5 runs of both algorithms, the average accuracy of the constrained model is 2.3% lower than that of the fully connected model. This shows that adopting a fully connected model with multi-level control structures indeed brings in extra modelling power for the chosen domain of soccer videos.

10.6 Conclusion

In this chapter we proposed algorithms for unsupervised discovery of structure from video sequences. We model the class of dense, stochastic structures in video using hierarchical hidden Markov models. The model parameters and model structure are learned using EM and Monte Carlo sampling techniques, and informative feature subsets are automatically selected from a large feature pool using an iterative filter-wrapper algorithm. When evaluated on TV soccer clips against manually labelled ground truth, we achieved results comparable to its supervised learning counterpart; when evaluated on baseball clips, the algorithm automatically selects two visual features, which agrees with our intuition that the status of a baseball game can be inferred from visual information only.

It is encouraging that in constrained domains such as sports, effective statistic models built over statistically optimized feature sets without human supervision have good correspondence with semantic events. We believe this success lends major credit to the correct choice in general model assumptions, and the selected test domain that matches this assumption. This unsupervised structure discovery framework leaves much room for generalizations and applications to many diverse domains. It also raises further theoretical issues that will enrich this framework if successfully addressed: modelling sparse events in domains such as surveillance videos; online model update using new data; novelty detection; automatic pattern association across multiple streams; a hierarchical modeling that automatically adapts to different temporal granularity; etc.

Appendix

Proposal probabilities for model adaptation.

$$p_{sp}(d) = c^* \cdot \min\{1, \rho/(k+1)\}; \tag{10.A.1}$$
$$p_{me}(d) = c^* \cdot \min\{1, (k-1)/\rho\}; \tag{10.A.2}$$
$$p_{sw}(d) = c^*; \tag{10.A.3}$$
$$p_{em} = 1 - \Sigma_{d=1}^{D}[p_{sp}(d) + p_{me}(d) + p_{sw}(d)]. \tag{10.A.4}$$

Here c^* is a simulation parameter, k is the current number of states; and ρ is the hyper-parameter for the truncated Poisson prior of the number of states [Andrieu et al., 2003], i.e. ρ would be the expected mean of the number of states if the maximum state size is allowed to be $+\infty$, and the scaling factor that multiplies c^* modulates the proposal probability using the resulting state-space size $k \pm 1$ and ρ.

Computing different moves in RJ-MCMC. EM is one regular hill-climbing iteration as described in Section 10.2; and once a move type other than EM is selected, one (or two) states at a certain level are selected at random for swap/split/merge, and the parameters are modified accordingly:

- Swap the association of two states:
  Choose two states from the same level, each belongs to a different higher-level state, swap their higher-level association.

- Split a state:
  Choose a state at random, the split strategy differs when this state is at different position in the hierarchy.

- When this is a state at the lowest level ($d = D$), perturb the mean of its associated Gaussian observation distribution as follows

$$\begin{aligned} \mu_1 &= \mu_0 + u_s \eta \\ \mu_2 &= \mu_0 - u_s \eta \end{aligned} \tag{10.A.5}$$

  where $u_s \sim U[0,1]$, and η is a simulation parameter that ensures reversibility between split moves and merge moves.

- When this is a state at $d = 1, \ldots, D-1$ with more than one children states, split its children into two disjoint sets at random, generate a new *sibling* state at level d associated with the same parent as the selected state. Update the corresponding multi-level Markov chain parameters accordingly.

■ Merge two states:
Select two *sibling* states at level d, merge the observation probabilities or the

- When $d = D$, merge the Gaussian observation probabilities by making the new mean as the average of the two.

$$\mu_0 = \frac{\mu_1 + \mu_2}{2}, \quad \text{if } |\mu_1 - \mu_2| \leq 2\eta \tag{10.A.6}$$

  here η is the same simulation parameter as in Eq. 10.A.5.

- When $d = 1, \ldots, D-1$, merge the two states by making all the children of these two states the children of the merged state, and modify the multi-level transition probabilities accordingly.

The acceptance ratio for different moves in RJ-MCMC. The acceptance ratio for *Swap* simplifies into the posterior ratio since the dimension of the space does not change. Denote Θ as the old model and $\hat{\Theta}$ as the new model :

$$r \triangleq (\text{posterior ratio}) = \frac{P(x|\hat{\Theta})}{P(x|\Theta)} = \frac{exp(\widehat{BIC})}{exp(BIC)} \tag{10.A.7}$$

When moves are proposed to a parameter space with different dimension, such as split or merge, we also need a proposal ratio term and a Jacobian term to align the spaces in order to ensure detailed balance [Green, 1995], as shown in Equations (10.A.8)–(10.A.11).

$$r_k \triangleq (\text{posterior ratio}) \cdot (\text{proposal ratio}) \cdot (\text{Jacobian}) \tag{10.A.8}$$

$$r_{split} = \frac{P(k+1, \Theta_{k+1}|x)}{P(k, \Theta_k|x)} \cdot \frac{m_{k+1}/(k+1)}{p(u_s)s_k/k} \cdot J \tag{10.A.9}$$

$$r_{merge} = \frac{P(k, \Theta_k|x)}{P(k+1, \Theta_{k+1}|x)} \cdot \frac{p(u_s)s_{k-1}/(k-1)}{m_k/k} \cdot J^{-1} \tag{10.A.10}$$

$$J = \left| \frac{\partial(\mu_1, \mu_2)}{\partial(\mu_0, u_s)} \right| = \left| \begin{array}{cc} 1 & \eta \\ 1 & -\eta \end{array} \right| = 2\eta \tag{10.A.11}$$

References

Andrieu, C., de Freitas, N., and Doucet, A. (2001). Robust full bayesian learning for radial basis networks. *Neural Computation*, 13:2359–2407.

Andrieu, C., de Freitas, N., Doucet, A., and Jordan, M. I. (2003). An introduction to MCMC for machine learning. *Machine Learning*, special issue on MCMC for Machine Learning.

Brand, M. (1999). Structure learning in conditional probability models via an entropic prior and parameter extinction. *Neural Computation*, 11(5):1155–1182.

Clarkson, B. and Pentland, A. (1999). Unsupervised clustering of ambulatory audio and video. In *International Conference on Acoustic, Speech and Signal Processing (ICASSP)*.

Cover, T. M. and Thomas, J. A. (1991). *Elements of Information Theory*. Wiley, New York.

Dempster, A. P., Laird, N. M., and Rubin, D. B. (1977). Maximum liklihood from incomplete data via the EM algorithm. *Journal of the Royal Statistical Society B*, 39:1–38.

Doucet, A. and Andrieu, C. (2001). Iterative algorithms for optimal state estimation of jump Markov linear systems. *IEEE Transactions of Signal Processing*, 49:1216–1227.

Dy, J. G. and Brodley, C. E. (2000). Feature subset selection and order identification for unsupervised learning. In *Proc. 17th International Conf. on Machine Learning*, pages 247–254. Morgan Kaufmann, San Francisco, CA.

Fine, S., Singer, Y., and Tishby, N. (1998). The hierarchical hidden Markov model: Analysis and applications. *Machine Learning*, 32(1):41–62.

Green, P. J. (1995). Reversible jump Markov chain Monte Carlo computation and Bayesian model determination. *Biometrika*, 82:711–732.

Hu, M., Ingram, C., Sirski, M., Pal, C., Swamy, S., and Patten, C. (2000). A hierarchical HMM implementation for vertebrate gene splice site prediction. Technical report, Dept. of Computer Science, University of Waterloo.

Ivanov, Y. A. and Bobick, A. F. (2000). Recognition of visual activities and interactions by stochastic parsing. *IEEE Transaction of Pattern Recognition and Machines Intelligence*, 22(8):852–872.

Iyengar, A., Squillante, M. S., and Zhang, L. (1999). Analysis and characterization of large-scale web server access patterns and performance. *World Wide Web*, 2(1-2):85–100.

Jain, A. K., Murty, M. N., and Flynn, P. J. (1999). Data clustering: a review. *ACM Computing Surveys*, 31(3):264–323.

Koller, D. and Sahami, M. (1996). Toward optimal feature selection. In *International Conference on Machine Learning*, pages 284–292.

Lawrence, C. E., Altschul, S. F., Boguski, M. S., Liu, J. S., Neuwald, A. F., and Wootton, J. C. (October 1993). Detecting subtle sequence signals: a Gibbs sampling strategy for multiple alignment. *Science*, 8(262):208–14.

Murphy, K. (2001). Representing and learning hierarchical structure in sequential data.

Murphy, K. and Paskin, M. (2001). Linear time inference in hierarchical HMMs. In *Proceedings of Neural Information Processing Systems*, Vancouver, Canada.

Naphade, M. and Huang, T. (2002). Discovering recurrent events in video using unsupervised methods. In *Proc. Intl. Conf. Image Processing*, Rochester, NY.

Rabiner, L. R. (1989). A tutorial on hidden Markov models and selected applications in speech recognition. *Proceedings of the IEEE*, 77(2):257–285.

Sahouria, E. and Zakhor, A. (1999). Content anlaysis of video using principal components. *IEEE Transactions on Circuits and Systems for Video Technology*, 9(9):1290–1298.

Schwarz, G. (1978). Estimating the dimension of a model. *The Annals of Statistics*, 7:461–464.

The HTK Team (2000). Hidden Markov model toolkit (HTK3). http://htk.eng.cam.ac.uk/.

Wang, Y., Liu, Z., and Huang, J. (2000). Multimedia content analysis using both audio and visual clues. *IEEE Signal Processing Magazine*, 17(6):12–36.

Xie, L., Chang, S.-F., Divakaran, A., and Sun, H. (2002a). Learning hierarchical hidden Markov models for video structure discovery. Technical Report ADVENT-2002-006, Dept. Electrical Engineering, Columbia Univ., http://www.ee.columbia.edu/~xlx/research/.

Xie, L., Chang, S.-F., Divakaran, A., and Sun, H. (2002b). Structure analysis of soccer video with hidden Markov models. In *Proc. Interational Conference on Acoustic, Speech and Signal Processing (ICASSP)*, Orlando, FL.

Xing, E. P. and Karp, R. M. (2001). Cliff: Clustering of high-dimensional microarray data via iterative feature filtering using normalized cuts. In *Proceedings of the Ninth International Conference on Intelligence Systems for Molecular Biology (ISMB)*, pages 1–9.

Xu, P., Xie, L., Chang, S.-F., Divakaran, A., Vetro, A., and Sun, H. (2001). Algorithms and systems for segmentation and structure analysis in soccer video. In *Proc. IEEE International Conference on Multimedia and Expo (ICME)*, Tokyo, Japan.

Yeung, M. and Yeo, B.-L. (1996). Time-constrained clustering for segmentation of video into story units. In *International Conference on Pattern Recognition (ICPR)*, Vienna, Austria.

Chapter 11

PSEUDO-RELEVANCE FEEDBACK FOR MULTIMEDIA RETRIEVAL

Rong Yan, Alexander G. Hauptmann and Rong Jin
School of Computer Science
Carnegie Mellon University
yanrong@cs.cmu.edu

Abstract Video information retrieval requires a system to find information relevant to a query which may be presented simultaneously in different ways through a text description, audio, still images and/or video sequences. The actual search also takes place in the text, audio, image or video domain. We present an approach that uses pseudo-relevance feedback from retrieved items that are NOT similar to the relevant items. An evaluation on the 2002 TREC Video Track queries shows that this technique can substantially improve image and video retrieval performance on a real collection. We believe that negative pseudo-relevance feedback shows great promise for very difficult multimedia retrieval tasks that involve image or video similarity, especially when combined with other retrieval agents.

Keywords: Content-based access to multimedia, video information retrieval, TREC Video Track, pseudo-relevance feedback, digital video search, evaluation, audio domain, relevance, retrieval performance, negative feedback, negative training examples, video similarity, mean average precision (MAP), SVM classifier, multiple modalities.

Introduction

Recent improvement in processor technology and speed, network systems and the availability of massive but cheap digital storage has led to a growing demand for content-based access to video information. Content-based access implies that we can search not only through manually indexed terms, but that the system can also directly evaluate if the video content, as represented by the images and the audio, is similar to what was specified in a query. Many large archives, such as the video archives from television networks [Evans, 2002], or still image archives [Porter, 2002] can become much more valuable if desired content can be found quickly and efficiently. This desire for content-based access is motivating researchers to develop approaches that allow users to query and retrieve based on the audio information and the imagery in video, which has become known as content-based video retrieval (CBVR) or content-based multimedia retrieval [Lew, 2002].

Pattern recognition techniques have been widely applied in visual information retrieval systems. In visual information retrieval, there are two fundamental problems to be addressed. One is the representation of the visual features and the other the design of a similarity metric which determines the "distance" between two example images. Among the most common distance metrics are generic metrics such as Euclidean distance and Mahalanobis distance [Antania et al., 2002]. However, CBVR systems that simply rely on a pre-defined generic similarity metric will be affected by two major difficulties. The first is that all visual feature representations are limited to capturing fairly low-level physical features of a generic nature (such as color, texture or shape), which makes it extremely difficult to determine appropriate similarity metrics without specific query information or in very restricted domains. The second difficulty is that different query scenarios may require different similarity metrics to model the distribution of relevant examples. For example, many animals can be discriminated by shape, but it is a probably a better choice to identify "sky" and "water" by color. Therefore, we should attempt to make the similarity metric adaptive with respect to different queries.

This goal of adapting the similarity metric to specific queries requires approaches that are able to automatically discover the discriminating feature subspace once the queries are provided. A natural solution is to cast this formulation of retrieval as a classification problem, where relevant examples are the positive instances and non-relevant examples are the negative instances of a class. Recent work [Tong and Chang, 2001; Tieu and Viola, 2001] has suggested that margin-based classifiers such as support vector machines (SVMs) and Adaboosting can yield high generalization performance and automatically emphasize the useful features by learning the maximal margin hyperplane in the embedding space. However, it is generally a characteristic of information retrieval that the user's query only provides a small amount of positive data and no explicit negative training data at all.

Thus, if we want to make use of margin-based learning algorithms for multimedia information retrieval, methods have to be devised which can provide more training data, especially some negative examples. One might consider random sampling to obtain negative examples [Tieu and Viola, 2001], since for any given query, there are usually on a very small number of positive, relevant data items in a collection. However, this random sampling policy is somewhat risky because it is possible that positive examples might be included as (false) negatives [Zhou and Huang, 2001], which would be extremely detrimental to a margin-based classifier. Standard relevance feedback addresses this issue in an interactive fashion, in which the system iteratively asks users to label more training examples as relevant/non-relevant for the learning algorithms [Picard et al., 1996; Rui et al., 1998]. However, it is tedious to hand pick negative examples and subjectively quite difficult to provide a good negative sample [Yu et al., 2002] that clearly shows the distinction to the positive examples, since negative instances are less well-defined as a coherent subset. After the interactive relevance feedback, the system must then re-build a new classifier, obtain a new similarity metric and provide an improved score to all items in the collection, all while the user is waiting. If one is looking at retrieval over very large collections, and for measurable, comparable performance evaluations not affected by individual, human differences, automatic retrieval is essential [TREC, 2002].

Instead of relying on the relevance feedback judgments of real users, it is worthwhile to consider the idea of obtaining additional relevant/non-relevant training examples via automatic relevance feedback based on a universal/generic similarity metric, which is not tailored to the specific queries. The retrieval results from this metric can provide guidance in selecting good potential training data for a margin-based learning algorithm. However, it quickly becomes apparent that it is inappropriate to consider the top-ranked examples from the generic similarity metric for positive feedback due to the poor performance of current visual information retrieval algorithms in general applications. We found that it is more reasonable to sample the bottom-ranked examples for negative feedback. For example, the bottom-ranked examples for the concept "cars" are more likely to be different from query images in shape. By feeding back these negative data instances, the learning algorithm can automatically re-weight and refine the discriminating feature subspace. Consequently, the similarity metric produced by the learning algorithm is expected to be better adapted to the current queries than a universal/generic similarity metric that is used in all query situations.

The purpose of this learning process is to discover a better similarity metric by finding the most discriminating subspace between positive and negative examples. However, we cannot expect to produce a fully accurate classification prediction in this context, because the sample size of the initial training data is often much too small to accurately represent the true distributions. Thus the class distribution of especially the negative distribution is not reliably modeled. Therefore, it is possible that the feedback will be based on an incorrect estimate of the difference between relevant and non-relevant distributions, resulting in a higher false alarm rate, which will greatly degrade overall system retrieval performance. To address this, one effective solution is to combine the relevance feedback results with results obtained from a generic similarity metric, which can filter out most of the false alarms while hopefully keeping the advantages of an updated and improved similarity metric. For video retrieval, the combination of results with retrieval scores obtained by search in other modalities can further reduce the false alarm rate.

In this chapter, we propose a novel automatic retrieval technique for multimedia data called pseudo-relevance feedback (PRF). It attempts to learn an adaptive similarity space by automatically feeding back the training data which are identified based on a generic similarity metric. We also discuss the combination strategy for different retrieval algorithms, which is essential in the face of unreliability in the PRF approach. The experimental results discussed later in the chapter confirm the effectiveness of the proposed approach.

11.1 Related Work

In the following paragraphs, we will briefly discuss some of the features of complete systems for video retrieval, using the Informedia Digital Video Library [Wactlar et al., 1999] system as a prototype. Similar features, components and interfaces in video retrieval systems have been described by [Smeaton et al., 2001] and [J.R. Smith and Tseng, 2002]. Commercial versions of similar video analysis software are marketed by companies such as Virage [Virage, 2003] and Sonic Foundry [SonicFoundry, 2003].

11.1.1 The Informedia Digital Video Library System

The Informedia Digital Video Library [Wactlar et al., 1999] project focuses specifically on information extraction from video and audio content. Over two terabytes of online data have been collected in MPEG-1 format, with metadata automatically generated and indexed for retrieving videos from this library. The architecture for the project is based on the premise that real-time constraints on library and associated metadata creation could be relaxed in order to realize increased automation and deeper parsing and indexing for identifying the library contents and breaking it into segments. Library creation is an offline activity, with library exploration by users occurring online and making use of the generated metadata and segmentation.

The Informedia research challenge is how the information contained in video and audio can be analyzed automatically and then made useful for a user. Broadly speaking, the Informedia project wants to enable search and discovery in the video medium, similar to what is currently widely available for text. One prerequisite for achieving this goal is the automated information extraction and metadata creation from digitized video. Once the metadata has been extracted, the system enables full-content search and retrieval from spoken language and visual documents. The approach that has been most successful to date involves the integration of speech, image and natural language understanding for library creation and retrieval.

The Informedia interface provides multiple levels of summaries and abstractions for users:

1 Visual icons with relevance measure. When the video story segment results are returned for a query, each keyframe has been selected to be representative for the story as it relates to the query. The keyframe has a little thermometer on the side, which indicates the relevance of this video story to the user query. Different colors in the thermometer bar correspond to different query words, so that the user can tell the contribution of each query word to the relevance of this clip. This allows a user to immediately see which query words matched in a story and which query words are dominant in this story [Christel and Martin, 1998].

2 Short titles or headlines. Moving the mouse over the keyframe brings up an automatically generated headline, which acts as a title for the story to summarize in text what this story is about. [Jin and Hauptmann, 2001]

3 Topic identification of stories. In addition to titles, stories can also be assigned to topics, allowing a better categorization of the results [Hauptmann and Lee, 1998].

4 Filmstrip (storyboard) views. A story keyframe can be expanded into a complete storyboard of images, one per shot, which summarize the complete video story at a glance [Christel et al., 1999]. This allows quick visual verification of often inaccurate results.

5 Transcript following, even when the speech recognition is errorful. While the video is playing, a transcription of the audio is visually synchronized by highlighting the words as they are spoken in the audio track [Hauptmann and Witbrock, 1996].

6 Dynamic maps. Sometimes the information in the video can best be summarized in a dynamic and interactive map, which dynamically shows the locations referenced in the video and allows the user to search geographically for video related to a given area [Christel et al., 2000].

7 Active video skims allow the user to quickly skim through a video play playing interesting excerpts [Christel et al., 1999].

8 Face detection and recognition allow a user to search for faces that are similar to the one specified and to associate faces with names. [Houghton, 1999]

9 Image retrieval. Images similar to a query can be retrieved based on color, texture and shape [QBIC, 2003; Alexander G. Hauptmann, 2002]

Once a relevant video clip has been found, a user might want to make annotations for herself or others to later reuse what was learned. The Informedia system provides a simple mechanism that allows a user to type or speak any comment that applies to a user-selected portion of the video. To this end, the indexing mechanism was modified to allow dynamic, incremental additions and deletions to the index. Finally, a fielded search capability enables the user to search only on selected fields, for example searching only the user annotation field, either through a statistical search based on the OKAPI BM-25 formula or with a classic Boolean search expression. Further re-use of relevant video clips is enabled through a cut-and-paste mechanism that allows a selected clip to be extracted from the library and imported into PowerPoint slide presentations or MS Word text documents.

11.1.2 Relevance and Pseudo-Relevance Feedback in Information Retrieval

Relevance feedback [Picard et al., 1996; Rui et al., 1998; Cox et al., 1998; Tong and Chang, 2001; Zhou and Huang, 2001] and pseudo-relevance feedback [Carbonell et al., 1997] are two common retrieval technique which both originated from document retrieval. For the last decade, relevance feedback has been rapidly embraced in the multimedia domain, as an effective way to improve the retrieval performance by gathering user information in an iterative fashion. Most of the early work [Picard et al., 1996; Rui et al., 1998] was based on the adjustment of feature weighting in an independent feature space, which attempts to associate more important features with higher weights. More recent work has begun to formulate the relevance feedback within a machine learning framework from multiple viewpoints, including distance optimization [Ishikawa et al., 1998], Bayesian learning [Cox et al., 1998], density estimation [Chen et al., 2001], active learning [Tong and Chang, 2001] and discriminating transformation [Zhou and Huang, 2001]. Tieu and Viola [Tieu and Viola, 2001] used a boosting technique to train a learning algorithm in the space of 45000 "highly selective features". This work demonstrates that the margin-based classifiers can generalize well even in a high dimensional feature space. Pseudo-relevance feedback is an automatic retrieval approach without any user intervention. In this approach, a small number of top-ranked documents are assumed to be the relevant examples and used in an relevance feedback process to construct an expanded query. Although pseudo-relevance feedback has gained its success in the document retrieval, there are few studies carried out in the domain of multimedia retrieval. One of the partial explanations is that we can no longer assume the top ranked documents are always relevant due to relatively poor performance of today's visual retrieval systems.

From the viewpoint of machine learning, our approach is closely related to a learning framework called positive example based learning or partially supervised learning [Yu et al., 2002]. Similar to what we proposed, these approaches begin with a small number of positive examples and no negative examples. The "strong negative" instances are extracted from unlabeled data and used to train a classifier [Yu et al., 2002]. This work has demonstrated that accurate classifiers can be built with sufficient positive and unlabeled data without any negative data. This provides a justification for a positive-only learning framework. However, the goal of these learning algorithms which is to associate all examples in a collection with one of the given categories, is different from our objectives of producing a ranked list of the examples. Another area of related work can be found in semi-supervised learning, which constructs a classifier using a training set of labeled data and a working set of unlabeled data. Transductive learning and co-training are two of the paradigms to utilize the information of unlabeled data. Transductive learning has recently been successfully applied in image retrieval [Wu and Huang, 2000], however, the computation of transductive learning is frequently too expensive to make the efficient retrieval feasible in large collections. Co-training [Blum and Mitchell, 1998] has also been suggested as a way

Input:
Query examples $q_1, ..., q_n$
Target examples in the video collection $t_1, ..., t_m$
Output:
Final retrieval score s_i^f for target example t_i
Algorithm:
1) For every i, $1 \leq i \leq m$,
compute base retrieval score $s_i^0 = f_b(t_i, q_1, .., q_n)$
where f_b is the base similarity metric.
2) Iteratively, k from 0 to max
a) Given the retrieval scores s_i^k,
sample the positive examples pos_i^k
and negative examples neg_i^k
using some sampling strategy p
b) Compute the updated retrieval score $s_i^{k+1} = f_l(t_i)$
where the learning algorithm f_l is trained by
pos_i^k and neg_i^k.
3) Combine all the retrieval scores into final score
$s_i^f = g(s_i^0, .., s_i^{max+1})$

Figure 11.1. Basic Algorithm for Peudo-Relevance Feedback

to handle the multimedia retrieval problem since redundant information is available from different modalities.

11.2 Pseudo-Relevance Feedback

In the task of content-based video retrieval, a query typically consists of a text description plus audio, images or video. This query is posed against a video collection. The job of the video retrieval algorithm is to retrieve a set of relevant video shots from a given data collection. Let T be the target video collection, and Q be the user queries. The retrieval algorithm should provide a permutation of the video shots t_i in T, which is sorted by their similarity to the user queries q_i in Q. Most current systems represent the video information as a set of features. The difference between two video segments is measured through a similarity metric between their feature vectors. Formally, a distance measure $d(f(t_i), f(q_i))$ is usually computed to sort each video shot $t_i \in \mathcal{T}$, where $f(x)$ stands for the features of x. Typically, the video features can be constructed from information of multiple perspectives, such as speech transcript, audio, camera motion and video frames. When retrieval is thought of as a classification problem, video data collection can be sepa-

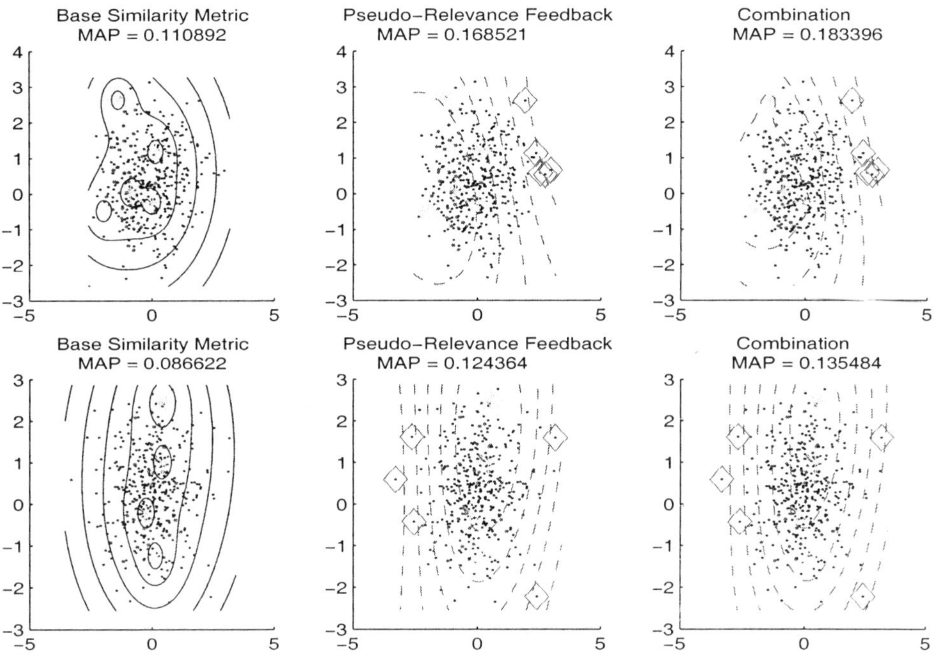

Figure 11.2. Comparison between different algorithms on a synthetic data set, cross are negative, dot are positive, dash-line is the decision boundary, green is query point/initial positive data, diamond is the negative points being selected

rated into two parts for each query, where positive examples T^+ are the relevant examples and negative examples T^- are non-relevant ones.

Precision and recall are two common performance measures for retrieval systems. However, it is well known that these two measure do not take the rank of retrieved examples into consideration. As a better alternative, we adopted mean average precision [TREC, 2002] as our performance measure, which corresponds to the area under an ideal recall/precision curve. To compute average precision, the precision after every retrieved relevant shot is computed, and these precisions are averaged over the total number of retrieved relevant shots in the collection. Mean average precision is the average of these average precision over all topics.

Figure 11.2 summarizes the pseudo-relevance feedback algorithm, which is similar to a relevance feedback process except that users' judgment is replaced by the output of a base similarity metric. This algorithm consists of four major components, that is, the base similarity metric f_b, the sampling strategy p, the learning algorithm f_l and the combination strategy g. It starts by computing the retrieval scores using the base similarity metric f_b for every target example t_i in the video collection. Next, it iteratively identifies new training examples and computes an updated retrieval score. For each run k, the sampling strategy p is used to extract positive and negative examples from the video collection. These positive and negative data are combined to train a learning algorithm f_l. The output s_i^{k+1} of f_l can be interpreted as an updated retrieval score for each target example t_i. Finally, the retrieval scores are fused into a set of final results via a combination strategy g. In our implementation, the positive examples are the query examples and the negative data are obtained from the strongest negative examples. Due to computational issues, the feedback process repeats for only one iteration. For the sake of simplicity, we call s_i^0 the base similarity metric, s_i^1 the PRF metric and s_i^{final} the combination metric. More details of our implementation will be discussed in Section 11.4.

The key idea for this multimedia PRF approach is to automatically feed back the training data which are identified based on a generic similarity metric, so as to learn an adaptive similarity metric and generalize the discriminating subspace for various queries. It is interesting to examine why this approach boosts retrieval performance using examples provided by a generic metric. One explanation relies on the good generalization ability of margin-based learning algorithms. For many retrieval systems, the assumption of isotropic data distribution is generally invalid and undesirable [Hastie and Tibshirani, 1996]. The distribution of relevant examples, especially in a high dimensional feature space, are more likely to cluster in specific directions. More importantly, these discriminative directions vary with different queries and topics. For example, the positive examples of the concept "sky" are similar in color space while those of "cars" are similar in shape space. The PRF approach can provide a better similarity metric than a generic similarity metric in this case, which will be illustrated by the following example.

To visualize the effect of PRF approach, let us consider figure 11.2 which depicts an example based on a synthetic data set. We randomly generated 47 positive examples and 357 negative examples, both sampled from a two-dimensional Gaussian distribution. The negative data has the same variance along both dimensions, while the positive data was designed to cluster along the vertical direction. Figure 11.2 depicts two cases where the positive data lie along the edge of the data collection and in the center of the data collection. Three different metrics are evaluated and their contours are also plotted. All the algorithms begin with five positive examples which were randomly chosen as the initial query. The base similarity metric is the same as discussed in Section 11.4. After the base metric is computed, the 5 strongest negative instances are selected as the negative examples for next run. The PRF metric is provided by an RBF-kernel SVM built on the positive and negative feedback. We also compute the combination metric, which is a 50%-50% linear combination of PRF metric and base similarity metric.

In both cases, the PRF metric turns out to be superior to the base similarity metric in terms of average precision. As noted before, the base similarity metric is typically a generic metric which can not be modified across queries. However, severe bias can be introduced along different dimensions, especially in a high dimensional space with finite samples. Compared with a generic metric, the PRF metric has the advantage that it can be adapted based on the global data distribution and the training examples. Figure 11.2 shows that PRF approach can learn a near optimal decision boundary by feeding back the negative examples. Another effect of the PRF metric is to associate higher scores with the examples which are farther away from the negative data. This might help the performance when the positive data are always near the margin of data collection, which is common in high dimensional spaces. However, this effect also has a downside. Some negative outliers will be even assigned a higher score than any positive data, which possibly leads to more false alarms. One solution for this is to combine the effects of the base metric and the PRF metric, which can be expected to smooth out most of the outliers. From Figure 11.2, we can observe that a simple linear combination of these two approaches leads to a reasonable trade-off between local classification behavior and global discriminating ability.

11.3 Analysis

The previous discussion using synthetic data collections can help us to understand graphically the advantages and drawbacks of the PRF metric in multimedia retrieval. In this section, we would like to provide in-depth discussion on how the PRF approach works when applied to an actual video collection. Our analysis is based on a statistical model of average precision. Moreover, we also present several metric fusion paradigms by transforming different types of similarity metrics into probabilistic outputs.

For each example t_i, the retrieval scores s_i^k provided by different similarity metrics can be treated as the distance between a target t_i and the queries in the corresponding metric space. In this context, we can define the positive distance d^+ as the distance between the positive data T^+ and the queries. The negative distance d^- is similarly defined. As suggested in previous studies [Tarel and Boughorbel, 2002; Platt, 1999], it is reasonable to assume that d^+ and d^- are both approximately Gaussian distributed. Tarel et al [Tarel and Boughorbel, 2002] show that if a similarity metric can be represented by the sum of the similarity metrics of its components, the positive distance d^+ and negative distance d^- will converge towards a Gaussian distribution when the number of examples goes to infinity. Therefore, the probability density function (pdf) p(x) for both distance functions are in the form of,

$$p(x) = \frac{1}{\sqrt{2\pi}\sigma} \exp(-\frac{(x-\mu)^2}{2\sigma^2}) \tag{11.1}$$

where μ is the mean and σ is the variance. The corresponding cumulative density function (cdf) are defined as

$$P(x) = erf(x) = \int_0^x p(t)dt \tag{11.2}$$

which sometimes is also called the error function $erf(x)$. For the sake of analytical simplicity, the erf function can be modeled as a sigmoid function [Platt, 1999] by equating their derivative when $x = \mu$,

$$P(x) = \frac{1}{1 + \exp(C\sigma^{-1}(x-\mu))} \tag{11.3}$$

where $C = -4/\sqrt{2\pi}$

11.3.1 A statistical Model for Average Precision

ID	Type	μ^+	σ^+	μ^-	σ^-	σ^+/σ^-	$\frac{\mu^- - \mu^+}{\sigma^-}$	Approx.	Actual
0	Base	0.63	0.28	1.47	0.44	0.64	1.90	0.0066	0.0070
0	PRF	0.61	0.11	1.18	0.46	0.25	1.25	0.0047	0.0052
14	Base	4.99	1.27	4.65	1.58	0.80	-0.22	0.0006	0.0006
14	PRF	0.06	0.04	0.08	0.04	0.84	0.44	0.0008	0.0012

Table 11.1. Results for average precision analysis

To provide more insights for the PRF approach, it will be useful to study the relationship between the probabilistic distribution of distance functions and our performance criterion, i.e. average precision. In the following discussion, let $p(t)$ be the probability density of T for the data distribution, $p^+(t)$ be that for the positive distribution and $p^-(t)$ be that for the negative distribution. The unconditional (prior) probabilities of

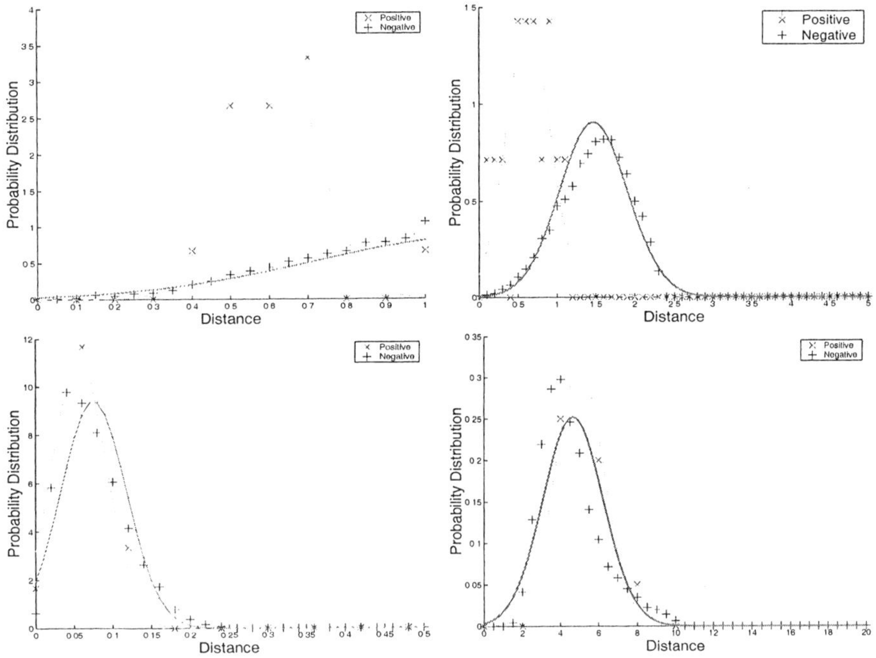

Figure 11.3. Distance function for Query 75 and Query 89 in TREC02 data

the positive class and negative class are denoted by π^+ and π^- respectively. We also denote the mean and variance for the d^+ are μ^+, σ^+. The mean and variance for d^- are μ^-, σ^-. The data collection has a total of N^+ positive examples and N^- negative examples.

Given an example e with distance t_e, the number of positive examples which have lower ranks than e is $N^+(t < t_e) = N^+P^+(t_e)$. Similarly, the number of negative examples which have lower ranks than e is $N^-(t < t_e) = N^-P^-(t_e)$. Therefore, the rank for example e can be translated into the number of examples which have lower ranks, that is, $Rank(e) = N^+(t < t_e) + N^-(t < t_e)$. Thus, the average precision can be defined as,

$$AP = \sum_{e \in T^+} \frac{N^+(t < t_e)}{N^+(t < t_e) + N^-(t < t_e)} \tag{11.4}$$

or we have,

$$AP = \int_0^{t_{max}} \frac{P^+(t)}{P^+(t) + KP^-(t)} p^+(t)dt \tag{11.5}$$

where t_{max} is the largest distance of positive examples, K is the constant N^-/N^+.

Because N^- is always much larger than N^+, we can ignore $P^+(t)$ in the denominator of (11.5),

$$AP = \int_0^{t_{max}} \frac{P^+(t)}{KP^-(t)} p^+(t)dt \tag{11.6}$$

Applying the approximation of (11.3) with some mathematical manipulation, we can get

$$P^-(t) = \left[1 + \left(\frac{P^+(t)}{1-P^+(t)}\right)^{-\frac{\sigma^+}{\sigma^-}} e^{-C\frac{\mu_- - \mu^+}{\sigma^-}}\right]^{-1} \tag{11.7}$$

Substituting (11.7) into (11.6), we can change the integral variable to be $P^+(t)$, with upper limit 1 and lower limit 0,

$$\begin{aligned} AP &= \frac{1}{K}\int_0^1 y\left[1 + \left(\frac{y}{1-y}\right)^{-\frac{\sigma^+}{\sigma^-}} A\right] dy \\ &= \frac{1}{K}\int_0^1 \left[y + y\left(\frac{y}{1-y}\right)^{-\frac{\sigma^+}{\sigma^-}} A\right] dy \end{aligned} \tag{11.8}$$

where $A = e^{-C\frac{\mu^- - \mu^+}{\sigma^-}}$

Obviously, (11.8) can be decomposed into two parts. The first part yields

$$\frac{1}{K}\int_0^1 ydy = \frac{1}{2K} \tag{11.9}$$

The second part can be simplified using a beta function,

$$\begin{aligned} & \frac{A}{K}\int_0^1 y\left(\frac{y}{1-y}\right)^{-\frac{\sigma^+}{\sigma^-}} dy \\ = & \frac{A}{K}\int_0^1 y^{1-\frac{\sigma^+}{\sigma^-}}(1-y)^{\frac{\sigma^+}{\sigma^-}} dy \\ = & \frac{A}{K} Beta(2 - \frac{\sigma^+}{\sigma^-}, 1 + \frac{\sigma^+}{\sigma^-}) \end{aligned} \tag{11.10}$$

Substituting both (11.9) and (11.10) into (11.8), the MAP can be simplified to

$$AP = \frac{1}{K}\left(\frac{1}{2} + A \cdot Beta(2 - \frac{\sigma^+}{\sigma^-}, 1 + \frac{\sigma^+}{\sigma^-})\right) \tag{11.11}$$

The formula (11.11) allows us to analyze the behavior of average precision more deeply. Since few assumption are valid for the data distribution in CBVR, it is difficult to estimate the parameters for data distribution. However, (11.10) shows that the computation of average precision is only related to two components: the variance ratio $\frac{\sigma^+}{\sigma^-}$ and the normalized mean distance $\frac{\mu^- - \mu^+}{\sigma^-}$. Average precision increases with higher normalized mean distance, and it varies with different variance ratios. This property suggests that we further analyze the performance based on the similarity distance distribution.

Figure 11.3 plots the distance distribution of the base similarity metric and the PRF metric for two actual queries in the TREC02 search tasks, using the setting discussed in Section 11.4. As expected, the negative distances can be perfectly fit to the Gaussian model, however, the positive distances are only approximately Gaussian distributed due to the small sampling size. Table 11.1 lists all the statistics of score distribution, the normalized mean distance, the variance ratio, as well as the comparison between predicted average precision and actual average precision. The results on these queries successfully demonstrate that our approximation shown in (11.10) is a reasonable estimate for the actual average precision. Typically, the major performance improvement of the PRF approach is due to the reduction of the normalized mean distance. In query 89, the base similarity metric has an even higher positive mean μ^+ than negative mean μ^-, indicating a poor performance for some non-visual queries. Unsurprisingly, PRF can greatly reduce the normalized positive mean distance μ^+/σ^-, since the PRF approach has the ability to adapt the metric space across queries. However, as a tradeoff, it will overemphasize the "false positives" which are far away from the negative training examples. In query 75, the base similarity metric does achieve a high average precision and leaves no room for the PRF to improve rankings further. Therefore, the normalized positive mean distance μ^+/σ^- does not change too much itself. Unfortunately, Figure 11.3 shows that many more false positives have been assigned a lower distance or a higher retrieval scores than true positive examples. This over-scoring problem greatly degrades the performance of the PRF approach. To address this, different retrieval scores can be combined to boost the retrieval performance, which will be discussed in next section.

11.3.2 Probabilistic Output and Combination

Fusion of different retrieval algorithms is an effective way to address the "false positive" problem in PRF. Combining the base metric and the PRF metric might offer a reasonable trade-off between these two algorithms. More interestingly, it has been found that combination of retrieval scores from different modalities can recover most of the performance hurt since most false positives can be filtered out by additional information. These combinations can also reduce the prediction variance and offer more stable results. In this section, we study how to combine different retrieval algorithms into one and present our combination schemes via the estimation of posterior probabilities.

As we know, the retrieval scores can encode the confidence of the different search algorithms. However, these confidences are relative numbers whose value will vary with the retrieval approaches used. To address the issue of comparability between different types of retrieval scores, the retrieval scores have to be normalized which allows them to be combined with each other and to produce the final ranking decisions. To linearly normalize the scores to some interval, such as $[-1, 1]$, is inappropriate since this does not take into account the score distribution. A more reasonable way is to calibrate the scores to positive posterior probabilities. Given that both negative and positive scores are Gaussian distributed, we can obtain the following form of posterior probability by applying Bayes rule,

$$\begin{aligned} p(+|t) &= \frac{p^+(t)\pi^+}{p^-(t)\pi^- + p^+(t)\pi^-} \\ &= \frac{1}{1 + exp(at^2 + bt + c)} \end{aligned} \quad (11.12)$$

where $p(+|t)$ is the posterior probability, as well as π^+ and π^- are the unconditional(prior) probabilities of the positive class and negative class respectively. However, the posterior estimation derived from two Gaussian assumption violates the monotonicity between scores and posterior probabilities. Platt et al [Platt, 1999] suggest using a parametric sigmoid model to fit the posterior directly,

$$p(+|t) = \frac{1}{1 + exp(At + B)} \quad (11.13)$$

However, it is impractical to fit parameters directly with maximum likelihood estimation due to the computational complexity. Note that it is not necessary to accurately model the posterior probability in our case. Therefore, we prefer an approximation which leads to reasonable prediction effectiveness with less computational effort.

One solution is to set the parameters manually based on empirical testing. Especially when the output of the retrieval algorithm is bounded by some interval [min, max], one can always set the parameters to make

$p(+|min)$ close to 0 and $p(+|max)$ close to 1. Experimental results show that this ad-hoc parameter setting can lead to reasonable performance.

Another form of approximation can be derived from the rank distribution. As mentioned before, the sigmoid function can be approximately modeled by the cumulative density function of a Gaussian distribution. In this case, if we assume that for any number Δt, $p(+|t = t_0) - p(+|t = t_0 + \Delta t) \propto Pr(t_0 < t < t_0 + \Delta t)$, the posterior probability for example e has a simplified form of,

$$\begin{aligned} p(+|t = t_0) &= Pr(t_0 < t < \infty) \\ &= 1 - P(t_0) = 1 - \frac{Rank(e)}{N} \end{aligned} \tag{11.14}$$

where N is the number of all the examples in the collection. This approximation allows a simpler of probability estimation without any parameter tweaking.

After the retrieval scores are calibrated to posterior probabilistic outputs, they can be linearly combined into the final score

$$s^{final} = \sum_i \lambda_i p_i(+|t = t_0) \tag{11.15}$$

However, we still don't know how to choose the best weights to combine these probabilistic outputs. As future work, the statistical model for average precision may be useful to understand how to estimate the best parameters in this case.

	Null	+Transcript	+Transcript +Video Summary
Null	0	0.0971	0.1316
Color(CUM)	0.0245	0.1161	0.1415
Texture1	0.0008	0.089	0.1081
Texture2	0.0017	0.0788	0.1048

Table 11.2. Baseline Performance for Nearest Neighbor Search with different features and combination with text information

11.4 Algorithm Details

Although the PRF approach can be applied to various retrieval tasks such as text retrieval and audio retrieval, in our initial work we have mainly employed it in image retrieval. In this section, we present more details for the PRF approach in image retrieval, discuss how we determine its major components and describe our retrieval algorithms for other modalities as well as the combination schemes.

11.4.1 Base Similarity Metric

The base similarity metric is used to generate the base retrieval scores s_i^0 and also a criterion to select the feedback examples. Given multiple examples as queries, the retrieval algorithm has to learn a more complex distance function than for single-point query. A number of systems have been developed to handle the multiple example-based query. Among them, the aggregate dissimilarity model proposed in [Wu et al., 2001] was able to learn disjunctive models within any metric space. In their model, the aggregate dissimilarity for example x to the query $q_1, ..., q_n$ is expressed by

$$d(x)^\alpha = \begin{cases} 0 & if(\alpha < 0) \wedge d(x, q_i) = 0 \\ \frac{1}{n}\sum_{i=1}^{n} d(x, q_i)^\alpha & otherwise \end{cases} \qquad (11.16)$$

Due to this model's ability to handle multiple examples in arbitrary metric spaces, we adopted this model as our base similarity metric in our experiments, where α is set to -1. The distance $d(x, q_i)$ is computed using the Euclidean distance function. Finally for each target shot t_i, the probabilistic output s_i^0 of the base similarity metric will be modeled by (11.14).

Note that in image retrieval, the retrieval algorithms typically assign a score for each video frame. But the basic unit for video retrieval is a video shot (comprised of multiple frames) instead of an individual frame. Therefore, we choose the maximal retrieval score of a frame within one video shot as the video shot's retrieval score.

11.4.2 Sampling Strategies

After the initial search results are available, a number of feedback training examples will be sampled as the input to a learning algorithm. In our work, the query examples are considered the only positive examples. A subset of the examples that are most dissimilar to the queries will be considered as the negative examples. Since we suspected that different amounts of negative data and different ranges of negativity would affect performance of the PRF approach, we empirically examined how the results were affected by these parameters.

A variety of different sampling strategies can be used and further investigation was necessary to assess their performance. We also examined how retrieval results from other modalities could improve the final rankings. For example, the top ranked examples retrieved from speech transcripts can be positive feedback for image retrieval due to the high precision of textual retrieval. Retrieval results for web search engines could be another source to expand the pool of feedback positive examples, but this was not further investigated.

11.4.3 Classification Algorithm

The training examples provided by sampling strategies are fed back to train a margin-based classifier. In our experiments, support vector machines (SVMs) were used since SVMs are known to yield good generalization performance especially in high dimensional data. Their decision function is of the form

$$y = sign\left(\sum_{i=1}^{N} y_i \alpha_i K(x, x_i) + b\right) \tag{11.17}$$

where x is the d-dimensional vector of a test example, $y \in \{-1, 1\}$ is a class label, x_i is the vector for the ith training example, N is the number of training examples, $K(x, x_i)$ is a kernel function, $\alpha = \{\alpha_1, ..., \alpha_N\}$ and b are the parameters of the model. These α_i can be learned by solving the following quadratic programming (QP) problem,

$$\min Q(\alpha) = -\sum_{i=1}^{N} \alpha_i + \frac{1}{2}\sum_{i=1}^{N}\sum_{j=1}^{N} \alpha_i \alpha_j y_i y_j K(x, x_i) \tag{11.18}$$

subject to $\sum_{i=1}^{N} \alpha_i y_i = 0$ and $0 \leq \alpha_i \leq C, \forall i$

(11.13) is employed to generate the posterior probabilistic output with the parameters (A, B) manually set, which serves as PRF metric s_i^1 for t_i. Ultimately, the posterior probabilities for the base similarity metric and the PRF metric will be linearly combined, i.e. $s_i^I = g(s_i^0, s_i^1) = s_i^1 + \lambda_b s_i^0$ as the final score of the image retrieval. We call λ_b the combination factor of the base similarity metric in the following discussion.

11.4.4 Combination with Text Retrieval

Apart from the image retrieval, we also extract information from other modalities, especially text information like speech and video OCR transcripts, movie titles and external video summaries. This text information can sometimes be helpful in answering high-level semantic queries, which constitutes a good complement to image retrieval. For example, the query "Show me a picture of George Washington" is difficult for image retrieval due to ill-defined visual features of a specific person. A text search based on speech transcripts, however, will give strong hints for where to find relevant shots.

The first type of textual information is from speech and video OCR transcripts. The retrieval of these transcripts is done using the OKAPI

BM-25 formula. The exact formula for the Okapi method is shown in the following equation

$$sim(Q,D) = \sum_{qw \in Q} \left(\frac{tf(qw,D)log(\frac{N-df(qw)+0.5}{df(qw)+0.5})}{0.5 + 1.5\frac{|D|}{avg\_dl} + df(qw)} \right) \tag{11.19}$$

where *tf(qw,D)* is the term frequency of word *qw* in document *D*, *df(qw)* is the document frequency for the word *qw* and *avg_dl* is the average document length for all the documents in the collection. No relevance feedback at the text level was used.

Externally provided video summaries are another source of textual information. If a video abstract contains any keyword of a query, all shots in that movie are believed to have a higher chance to be relevant. Thus, for each query, the posterior probability of a video shot is set to 1 if any keyword of the query can be found in the video abstract for the corresponding movie, otherwise the posterior probability is set to 0.

Again, all the output scores from the various modalities will be converted into the probabilities using equation (11.14). Although retrieval scores are not Gaussian distributed for textual information, we still use (11.14) to compute the probabilistic output due to its simple form. Let s_i^{sp} and s_i^v be posterior probabilities of transcript retrieval and video summary retrieval for t_i. Finally, these posterior probabilities will be linearly combined, i.e. $s_i^f = s_i^I + \lambda_{sp} s_i^{sp} + \lambda_v s_i^v$.

11.5 Experimental Results

We present our experimental results on 2002 TREC Manual Video Retrieval Task [TREC, 2002] in order to demonstrate the effectiveness of the PRF approach. Note that in the following experiments, only 100 video shots are retrieved for each query in accordance with the requirements of the TREC'02 video track.

11.5.1 Experimental Setting

The video data came from the video collection provided by the TREC Video Retrieval Track. The definitive information about this collection can be found at the NIST TREC Video Track web site [TREC, 2002]. The Text REtrieval Conference evaluations are sponsored by the National Institute of Standards and Technology (NIST) with additional support from other U.S. government agencies. Their goal is to encourage research in information retrieval by providing a large test collection, creating typical queries that might be asked, pooling the relevant results returned from multiple approaches, and applying uniform scoring procedures to evaluate systems. Thus the TREC evaluations create a forum for organizations interested in comparing their results. The video retrieval evaluation centered around the shot as the unit of information retrieval (equivalent to a 'document' for text retrieval) rather than the scene (a related sequence of multiple shots) or story/segment as the video document to be retrieved.

The 2002 Video Collection for the Video TREC retrieval task consisted of 40 hours of MPEG-1 video in the search test collection. The data came from the Internet Archive [Archives, 2003] of videos, which is a collection of advertising, industrial, and amateur films from about 1930 - 1970. The films were generally produced by the US Government, corporations, non-profit organizations or trade groups. This corpus translated into 1160 segments as processed by Carnegie Mellon University or 14,524 shots where the boundaries were provided as the common shot reference of the Video TREC evaluation effort. The shots were composed of a total of 292,000 I-frames which we extracted directly from the MPEG-1 compressed video files.

The audio processing component of our video retrieval system splits the audio track from the MPEG-1 encoded video file, and decodes the audio and down-samples it to 16 kHz, 16 bit samples. These samples are then passed to a speech recognizer. The speech recognition system we used for these experiments is a state-of-the-art large vocabulary, speaker independent speech recognizer. For this evaluation, a 64000-word language model derived from a large corpus of broadcast news transcripts was used. Previous experiments had shown the word error rate on this type of mixed documentary-style data with frequent overlap of music and speech to be 35 - 40%.

On the image processing side, two types of low-level image features including color features and texture features were used in our system. The color feature is the cumulative color histogram for the HSV (Hue-Saturation-Value) color space [Chapelle et al., 1999]. The hue is quantized into 16 bins. Both saturation and value are quantized into 6 bins. The texture features are obtained from the convolution of the image pixels with various Gabor filters [Schiele and Crowley, 1996]. We compute a histogram for each filter which is quantized into 16 bins. Their central and second-order moments are generated as the texture feature. In our implementation, two versions of texture features are generated: one uses six filters in a 3×3 image tessellation, the other uses 12 filters for the whole image. As a preprocessing step, each dimension of the feature vectors is normalized by its covariance.

A typical query consists of video, image, audio and text descriptions. An example of an actual TREC video query as specified in the XML format of video topics is shown in Figure 11.4. The images referred to in the query are shown in Figure 11.5. Queries consisting of video clips will be decomposed into a sequence of images. None of the actual queries contained audio examples. Therefore, only text and images were available for each query. Figure 11.6 gives some examples for different queries in TREC02 video track, including the query keywords and sample of images. Examples for negative images chosen by sampling strategies are also provided. We used the SVM^{Light} implementation. The setting for the feedback learning algorithm is RBF kernel SVMs with parameter 0.05. For the probabilistic output, the parameters (A, B) are manually set to be $(-10, -2)$

<! DOCTYPE videoTopic SYSTEM "videoTopics.dtd">
<videoTopic num="083">
<textDescription text="Find shots of the Golden Gate Bridge" / >
<imageExample
src="http://www.nonprofitcenters.org/documents/images/golden-gate-view.jpg"
desc="clear side view" / >
<imageExample
src="http://www.onroute.com/guides/sanfrancisco/images/golden-gate-sun.gif"
desc="red tower and part of span from side" / >
<imageExample src="http://www.windscreen.tv/images/San Francisco Golden Gate Bridge/2006.12.00.jpg" desc="on bridge" / >
<imageExample src="http://www.wetterschau.de/wecke/Nebel/golden gate.jpg" desc="tower and span in fog" / >
< /videoTopic>

Figure 11.4. The XML specification for Query 83: *Find shots of the Golden Gate Bridge*

For the combination of the posterior probabilities, except when stated otherwise, the weights λ_b, λ_{sp} are set to 1 but the weight for video summary information λ_v was set as low as 0.2, because its prediction is based on the whole video movie as a unit and thus is too coarse to provide an accurate score for an individual video shot. Note that these weights are arbitrarily set which is not necessarily the best choice. In fact, the experimental results reported in next section shows that this weight setting is far from optimal.

Figure 11.5. The five sample images provided as part of query 83: *Find shots of the Golden Gate Bridge*

11.5.2 Results

The first series of experiments was designed to verify the performance of the base similarity metrics with three types of low-level image features.

	Precision	Recall	MAP
Base	0.1108	0.3000	0.1415
PRF	0.1320	0.3318	0.1522

Table 11.3. Comparison for base similarity metric and PRF approach

	λ_t=0.5	λ_t=0.8	λ_t=1	λ_t=1.2	λ_t=1.5
λ_v=0.1	0.1437	0.146	0.1459	0.144	0.1439
λ_v=0.2	0.1513	0.1516	0.1516	0.147	0.1469
λ_v=0.3	0.1548	0.1539	0.1538	0.1495	0.1471
λ_v=0.4	0.1568	0.1545	0.1548	0.1518	0.1469

Table 11.4. Different weights for the combination of text retrieval and image retrieval

ID	Key Words	Query Images		Negative Images	
75	Eddie Rickenbacker				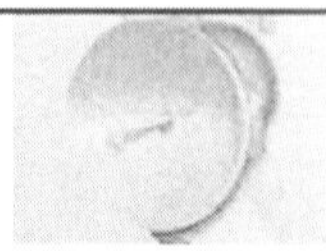
83	Golden Gate Bridge				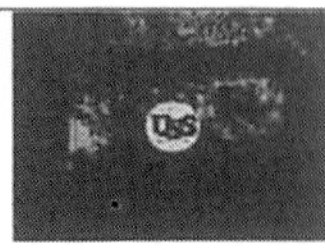
93	Dairy cattle, cows, bulls, cattle				

Figure 11.6. Examples for Queries

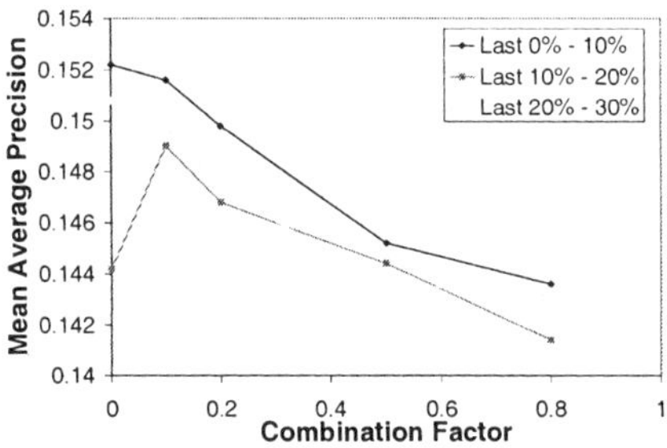

Figure 11.7. Various sampling range for negative examples

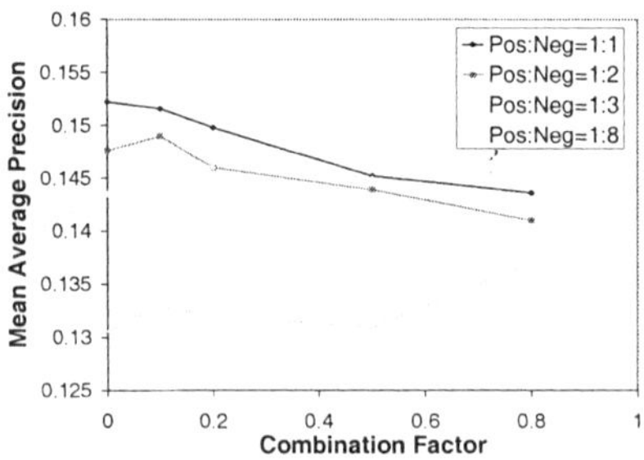

Figure 11.8. Various pos/neg ratios for negative examples

	# better	# worse	# equal
$\lambda_b = 0$	11	10	3
$\lambda_b = 0.2$	14	7	3
$\lambda_b = 0.5$	13	7	4
$\lambda_b = 0.8$	15	4	5

Table 11.5. Query Analysis for individual queries

Combinations with text retrieval are also studied, including transcript retrieval and video summary retrieval. Table 11.2 lists the mean average precisions (MAP) for all 12 possible combinations. It clearly shows that pure image retrieval without any text combination produces relatively poor performance, with the highest MAP only reaching 2%. By comparison, text retrieval based only on speech and VOCR transcripts can achieve much better results. Moreover, retrieval based on video summaries can push retrieval performance even higher. Finally, we get the highest MAP of 14.1% using the color based image retrieval combined with text retrieval. To avoid exponential explosion of combination in the following experiments, color features alone are used as base image features and we only report the retrieval results combined with transcript and video summary retrieval.

Next, we analyzed the performance of the PRF approach. Let us define the rank ratio for an example e as $1 - rank(e)/MaxRank$. The negative sampling range can be represented as a pair of rank ratios $[a, b]$, indicating that we only sample negative feedback from the examples

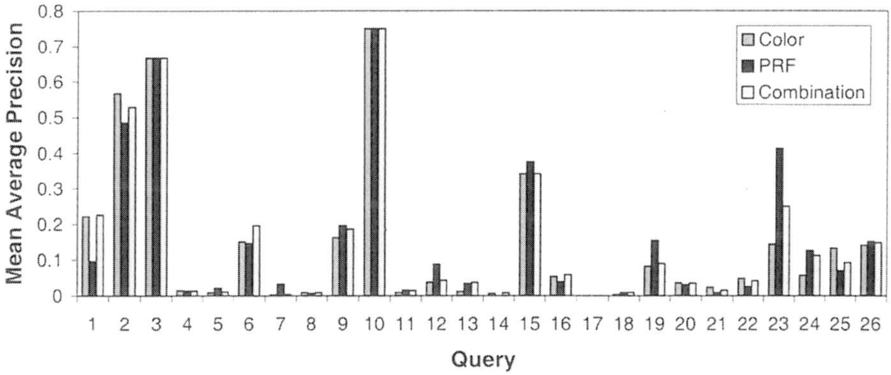

Figure 11.9. Comparison of MAP for individual queries

where the rank ratios are within $[a, b]$. We also define the feedback class ratio $Ratio_{Neg}$ as the ratio between positive feedback and negative feedback examples. The basic setting for our experiments is that the negative sampling range is set to [0%, 10%] and the feedback class ratio is 1. Table 11.3 lists the comparison between PRF and the base similarity metric in terms of precision, recall and mean average precision. As can be seen from the figure, PRF achieves a performance improvement over the base retrieval algorithm in all three performance measures. In terms of MAP, the result for the PRF approach can be as high as 15.2% which achieves about a 7% relative improvement beyond the base similarity metric.

To study the behavior of PRF in more detail, we evaluated the effects for various parameters including the negative sampling range and the feedback class ratio. The results are plotted in Figures 11.7 and 11.8. For each figure, a different combination factor λ_b with the base similarity metric are also reported. In both experiments, we test the case when $\lambda_b = 0, 0.1, 0.2, 0.5, 0.8$. $\lambda_b = 0$ indicates that the only the PRF score was used and the base similarity metric was ignored. Most of the performance curves go down with higher combination factors. This can be explained by the fact that the false positive problem in the PRF approach has largely been addressed through the combination with the text retrieval results. Therefore, the PRF approach can always benefit from a more adaptive metric space. However, $\lambda_b = 0$ does not work well in all cases. As an alternative, $\lambda_b = 0.1$ seems to be a fairly good trade-off combination factor.

In Figure 11.7, three cases are studied where the negative sampling range is [0%, 10%], [10%, 20%] and [20%, 30%]. It shows that the best case is when the negative examples are sampled from the strongest negative examples. This might partially be explained by the fact that the most dissimilar examples are more likely to be the most negative examples. Figure 11.8 indicates that the performance is lower when the feedback class ratio becomes higher. This is maybe caused by the fact that the classifier will produce poor probabilistic output for unbalanced training sets. However, further investigations are required to determine how these parameters can affect the results.

We evaluated the combination factor of the transcript retrieval λ_{sp} and video summaries retrieval λ_v. Table 11.4 shows the comparison with different combination factors. Basically, MAP becomes worse when λ_{sp} is higher and λ_v is lower. The highest performance has a MAP of 15.8% with λ_t 0.5 and λ_v 0.4. This implies that our default setting for combination factors was far from optimal. A better scheme should be developed to determine near optimal weighting for different retrieval algorithms.

So far, the experimental results presented above depict the average performance over all queries. However, it is still not clear whether the PRF approach benefits majority of the queries or only a small number of them. Our last experiment was designed to examine the effect of the PRF approach for individual queries. Figure 11.9 compares the mean average precision per query of the base similarity metric, the PRF approach and their combination when $\lambda_b = 0.5$. Compared to the base similarity metric, the PRF metric results in a large increase for queries 23, and 19 but mostly loses in queries 1, and 25. The combination of both achieves a fairly good trade-off between them. In table 11.5, we show how many queries are better, worse or equal to the MAP of the base similarity metric with different combination factors λ_b. As expected, only half of the queries can achieve a higher MAP with the PRF approach over the base retrieval algorithm, but their combination seems to benefit most of the queries. This again indicates the importance of the combination strategies.

11.6 Conclusion

This chapter presented a novel technique for improved multimedia information retrieval, negative pseudo-relevance feedback. After looking at content-based video retrieval, we found that the task can be framed as a concept classification task. Since learning algorithms for classification tasks have been extraordinarily successful in recent years, we were able to apply insights from machine learning theory to the video information retrieval task. Specifically, the multimedia query examples provide the positive training examples for the classifier. Negative training examples are obtained from an initial simple Euclidian similarity metric, which selects the worst matching images for pseudo-relevance feedback. An SVM classifier then learns to weight the discriminating features, resulting in improved retrieval performance. An analysis of the negative PRF technique shows that the benefit of the approach derives from the ability to separate the means of the (Gaussian) distributions of the negative and positive image examples, as well as reducing the variances of the distributions. Since extreme outliers in high-dimensional feature spaces can result in over-weighting of some dimensions, empirical results suggest that smoothing with the initial simple distance metric safeguards against egregious errors. Experiments on the data from the 2002 TREC Video track evaluations confirmed the effectiveness of the approach on a collection of over 14000 shots in 40 hours of video.

References

Alexander G. Hauptmann, N. P. (2002). Video-cuebik: adapting image search to video shots. In *JCDL*, pages 156–157.

Antania, S., Kasturi, R., and Jain, R. (2002). A survey on the use of pattern recognition methods for abstraction, indexing and retrieval of images and video. *Pattern Recognition*, 4:945–65.

Archives, I. (2003). Internet Archives, http://www.archive.org/movies/prelinger.php

Blum, A. and Mitchell, T. (1998). Combining labeled and unlabeled data with co-training. In *COLT: Proceedings of the Workshop on Computational Learning Theory.*

Carbonell, J. G., Yang, Y., Frederking, R. E., Brown, R. D., Geng, Y., and Lee, D. (1997). Translingual information retrieval: A comparative evaluation. In *IJCAI*, pages 708–715.

Chapelle, O., Haffner, P., and Vapnik, V. (1999). SVMs for histogram-based image classification. *IEEE Transactions on Neural Networks*, 10(5):1055–1065.

Chen, Y., Zhou, X., and Huang, T. (2001). One-class svm for learning in image retrieval. In *Proc. IEEE International Conf. on Image Processing*, Thessaloniki, Greece.

Christel, M. and Martin, D. (1998). Information visualization within a digital video library. *Journal of Intelligent Information Systems*, 11(3):235–257.

Christel, M., Olligschlaeger, A., and Huang, C. (2000). Interactive maps for a digital video library. *IEEE MultiMedia*, 7(1).

Christel, M. G., Hauptmann, A. G., Warmack, A., and Crosby, S. A. (1999). Adjustable filmstrips and skims as abstractions for a digital video library. In *Advances in Digital Libraries*, pages 98–104.

Cox, I. J., Miller, M., T.Minka, and Yianilos, P. (1998). An optimized interaction strategy for bayesian relevance feedback. In *IEEE Conf. on Computer Vision and Pattern Recognition*, pages 553–558, California.

Evans, J. (2002). Managing the digital television archive: the current and future role of the information professional. In *Online information.*

Hastie, T. and Tibshirani, R. (1996). Discriminant adaptive nearest neighbor classification and regression. In Touretzky, D. S., Mozer, M. C., and Hasselmo, M. E., editors, *Advances in Neural Information Processing Systems*, volume 8, pages 409–415. The MIT Press.

Hauptmann, A. and Witbrock, M. (1996). Informedia news on demand: Multimedia information acquisition and retrieval. In *Intelligent Multimedia Information Retrieval.* AAAI Press/MIT Press, Menlo Park, CA.

Hauptmann, A. G. and Lee, D. (1998). Topic labeling of broadcast news stories in the informedia digital video library. In *Proceedings of the third ACM conference on Digital libraries*, pages 287–288. ACM Press.

Houghton, R. (1999). Named faces: putting names to faces. *IEEE Intelligent Systems*, 14(5).

Ishikawa, Y., Subramanya, R., and Faloutsos, C. (1998). MindReader: Querying databases through multiple examples. In *24th International Conference on Very Large Data Bases, VLDB*, pages 218–227.

Jin, R. and Hauptmann, A. (2001). Headline generation using a training corpus. In *Second International Conference on Intelligent Text Processing and Computational Linguistics (CICLING01)*, pages 208–215, Mexico City, Mexico.

J.R. Smith, C.Y. Lin, M. N. P. N. and Tseng, B. (2002). Advanced methods for multimedia signal processing. In *International Workshop for Digital Communications IWDC*, Capri, Italy.

Lew, M., editor (2002). *International Conference on Image and Video Retrieval.*

Picard, R. W., Minka, T. P., and Szummer, M. (1996). Modeling user subjectivity in image libraries. In *IEEE International Conf. On Image Processing*, volume 2, pages 777–780, Lausanne, Switzerland.

Platt, J. (1999). Probabilistic outputs for support vector machines and comparison to regularized likelihood methods. In A. Smola, P. Bartlett, B. S. and Schuurmans, D., editors, *Advances in Large Margin Classiers.* MIT Press.

Porter, C. (2002). The challenges of video indexing and retrieval within a commercial environment. In *CIVR.*

QBIC (2003). IBM QBIC web site, http://wwwqbic.almaden.ibm.com.

Rui, Y., Huang, T. S., and Mehrotra, S. (1998). Relevance feedback: A power tool for interactive content-based image retrieval. *IEEE Trans. on Circuits and Systems for Video Technology*, 8:644–655.

Schiele, B. and Crowley, J. L. (1996). Object recognition using multidimensional receptive field histograms. In *ECCV.*

Smeaton, A., Murphy, N., O'Connor, N., Marlow, S., Lee, H., McDonald, K., Browne, P., and Ye, J. (2001). The fschlr digital video system: A digital library of broadcast tv programmes. In *Joint Conference on Digital Libraries*, pages 24–28, Roanoke, VA.

SonicFoundry (2003). SonicFoundry Inc. Website, http://sonicfoundry.com.

Tarel, J. P. and Boughorbel, S. (2002). On the choice of similarity measures for image retrieval by example. In *ACM International Conference on Multimedia*, pages 107–118.

Tieu, K. and Viola, P. (2001). Boosting image retrieval. In *International Conference on Computer Vision*, pages 228–235.

Tong, S. and Chang, E. (2001). Support vector machine active learning for image retrieval. In *ACM International Conference on Multimedia*, pages 107–118.

TREC (2002). TREC-2002 video track, http://www-nlpir.nist.gov/projects/t2002v/t2002v.html.

Virage (2003). Virage, Inc. Website, http://www.virage.com/.

Wactlar, H., Christel, M., Gong, Y., and Hauptmann, A. (1999). Lessons learned from the creation and deployment of a terabyte digital video library. *IEEE Computer*, 32(2):66–73.

Wu, L., Faloutsos, C., Sycara, K. P., and Payne, T. R. (2001). Multimedia queries by example and relevance feedback. *IEEE Data Engineering Bulletin*, 24(3).

Wu, Y. and Huang, T. S. (2000). Self-supervised learning for visual tracking and recognition of human hand. In *AAAI*, pages 243–248.

Yu, H., Han, J., and Chang, K. C. (2002). PEBL: Positive example based learning for web page classification using svm. In *Proceedings of the 2002 ACM SIGKDD Conference (KDD 2002)*, pages 239–248.

Zhou, X. S. and Huang, T. S. (2001). Comparing discriminating transformations and svm for learning during multimedia retrieval. In *ACM International Conference on Multimedia*.

Index